PEARSON EDEXCEL INTERNATIONAL GCSE (9–1)
HUMAN BIOLOGY
Student Book

Phil Bradfield
Steve Potter

Published by Pearson Education Limited, 80 Strand, London, WC2R 0RL.

www.pearsonglobalschools.com

Copies of official specifications for all Edexcel qualifications may be found on the website: https://qualifications.pearson.com

Text © Pearson Education Limited 2017
Edited by Stephanie White and Daniel Gill
Designed by Cobalt id
Typeset by Tech-Set Ltd, Gateshead, UK
Original illustrations © Pearson Education Limited 2017
Illustrated by Tech-Set Ltd, Gateshead, UK
Cover design by Pearson Education Limited
Picture research by Andreas Schindler
Cover photo/illustration © gettyimages.co.uk: Chris Raven/EyeEM

The rights of Phil Bradfield and Steve Potter to be identified as authors of this work have been asserted by them in accordance with the Copyright, Designs and Patents Act 1988.

First published May 2017

25
10 9 8

British Library Cataloguing in Publication Data
A catalogue record for this book is available from the British Library

ISBN 978 0 435 18498 8

Copyright notice
All rights reserved. No part of this publication may be reproduced in any form or by any means (including photocopying or storing it in any medium by electronic means and whether or not transiently or incidentally to some other use of this publication) without the written permission of the copyright owner, except in accordance with the provisions of the Copyright, Designs and Patents Act 1988 or under the terms of a licence issued by the Copyright Licensing Agency, 5th Floor, Shackleton House, 4 Battlebridge Lane, London, SE1 2HX (www.cla.co.uk). Applications for the copyright owner's written permission should be addressed to the publisher.

Printed and bound by CPI Group (UK) Ltd, Croydon, CR0 4YY

Acknowledgements
The author and publisher would like to thank the following individuals and organisations for permission to reproduce photographs:

(Key: b-bottom; c-centre; l-left; r-right; t-top)

Alamy Stock Photo: Blend Images 89bl, Blue Jean Images 109tl, BSIP SA 128cr, 235tl, David Cole 245cl, Cultura Creative (RF) 52, Cultura RM 234bc, domonabikebali 164cl, FLPA 213bl, Image Source 159br, imageBROKER 239cr, Images of Africa Photobank 56cl, Lenscap 204br, MBI 224tl, Mediscan 77br, Phanie 229cr (a), 241cl, Phanie 229cl (a), 241cl, Keith Bedford / Reuters 179bl, sciencephotos 146br, Trevor Smith 46tr, tdbp 245tl, Kevin Wheal 71cl
Fotolia.com: bit24 54cr
Getty Images: Anthony Bannister / Gallo Images 239tl, Nancy R. Cohen / DigitalVision 207tl, DR GOPAL MURTI / SCIENCE PHOTO LIBRARY 241tl, FatCamera 192, John P Kelly / Stockbyte 131cr, Klaus Vedfelt / Digital Vision 2, Klaus Vedfelt / Stone 124
Greg Jianas: 237bl
Pearson Education Ltd: Trevor Clifford 27cl

Science Photo Library Ltd: 36tc (a), 103br (b), 221br, MICHAEL ABBEY 229br (d), AJ PHOTO 87tl, 237br, Andrew Lambert Photography 36cr (d), ANIMATED HEALTHCARE LTD 193bl, JUERGEN BERGER 230bl, Biophoto Associates 58cl, 88cl, 102tr, 246cl, BSIP 229cr (b), James Cavallini 219br, 238tl, Martyn F. Chillmaid 36tr (b), 36cl (c), CNRI 0 (vi), 21bl, 99bl, MICHELLE DEL GUERCIO 105tl, Eye of Science 229bl (c), GASTROLAB 69cl, Steve Gschmeissner 12bl, 29tr, 126cl, 127cl, INNERSPACE IMAGING 95cr, ISM 240cr, KATERYNA KON 240cl, DENNIS KUNKEL MICROSCOPY 104br (b), NANCY MCKENNA 57cl, ASTRID & HANNS-FRIEDER MICHLER 167tr, Dr. Brad Mogen 0 (vi), 20cl, DR GOPAL MURTI 5tl, NATURAL HISTORY MUSEUM, LONDON 155cl, NIBSC 79tr, Susumu Nishinaga 29tl, ALFRED PASIEKA 34cr, Keith R. Porter 4bl, REVY, ISM 105tr, SATURN STILLS 44bl, DR K. SIKORA 113cr, Sputnik 181cr, VOLKER STEGER 107cl (a), Spencer Sutton 107bl (b), DR KEITH WHEELER 203tl
Shutterstock.com: Africa Studio 55br, areeya_ann 205cr, Diego Cervo 89cl, Boyan Dimitrov 160cl, Juan Gaertner 16cl, Mauricio Graiki 166cl, Melory 217bl, Jesada Sabai 154bl, schankz 64br, sciencepics 150tl, TRAIMAK 18bl, zulufoto 131tl, Zurijeta 89tl
Cover images: Front: **Getty Images:** Chris Raven / EyeEm
Inside front cover: **Shutterstock.com:** Dmitry Lobanov

All other images © Pearson Education

Select glossary terms have been taken from *The Longman Dictionary of Contemporary English Online*.

Endorsement Statement
In order to ensure that this resource offers high-quality support for the associated Pearson qualification, it has been through a review process by the awarding body. This process confirms that this resource fully covers the teaching and learning content of the specification or part of a specification at which it is aimed. It also confirms that it demonstrates an appropriate balance between the development of subject skills, knowledge and understanding, in addition to preparation for assessment.

Endorsement does not cover any guidance on assessment activities or processes (e.g. practice questions or advice on how to answer assessment questions), included in the resource nor does it prescribe any particular approach to the teaching or delivery of a related course.

While the publishers have made every attempt to ensure that advice on the qualification and its assessment is accurate, the official specification and associated assessment guidance materials are the only authoritative source of information and should always be referred to for definitive guidance.

Pearson examiners have not contributed to any sections in this resource relevant to examination papers for which they have responsibility.

Examiners will not use endorsed resources as a source of material for any assessment set by Pearson. Endorsement of a resource does not mean that the resource is required to achieve this Pearson qualification, nor does it mean that it is the only suitable material available to support the qualification, and any resource lists produced by the awarding body shall include this and other appropriate resources.

Disclaimer: **neither Pearson, Edexcel nor the authors take responsibility for the safety of any activity**. Before doing any practical activity you are legally required to carry out your own risk assessment. In particular, any local rules issued by your employer must be obeyed, regardless of what is recommended in this resource. Where students are required to write their own risk assessments they must always be checked by the teacher and revised, as necessary, to cover any issues the students may have overlooked. The teacher should always have the final control as to how the practical is conducted.

COURSE STRUCTURE	iv
ABOUT THIS BOOK	v
ASSESSMENT OVERVIEW	viii
UNIT 1	2
UNIT 2	52
UNIT 3	124
UNIT 4	192
APPENDICES	
APPENDIX A: A GUIDE TO EXAM QUESTIONS ON EXPERIMENTAL SKILLS	257
APPENDIX B: COMMAND WORDS	262
GLOSSARY	263
INDEX	271

COURSE STRUCTURE

UNIT 1

1. CELLS: 03
   1.1, 1.2, 1.3, 1.4, 1.5, 1.6, 1.8, 1.7, 1.10, 1.11, 1.14, 1.15, 1.16, 1.12, 1.13, 1.9

2. MOVEMENT OF SUBSTANCES INTO AND OUT OF CELLS: 24
   3.1, 3.2, 3.3

3. BIOLOGICAL MOLECULES: 32
   2.1, 2.2, 2.3, 2.6, 2.7, 2.8, 2.9, 2.10

UNIT 2

4. NUTRITION AND ENERGY: 53
   6.1, 6.3, 6.4, 2.4, 6.2, 2.5, 6.5, 6.10, 6.6, 6.7, 6.8, 6.9, 6.11, 6.12

5. RESPIRATION AND GAS EXCHANGE 74
   7.1, 7.3, 7.6, 7.5, 7.4, 8.12, 8.1, 8.2, 8.3, 7.2, 8.5, 8.4, 8.6, 8.13

6. INTERNAL TRANSPORT: 92
   9.9, 9.10, 1.15, 9.15, 8.10, 9.8, 8.9, 8.11, 9.3, 9.1, 9.2, 9.4, 9.6, 9.7, 9.5, 9.11, 9.12, 8.8, 8.7, 9.13, 9.14, 9.16, 9.17, 9.18, 9.19

UNIT 3

7. BONES, MUSCLES AND JOINTS: 125
   4.1, 1.15, 4.6, 4.3, 4.2, 4.4, 4.5

8. SENSORY RECEPTORS – THE EYE AND THE EAR: 136
   5.5, 5.7, 5.11, 5.12, 5.13, 5.15, 5.14

9. COORDINATION: 149
   5.2, 5.1, 5.8, 5.6, 5.4, 5.3, 5.19, 5.18, 5.16, 5.17, 5.10, 5.9, 10.7

10. HOMEOSTASIS AND EXCRETION: 171
    10.8, 10.2, 10.11, 10.3, 10.4, 10.5, 10.6, 10.9, 10.10, 10.1

UNIT 4

11. REPRODUCTION: 193
    11.14, 11.3, 1.16, 11.1, 11.17, 11.23, 11.2, 11.4, 11.6, 11.8, 11.7, 11.5, 11.10, 11.11, 11.12, 11.9

12. HEREDITY: 212
    11.13, 11.14, 11.19, 11.20, 11.18, 11.15, 11.16, 11.21, 11.22, 11.24

13. MICROORGANISMS: 228
    12.2, 12.5, 12.3, 12.1, 12.4, 12.8, 12.7, 12.9, 12.10, 12.11, 12.12, 12.14, 12.13, 12.15, 12.16, 12.6, 12.17, 12.18

APPENDICES 257

GLOSSARY 263

ABOUT THIS BOOK

This book is written for students following the Pearson Edexcel International GCSE (9–1) Human Biology specification. You will need to study all of the content in this book for your Human Biology examinations, except content in Extension boxes, which is meant to extend your learning.

In each unit of this book, there are concise explanations and numerous exercises that will help you build up confidence. The book also describes the methods for carrying out all of the required practicals.

The language throughout this textbook is graded for speakers of English as an additional language (EAL), with advanced Human Biology-specific terminology highlighted and defined in the glossary at the back of the book. A list of command words, also at the back of the book, will help you to learn the language you will need in your examinations.

You will find that questions in this book have Progression icons and Skills tags. The Progression icons refer to Pearson's Progression scale. This scale – from 1 to 12 – tells you what level you have reached in your learning and will help you to see what you need to do to progress to the next level. Furthermore, Edexcel has developed a Skills grid showing the skills you will practise throughout your time on the course. The skills in the grid have been matched to questions in this book to help you see which skills you are developing. You can find Pearson's Progression scale, along with guidelines on how to use it at www.pearsonglobalschools.com/igscienceprogression.

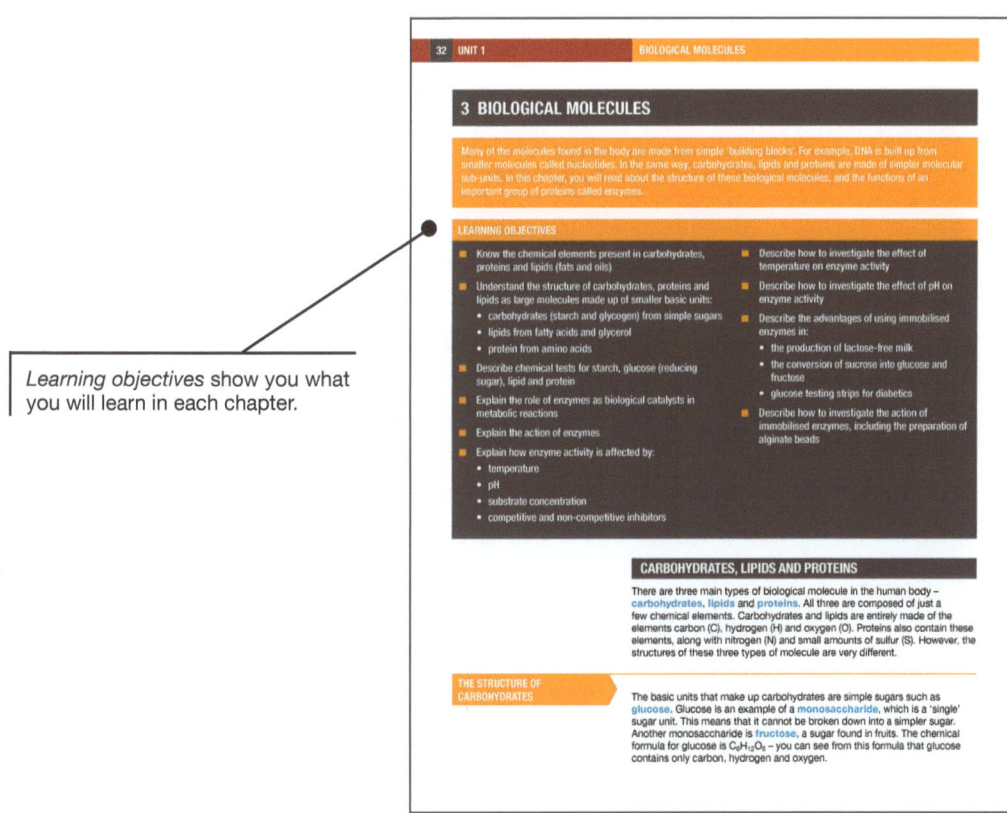

Learning objectives show you what you will learn in each chapter.

ABOUT THIS BOOK

Did You Know boxes give interesting facts about the topic you are studying.

Looking Ahead features tell you what you will learn if you continue your study of Human Biology to a higher level, such as International A Level.

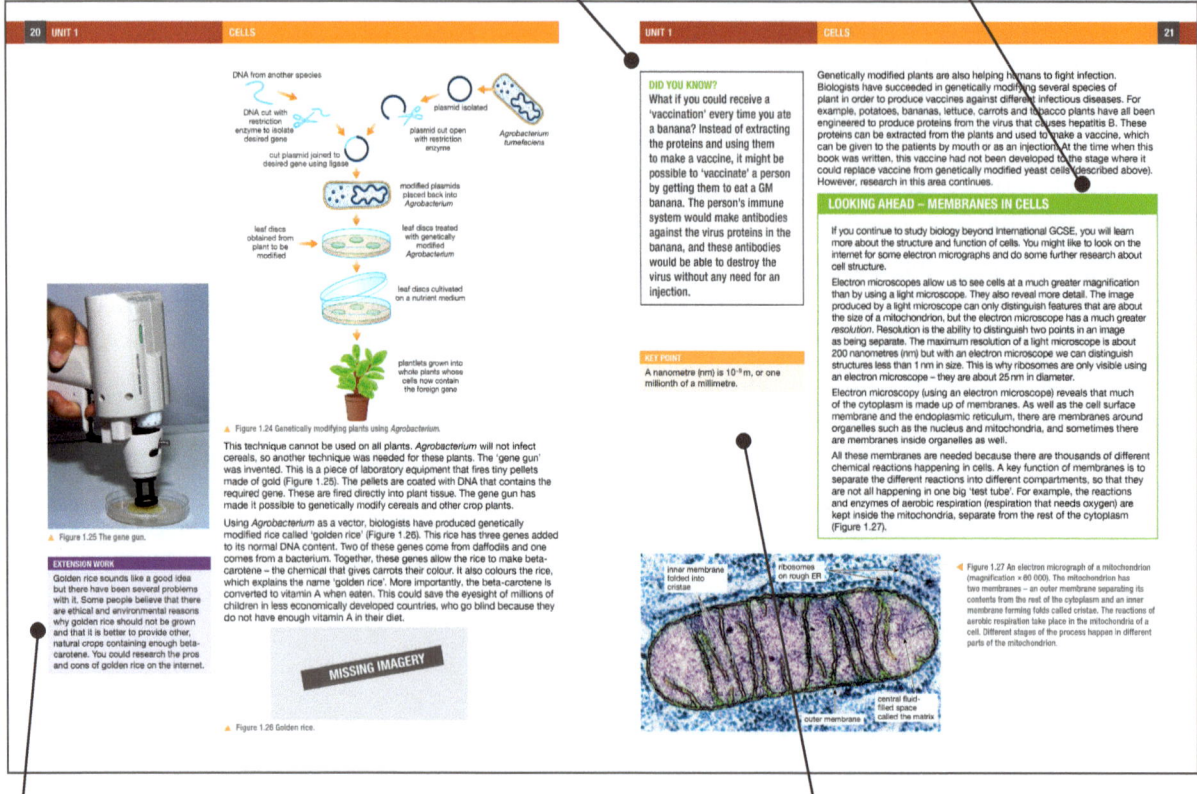

Extension boxes include content which is not on the specification and which you do not have to learn for your examination. However, it will help to extend your understanding of the topic.

Key Point boxes summarise the essentials.

Hint boxes give you tips on important points to remember in your examination.

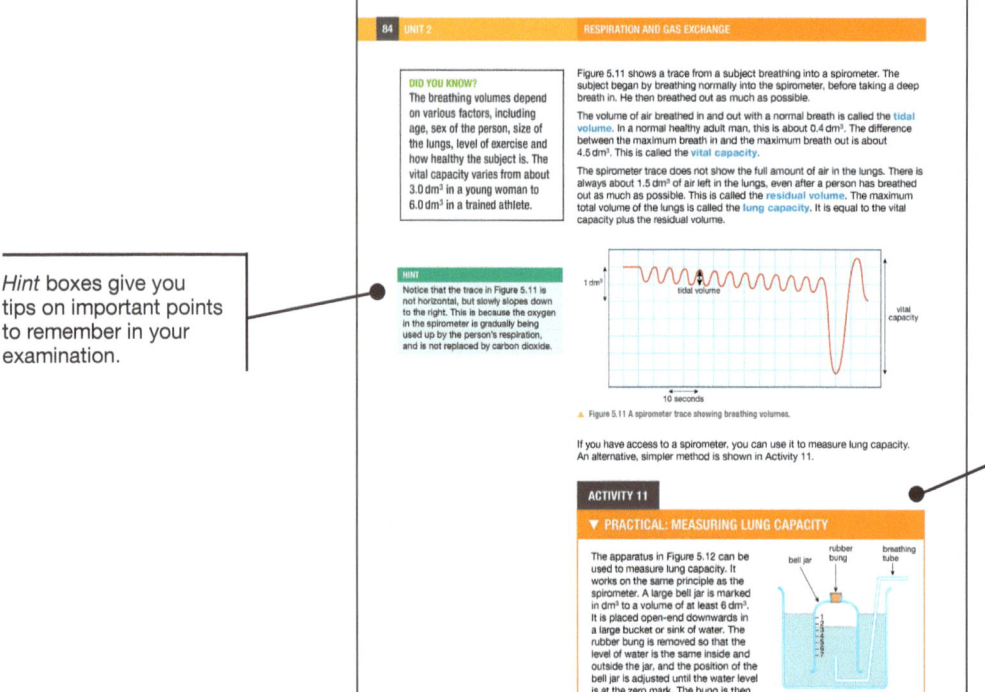

Practical activities describe the methods for carrying out all of the practicals you will need to know for your examination.

ABOUT THIS BOOK

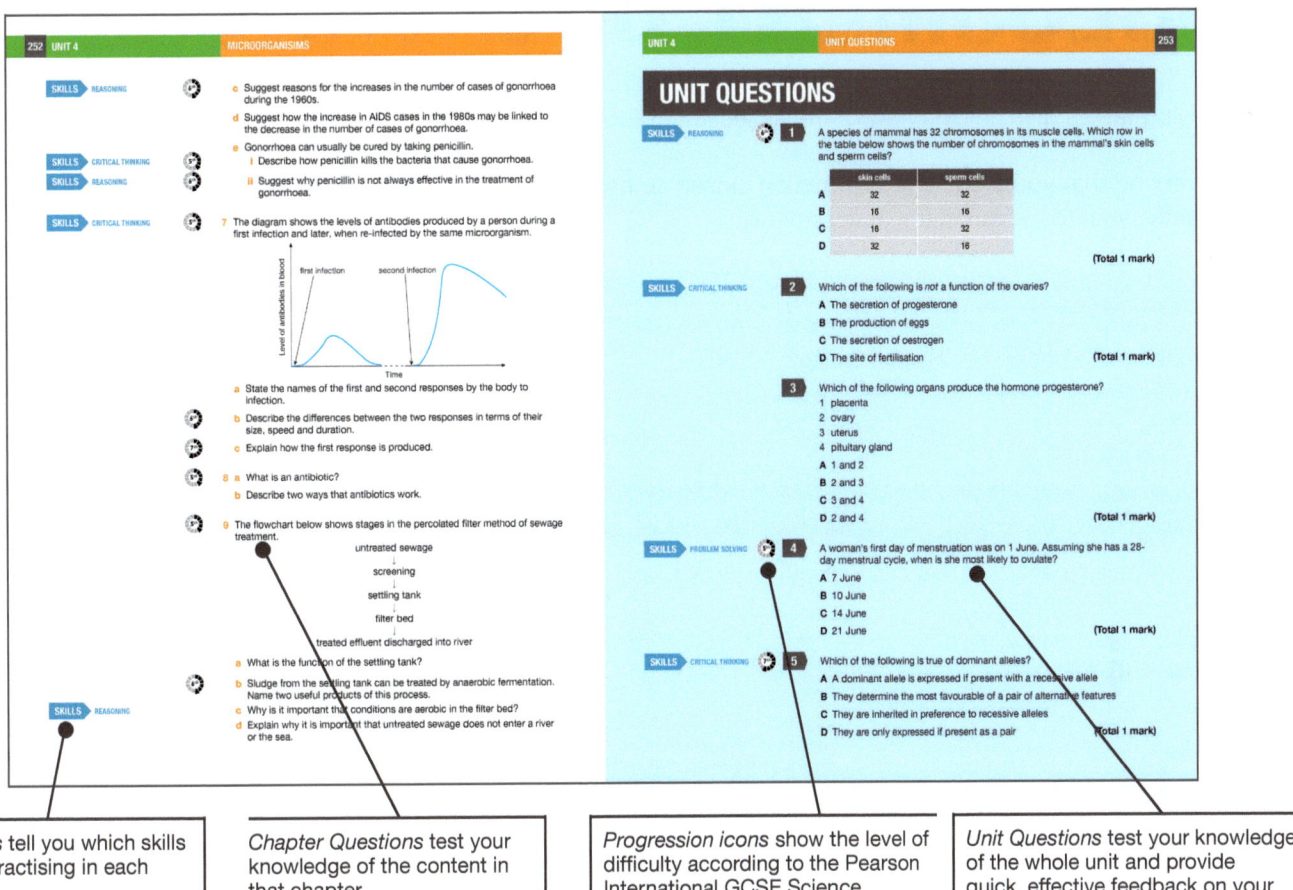

Skills tags tell you which skills you are practising in each question.

Chapter Questions test your knowledge of the content in that chapter.

Progression icons show the level of difficulty according to the Pearson International GCSE Science Progression Scale.

Unit Questions test your knowledge of the whole unit and provide quick, effective feedback on your progress.

ASSESSMENT OVERVIEW

The following tables give an overview of the assessment for this course.

We recommend that you study this information closely to help ensure that you are fully prepared for this course and know exactly what to expect in the assessment.

PAPER 1	SPECIFICATION	PERCENTAGE	MARK	TIME	AVAILABILITY
Written examination paper, externally set and assessed by Pearson Paper code 4HB1/01	Human Biology	50%	90	1 hour 45 mins	January and June examination series First assessment June 2019
PAPER 2	**SPECIFICATION**	**PERCENTAGE**	**MARK**	**TIME**	**AVAILABILITY**
Written examination paper, externally set and assessed by Pearson Paper code 4HB1/02	Human Biology	50%	90	1 hour 45 mins	January and June examination series First assessment June 2019

ASSESSMENT OBJECTIVES AND WEIGHTINGS

ASSESSMENT OBJECTIVE	DESCRIPTION	% IN INTERNATIONAL GCSE
AO1	Knowledge and understanding of human biology	38%–42%
AO2	Application of knowledge and understanding, analysis and evaluation of human biology	38%–42%
AO3	Experimental skills, analysis and evaluation of data and methods in human biology	19%–21%

EXPERIMENTAL SKILLS

In the assessment of experimental skills, students may be tested on their ability to:

- solve problems set in a practical context
- apply scientific knowledge and understanding in questions with a practical context
- devise and plan investigations, using scientific knowledge and understanding when selecting appropriate techniques
- demonstrate or describe appropriate experimental and investigative methods, including safe and skilful practical techniques
- make observations and measurements with appropriate precision, record these methodically and present them in appropriate ways
- identify independent, dependent and control variables
- use scientific knowledge and understanding to analyse and interpret data to draw conclusions from experimental activities that are consistent with the evidence
- communicate the findings from experimental activities, using appropriate technical language, relevant calculations and graphs
- assess the reliability of an experimental activity
- evaluate data and methods taking into account factors that affect accuracy and validity.

CALCULATORS

Students are expected to take a suitable calculator into the examinations. Calculators with QWERTY keyboards or that can retrieve text or formulae will not be permitted.

| CELLS 03 | MOVEMENT OF SUBSTANCES INTO AND OUT OF CELLS 24 | BIOLOGICAL MOLECULES 32 |

UNIT 1

Humans are composed of microscopic units known as cells, which are the 'building blocks' of life. Cells have a number of features in common, which allow them to grow, reproduce and generate more cells. In Chapter 1, you will start by looking at the structure and function of cells. You will also consider the role of DNA, which contains the genetic instructions for the development and functions of the body.

Cells need a supply of raw materials in order to function, and they produce other materials as waste products. In Chapter 2, you will learn about the ways in which these substances are exchanged between a cell and its surroundings. In Chapter 3, you will study the chemistry of cells – the structure and function of the different molecules that make up the human body.

1 CELLS

There are structural features that are common to the cells of all living organisms. In this chapter, you will find out about the structure and function of human cells, and how they are organised into tissues and organs. You will also learn about the role of the genetic material in cells – the DNA – and the principles of genetic engineering.

LEARNING OBJECTIVES

- Recognise cell structures as seen with a light microscope and transmission electron microscope, including the nucleus, chromosomes, cell membrane, mitochondria, endoplasmic reticulum and ribosomes
- Describe the functions of the nucleus, chromosomes, cell membrane, mitochondria, endoplasmic reticulum and ribosomes
- Describe the structure of a DNA molecule as two strands coiled to form a double helix, containing nucleotides, strands linked by complementary bases, and bases linked by hydrogen bonds
- Describe the process of DNA replication as the separation of DNA strands and the formation of a new strand by complementary base pairing of nucleotides, including the role of DNA polymerase
- Understand that a gene is a length of DNA containing a sequence of bases coding for a specific protein
- Know that RNA is a second type of nucleic acid that has the following features: single stranded, contains ribose, contains uracil; and that RNA is used to take information from DNA in the nucleus to the ribosomes for the synthesis of proteins
- Describe protein synthesis as:
  - transcription – the formation of mRNA in the nucleus and the transfer of mRNA to ribosomes in the cytoplasm
  - translation of the genetic code by tRNA from mRNA codons; the formation of a polypeptide chain using amino acids
- Understand that a DNA mutation involves a change in the sequence of bases that could lead to a change in the amino acid sequence and thus a change in the phenotype of an individual

- Understand that mitosis occurs during growth, repair, cloning and asexual reproduction
- Know the four main stages of mitosis – prophase, metaphase, anaphase and telophase – which result in the production of two genetically identical diploid daughter cells
- Understand that cells are grouped into tissues and tissues are organised into organs
- Describe the structure of bone, muscle (voluntary, involuntary and cardiac, as observed under a light microscope), blood, nervous tissue, and epithelium (squamous and ciliated, with reference to cells lining the cheek and trachea)*
- Describe the structure of cells specialised for reproduction (egg (ovum) and sperm) and relate their structure to their function*
- Know that there are different types of stem cell, including embryonic and adult stem cells, which have the ability to develop into other body cells
- Describe the advantages, disadvantages and ethics in the research and use of embryonic and adult stem cells
- Outline the principles of genetic engineering, including the production of genetically modified bacteria to produce human insulin, and the production of genetically modified plants to produce vaccines (e.g. hepatitis B) and to improve health (e.g. 'golden rice' to increase vitamin A in the diet)

* These specialised cells and tissues are described in more detail in later chapters.

CELL STRUCTURE

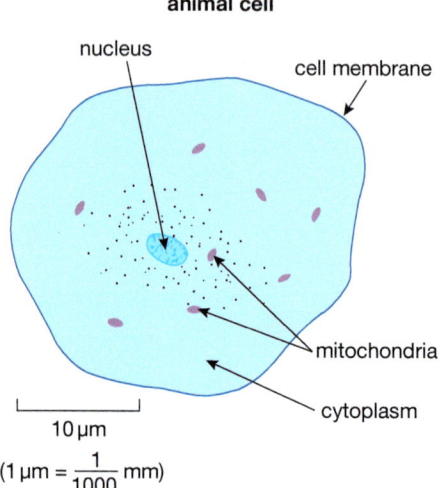

▲ Figure 1.1 The structure of a generalised animal cell, as seen through a light microscope.

The basic building block of living organisms is the **cell**. The human body is composed of countless millions of cells. There are many different types of cell, which are specialised so they can carry out particular functions in the body. Despite these differences, certain features are the same in most cells. Figure 1.1 shows some of the structures present in a typical animal cell.

The living material that makes up a cell is called **cytoplasm**. It has a texture rather like sloppy jelly, in other words, somewhere between a solid and a liquid. Unlike a jelly, it is not made of one substance; rather, it is a complex material that contains many different structures called **organelles**. You cannot see many of these structures under an ordinary light microscope. An electron microscope has a much higher magnification and can show the details of these parts of the cell (Figure 1.2).

The largest organelle in the cell is the **nucleus**. Nearly all cells have a nucleus, with a few exceptions, such as red blood cells. The nucleus controls the activities of the cell. It contains **chromosomes** (46 in human body cells) which carry the genetic material or **genes**. Genes control the activities in the cell by determining which proteins the cell can make (see below). One very important group of proteins found in cells is **enzymes** (see Chapter 3). Enzymes control chemical reactions that take place in the cytoplasm.

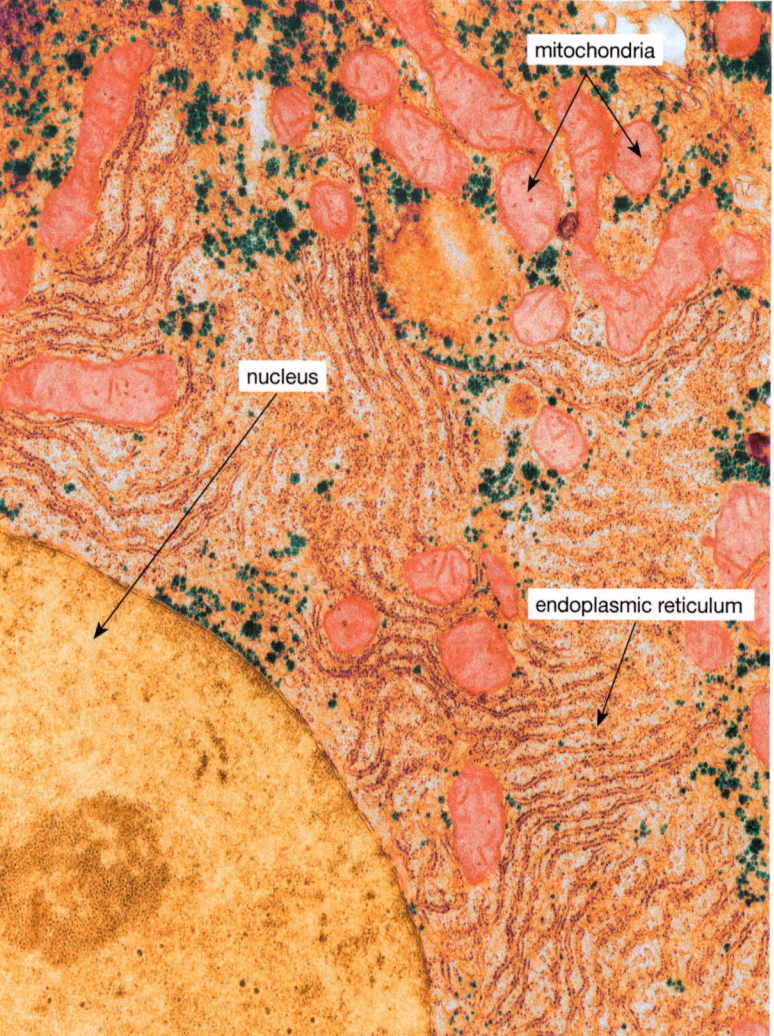

All cells are surrounded by a **cell membrane**, sometimes called the cell *surface* membrane to distinguish it from other membranes inside the cell. This is a thin layer like a 'skin' on the surface of the cell. It forms a boundary between the cytoplasm of the cell and the outside. However, it is not a complete barrier. Some chemicals can pass into the cell and others can pass out of it. We say that the membrane is **partially permeable**. In fact, as you will see, the membrane can actively control the movement of some substances. Because of this, it is also described as **selectively permeable**.

There are many other membranes inside a cell. Throughout the cytoplasm, there is a network of membranes called the **endoplasmic reticulum** (**ER**). In places, the endoplasmic reticulum is covered with tiny granules called **ribosomes**. These are the organelles where proteins are made or *synthesised* (see 'The stages of protein synthesis' later in this chapter). The spaces between the membranes of the endoplasmic reticulum act as a transportation system, sending protein to the part of the cell where it is needed.

◀ Figure 1.2 The higher magnification of the electron microscope lets us see more detail in the cell. This photograph shows a small part of a liver cell, with the nucleus in the bottom left-hand corner. The colours are not real – they have been added later to show up the different structures. You can just make out the ribosomes on the endoplasmic reticulum.

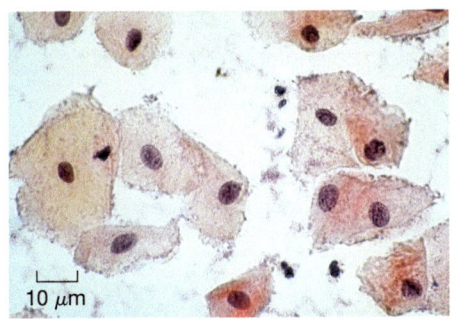

▲ Figure 1.3 Cells from the inside lining of a human cheek. The cells are stained to show up the details of structures such as the nucleus.

One organelle that is found in the cytoplasm of nearly all living cells is the **mitochondrion** (plural mitochondria). In cells that need a lot of energy, such as muscle or nerve cells, there are many mitochondria. This gives us a clue as to their role. They perform some of the reactions of **respiration**, releasing energy that the cell can use (see Chapter 5). Most of the energy from respiration is released in the mitochondria.

Figure 1.3 shows some cells from the lining of a human cheek. They were obtained by gently rubbing a cotton swab on the inside of a person's mouth and transferring the cells to a slide. They are stained with a dye to show them more clearly. How many different organelles can you identify?

CHROMOSOMES, GENES AND DNA

The chemical that is the basis of inheritance is **deoxyribonucleic acid** or **DNA**. DNA is usually found in the nucleus of a cell, in structures called chromosomes (Figure 1.4). A section of DNA that determines a particular feature is called a gene. Genes determine a person's characteristics by instructing cells to produce particular proteins (see below).

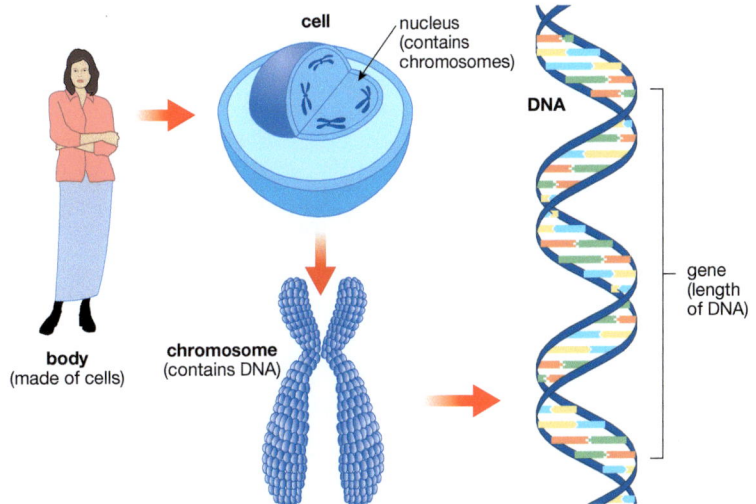

▲ Figure 1.4 Our genetic make-up.

Each chromosome contains one DNA molecule. The DNA is folded and coiled so that it can be packed into a small space. The DNA is coiled around proteins called **histones** (Figure 1.5).

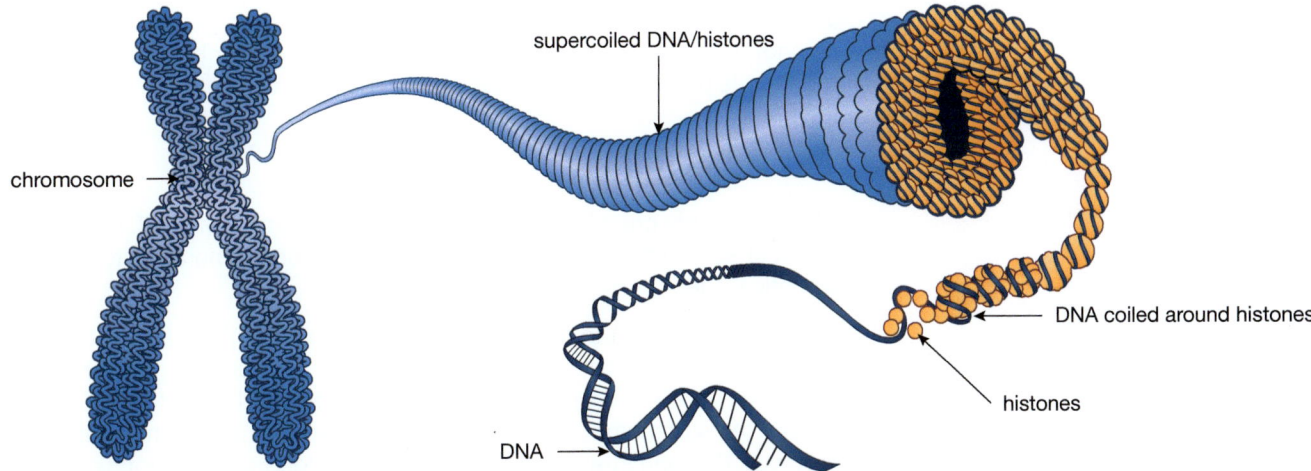

▲ Figure 1.5 The structure of a chromosome.

THE STRUCTURE OF DNA

A molecule of DNA is made from two strands of molecular groups called **nucleotides** (Figure 1.6).

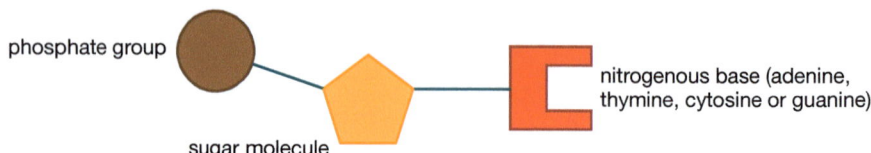

▲ Figure 1.6 The structure of a single nucleotide.

Each nucleotide contains a sugar called deoxyribose, a phosphate group, and a nitrogen-containing group called a **base**. There are four bases: adenine (A), thymine (T), cytosine (C) and guanine (G) (Figure 1.7).

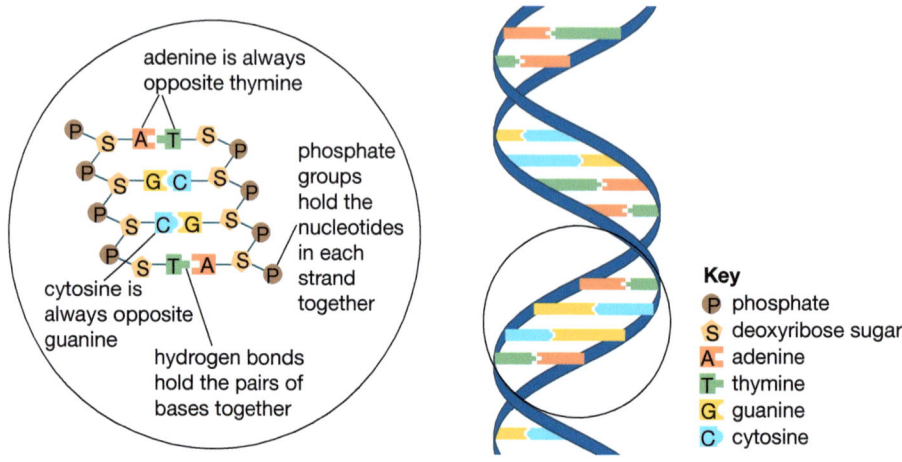

▲ Figure 1.7 Part of a molecule of DNA.

> **KEY POINT**
>
> One consequence of the base-pairing rule is that, in each molecule of DNA, the amounts of adenine and thymine are equal, as are the amounts of cytosine and guanine.

Notice that, in the two strands, nucleotides with adenine are always opposite nucleotides with thymine, while cytosine is always opposite guanine. Adenine and thymine are *complementary* bases, as are cytosine and guanine. Complementary bases always bind with each other and never with any other base. This is known as the *base-pairing rule*. The two strands are held together by hydrogen bonds between the complementary base pairs. These are weak bonds between hydrogen atoms on one base and oxygen or nitrogen atoms on another base. They are easily broken, allowing the chains to separate. This property is used when DNA makes a copy of itself.

DNA REPLICATION

DNA is the only chemical that can make exact copies of itself. Because of this, it is able to pass genetic information from one generation to the next as a 'genetic code'.

When a cell is about to divide (see 'Mitosis' later in this chapter) it must first make an exact copy of each DNA molecule in the nucleus. This process is called **replication**. As a result, each 'daughter cell' that is formed receives exactly the same amount and type of DNA. Figure 1.8 summarises this process. The new strands of DNA are assembled from nucleotides under the control of an enzyme called **DNA polymerase**.

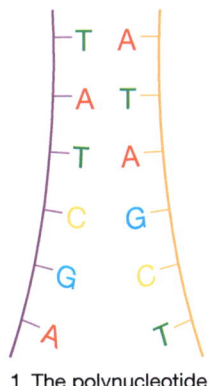 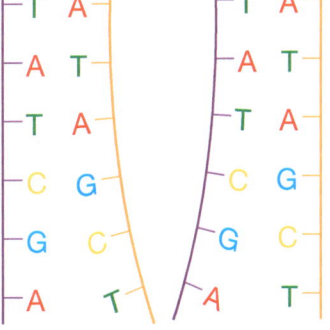 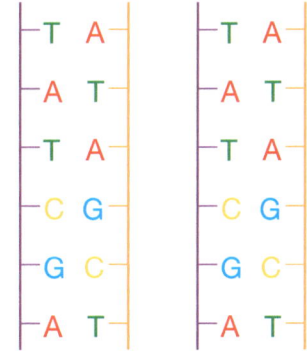

1. The polynucleotide strands of DNA separate.
2. Each strand acts as a template for the formation of a new strand of DNA.
3. DNA polymerase assembles nucleotides into two new strands according to the base-pairing rule.
4. Two identical DNA molecules are formed – each contains a strand from the parent DNA and a new complementary strand.

▲ Figure 1.8 How DNA replicates itself.

THE GENETIC CODE

Only one of the strands of a DNA molecule actually codes for the manufacture of proteins in a cell. This strand is called the **template strand**. The other strand is called the non-template strand.

> **DID YOU KNOW?**
> A 'template' is a pattern that can be used to make something. For example, a dress template is a paper pattern for cutting out the material of a dress.

Many of the proteins manufactured are enzymes, which go on to control processes within the cell. Some proteins are structural, for example, keratin in the skin or myosin in muscles. Other proteins have particular functions, such as haemoglobin and some hormones.

Proteins are made of chains of amino acids. A sequence of *three* bases in the template strand of the DNA codes for *one* amino acid. For example, the base sequence TGT codes for the amino acid cysteine. Because three bases are needed to code for one amino acid, the DNA code is a *triplet code*. The sequence of bases that codes for *all* the amino acids in a protein is a gene (Figure 1.9).

> **KEY POINT**
> A gene is a section of a molecule of DNA that codes for a specific protein.

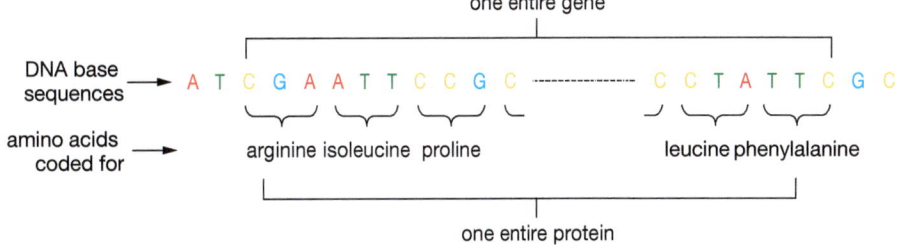

▲ Figure 1.9 The triplet code.

The triplets of bases that code for individual amino acids are the same in all organisms. The base sequence TGT codes for the amino acid cysteine in humans, bacteria, bananas, fish, or any other organism you can think of. The DNA code is a *universal* code.

THE STAGES OF PROTEIN SYNTHESIS

DNA stays in the nucleus but protein synthesis takes place in the cytoplasm. This means that, before proteins can be made, the genetic code must be copied and transferred out from the nucleus to the cytoplasm. This is carried out by a different kind of nucleic acid called **ribonucleic acid** (**RNA**).

DID YOU KNOW?
Ribose and deoxyribose are very similar in structure. Ribose contains an extra oxygen atom. Similarly, the bases uracil and thymine are very similar in structure.

DID YOU KNOW?
There is a third type of RNA called ribosomal RNA (rRNA). Ribosomes are made of RNA and protein.

There are three main differences between DNA and RNA:

- DNA is a double helix, RNA is a single strand
- DNA contains the sugar deoxyribose, RNA contains ribose
- RNA contains the base uracil (U) instead of thymine (T).

Two types of RNA take part in protein synthesis:

- **messenger RNA (mRNA):** forms a copy of the DNA code
- **transfer RNA (tRNA):** carries amino acids to the ribosomes to make the protein.

Protein synthesis takes place in two stages, called **transcription** and **translation**.

TRANSCRIPTION

Transcription happens in the nucleus. In a chromosome, part of the DNA double helix unwinds and 'unzips', so the two strands separate, exposing the bases along the template strand (Figure 1.10).

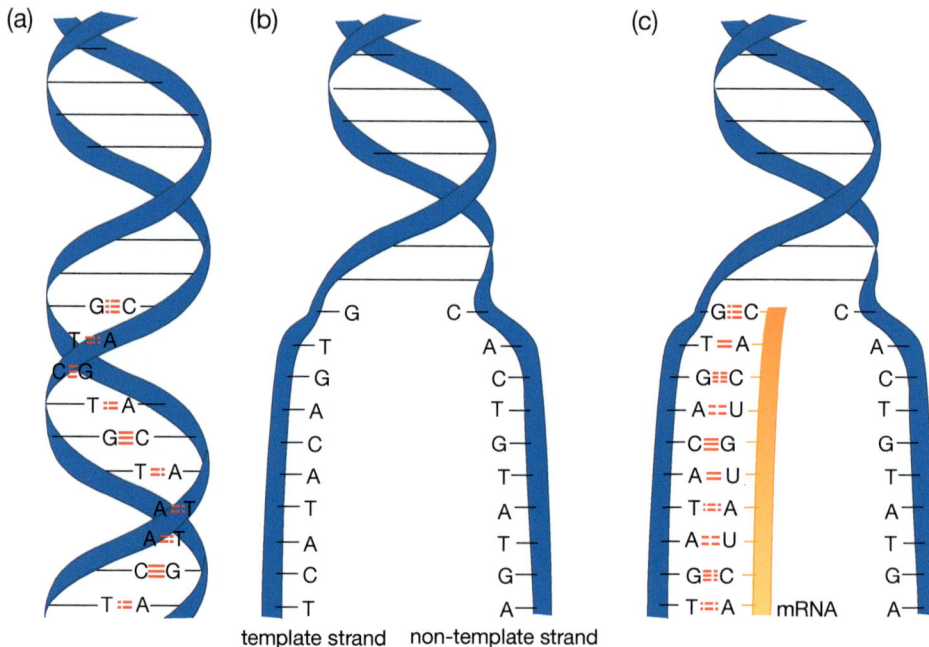

▲ Figure 1.10 Transcription. (a) The DNA double helix, showing some base pairs. (b) The two strands of the DNA have separated and unwound. (c) A short length of mRNA has formed. (The mRNA responsible for forming a whole protein would be much longer than this.)

The template strand of the DNA forms a framework upon which a molecule of mRNA is formed. The building blocks of the mRNA are RNA nucleotides. They line up alongside the template strand according to the complementary base-pairing rules (Table 1.1).

The RNA nucleotides link up one at a time to form an mRNA molecule. Bonds form between the ribose and phosphate groups, joining them together to make the sugar–phosphate backbone of the molecule. When a section of DNA corresponding to a protein (a gene) has been transcribed, the mRNA molecule leaves the DNA and passes out of the nucleus to the cytoplasm. It leaves through pores (holes) in the nuclear membrane. The DNA helix then 'zips up' again. Because of complementary base pairing, the triplet code of the DNA is converted into a triplet code in the mRNA.

KEY POINT
An RNA nucleotide consists of ribose sugar, a phosphate and one of four bases (C, G, A or U).

▼ Table 1.1 Base-pairing rules in transcription.

Base on DNA	Base on mRNA
G	C
C	G
T	A
A	U

TRANSLATION

Converting the code in the mRNA into a protein is called translation. This takes place at the ribosomes. By this stage, the code consists of sets of three bases in the mRNA (e.g. AUG, CCG, ACA). These triplets of bases are called **codons**. Each codon codes for a particular amino acid; for example, CCU codes for the amino acid proline, and AUG codes for methionine.

The mRNA molecule attaches to a ribosome. Now the tRNA molecules begin their part in the process. Each tRNA molecule has an **anticodon** of three bases at one end of the molecule. This is complementary to a particular codon on the mRNA. At the other end of the tRNA molecule is a site where a specific amino acid can attach (Figure 1.11). This means that there is a particular tRNA molecule for each type of amino acid. The tRNA molecule carries its amino acid to the ribosome, where its specific anticodon links up with the three bases of the corresponding mRNA codon.

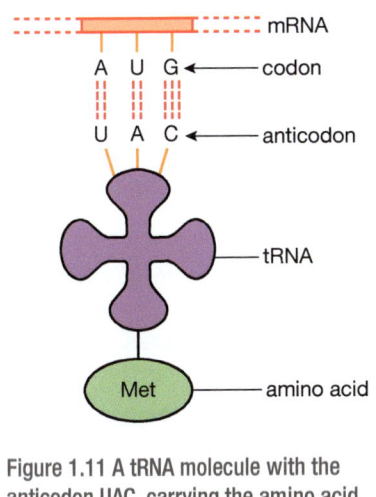

▲ Figure 1.11 A tRNA molecule with the anticodon UAC, carrying the amino acid methionine. The anticodon is complementary to the codon AUG on the mRNA.

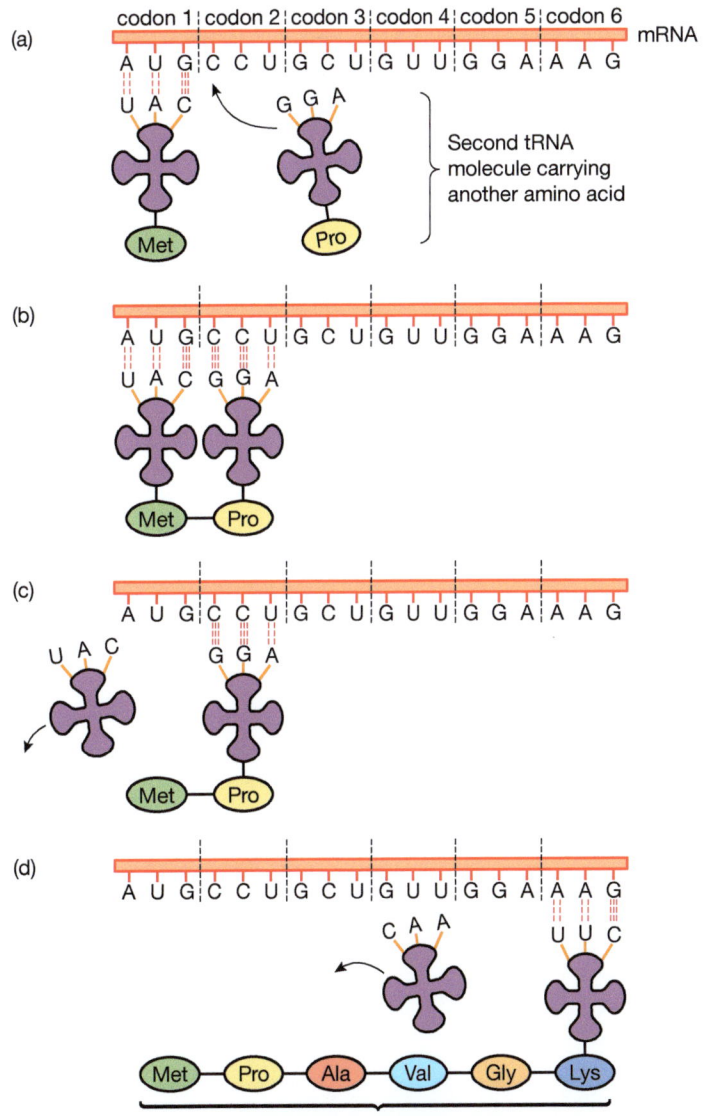

▲ Figure 1.12 Translation. (a) A tRNA molecule carrying the first amino acid is attached to the first codon on the mRNA. A second tRNA is arriving, carrying the second amino acid. (b) Two tRNA molecules are attached to the mRNA. A bond is formed between the two amino acids. (c) The first tRNA molecule is released. The process continues as more tRNA molecules bring amino acids to the ribosome. (d) The situation after six tRNA molecules have brought amino acids. A chain of amino acids called a polypeptide is starting to form – this is the beginning of a protein molecule.

> **DID YOU KNOW?**
> Protein synthesis is a process that uses up a lot of the chemical energy produced in a cell.

> **KEY POINT**
> Summary of protein synthesis:
> The order of bases in the template strand of the DNA forms the genetic code. The code is converted into the sequence of bases in the mRNA. In the cytoplasm, the sequence of mRNA bases is used to determine the position of amino acids in a protein.

This interaction between mRNA and tRNA is the basis of translation. The process is shown in Figure 1.12.

Translation takes place as follows:

- The first tRNA to bind at the mRNA does so at the 'start codon', which always has the base sequence AUG. This codes for the amino acid methionine.
- Another tRNA brings along a second amino acid. The anticodon of the second tRNA binds to the next codon on the mRNA.
- A bond forms between the methionine and the second amino acid.
- The first tRNA molecule is released and goes off to collect another amino acid.
- More tRNA molecules arrive at the mRNA and add their amino acids to the growing chain, forming a polypeptide.

At the end of the chain, a 'stop codon' tells the 'translation machinery' that the protein is complete, and it is released.

There are 20 different amino acids, so there must be at least 20 different codons (and 20 different anticodons). In fact, there are more than this, because some amino acids use more than one triplet code. For example, the mRNA codons GGU, GGC, GGA and GGG all code for the amino acid glycine.

GENE MUTATIONS – WHEN DNA MAKES MISTAKES

A **mutation** is a random change in the DNA of a cell. Sometimes, when DNA is replicating, mistakes are made and the wrong nucleotide is used. The result is a gene mutation, which can change the sequence of the bases in a gene. In turn, this can lead to the gene coding for the wrong amino acid in a protein. There are several ways in which gene mutations can occur (Figure 1.13).

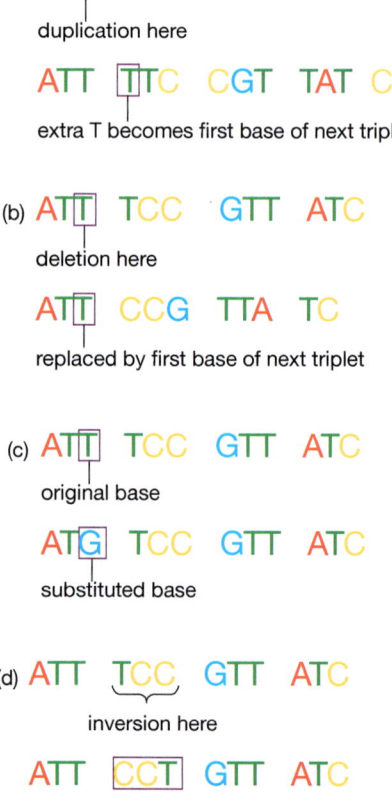

▲ Figure 1.13 Gene mutations: (a) duplication; (b) deletion; (c) substitution; (d) inversion.

- **Duplication:** In duplication, Figure 1.13(a), the nucleotide is inserted twice instead of once. This means that the entire base sequence is altered, because each triplet after the point where the mutation occurs is changed. The whole gene is different and will code for an entirely different protein.
- **Deletion:** In deletion, Figure 1.13(b), a nucleotide is missed out. Again, the entire base sequence is altered. Each triplet after the mutation is changed and the whole gene is different. As with duplication, the gene will now code for an entirely different protein.
- **Substitution:** In substitution, Figure 1.13(c), a different nucleotide is used. The triplet of bases in which the mutation occurs is changed and it *may* code for a different amino acid. If it does, the structure of the protein molecule will be different. This may be enough to produce a significant change in the functioning of the protein, or it may mean that the protein does not function at all. However, most amino acids have more than one code, so the new triplet may not code for a different amino acid. If this is the case, the protein will have its normal structure and function.
- **Inversion:** In inversion, Figure 1.13(d), the sequence of the bases in a triplet is reversed. The effects are similar to substitution. Only one triplet is affected and this may or may not result in a different amino acid and altered protein structure.

Mutations that occur in body cells, such as those in the heart, intestines or skin, will only affect the particular cell in which they occur. If the mutation is very harmful, the cell will die and the mutation will be lost. If the mutation does not significantly affect the functioning of the cell, the cell may not die. If the cell then divides, a group of cells containing the mutant gene will be formed. When the person dies, however, the mutation will be lost – it will not be passed to their children. Only mutations in the sex cells (**gametes**), or in the cells that divide to form gametes, can be passed on to the next generation. This is how genetic diseases begin.

CELL DIVISION

There are two kinds of cell division – **mitosis** and **meiosis**.

In most parts of the body, cells need to divide so that organisms can grow and replace old or damaged cells. The cells that are produced by this type of cell division should be exactly the same as the cells they are replacing. This is the most common form of cell division and is called mitosis. Mitosis forms all the cells in our bodies except the sex cells.

Only in the sex organs is cell division different. Here, some cells divide to produce sex cells or gametes, which contain only half the original number of chromosomes. When male and female gametes join together at **fertilisation**, the resulting cell (called a **zygote**) will contain the full set of chromosomes and can then divide and grow into a new individual. This type of cell division is called meiosis and is described in Chapter 11.

Human body cells have 46 chromosomes in 23 pairs called **homologous pairs**. These body cells are **diploid** cells – they have *two* copies of each chromosome. The sex cells have 23 chromosomes (only one copy of each chromosome); they are **haploid** cells.

MITOSIS

When a 'parent' cell divides, it produces *daughter cells*. Mitosis produces two daughter cells that are genetically identical to the parent cell – both daughter cells have the same number and type of chromosomes as the parent cell.

To achieve this, the dividing cell must copy each chromosome before it divides. The DNA replicates and more proteins are added to the structure. Each daughter cell can then receive a copy of each chromosome (and each molecule of DNA) when the cell divides. If it does not do this, the daughter cells will not contain all the genes.

A number of stages occur when a cell divides by mitosis. These are shown in Figure 1.14.

(a) prophase

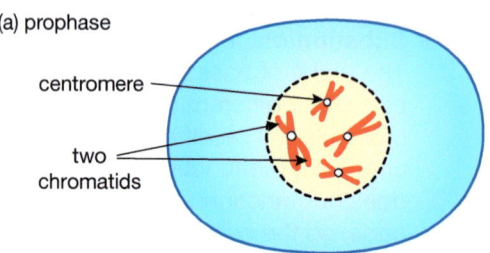

Before mitosis the DNA replicates and the chromosomes form two exact copies called chromatids. During the first stage of mitosis (prophase) the chromatids become visible, joined at a centromere. The nuclear membrane breaks down.

(b) metaphase

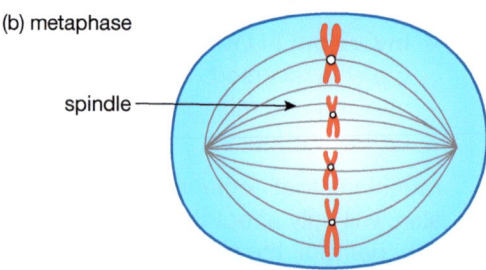

During metaphase a structure called the spindle forms. The chromosomes line up at the 'equator' of the spindle, attached to it by their centromeres.

(c) anaphase

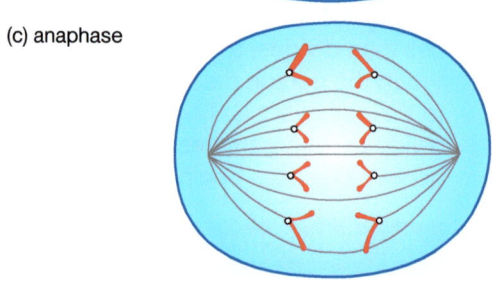

During anaphase, the spindle fibres shorten and pull the chromatids to the opposite ends ('poles') of the cell. The chromatids separate to become the chromosomes of the two daughter cells.

(d) telophase

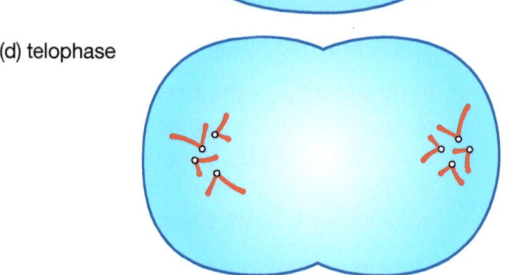

In the last stage (telophase) two new nuclei form at the poles of the cell. The cytoplasm starts to divide to produce two daughter cells. Both daughter cells have a copy of each chromosome from the parent cell.

▲ Figure 1.14 The stages of mitosis. For simplicity, the cell shown contains only two homologous pairs of chromosomes (one long pair, one short).

It is easiest to see mitosis in plant cells, because these are usually large and well defined by their cell walls. Figure 1.15 is a photograph of some cells from the root tip of an onion. Cells in this region divide by mitosis as the root grows. Although these are plant cells, the process is very similar in human cells.

Each daughter cell formed by mitosis is diploid, receiving a copy of every chromosome (and therefore every gene) from the parent cell. Each daughter cell is genetically identical to the others. A group of genetically identical cells produced by mitosis is called a **clone**. Apart from the sex cells, all the cells in our body are clones. They are formed by mitosis and contain copies of all of our chromosomes and genes.

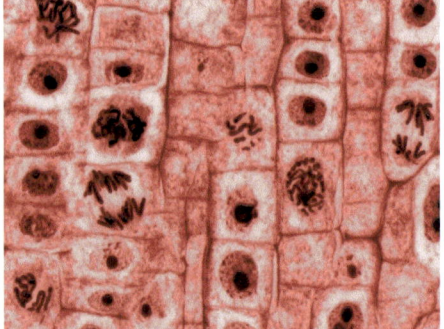

▲ Figure 1.15 Cells in the root tip of an onion dividing by mitosis. Can you identify any of the stages shown in Figure 1.14?

> **DID YOU KNOW?**
> You may have heard the words 'clone' and 'cloning' used to describe the production of entire organisms (animals and plants) from body cells, by mitosis. Many plants reproduce naturally from parts of leaves or roots broken off from the parent plant. This is an example of cloning, and is used commercially to grow plants. Cloning animals is more difficult, but several species have been grown artificially in the laboratory by mitosis from body cells. Since this type of reproduction does not involve sex cells, it is called asexual reproduction.

DIFFERENTIATION OF CELLS

A human begins life as a zygote, which divides by mitosis to form two cells, then four, then eight and so on, until an embryo is formed, containing many millions of cells (Figure 1.16).

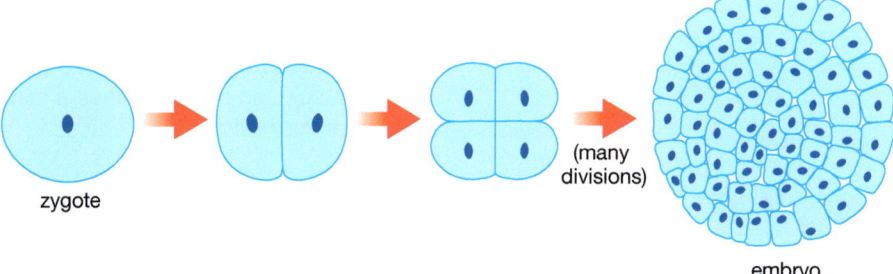

▲ Figure 1.16 A human embryo grows by mitosis.

As the developing embryo grows, cells become specialised to perform particular roles. This specialisation is also under the control of the genes, and is called differentiation. Different kinds of cell develop depending on where they are located in the embryo, for example, a nerve cell in the spinal cord, or an epidermal cell in the outer layer of the skin (Figure 1.17).

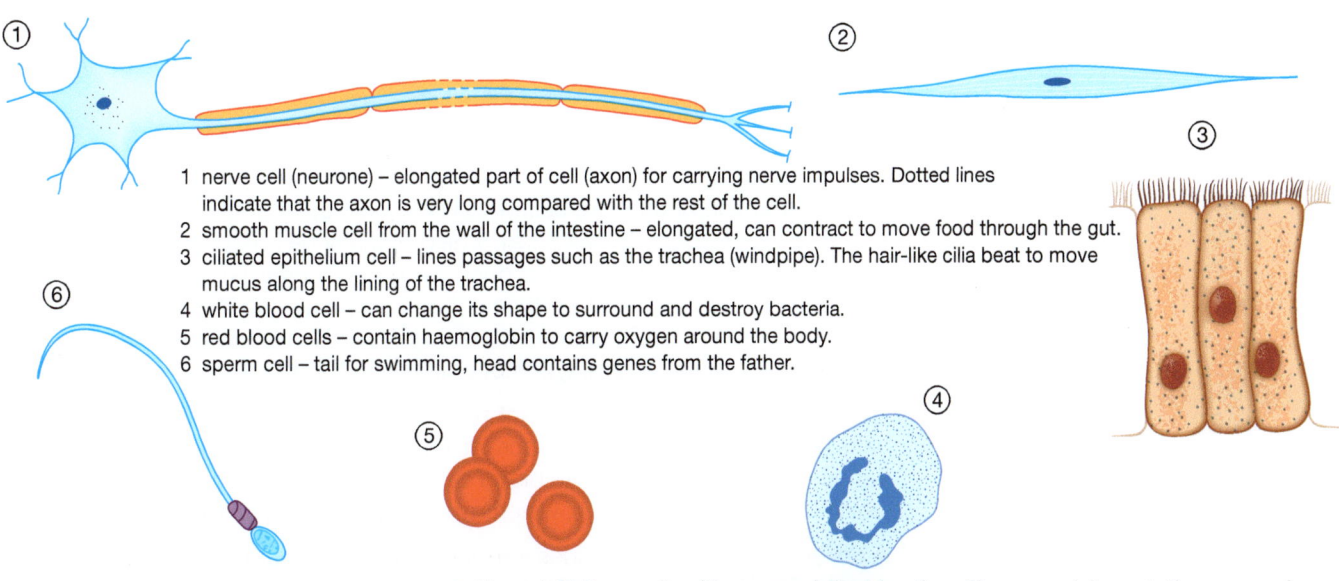

1. nerve cell (neurone) – elongated part of cell (axon) for carrying nerve impulses. Dotted lines indicate that the axon is very long compared with the rest of the cell.
2. smooth muscle cell from the wall of the intestine – elongated, can contract to move food through the gut.
3. ciliated epithelium cell – lines passages such as the trachea (windpipe). The hair-like cilia beat to move mucus along the lining of the trachea.
4. white blood cell – can change its shape to surround and destroy bacteria.
5. red blood cells – contain haemoglobin to carry oxygen around the body.
6. sperm cell – tail for swimming, head contains genes from the father.

▲ Figure 1.17 Some cells with very specialised functions. They are not drawn to the same scale.

> **DID YOU KNOW?**
> Throughout this book, you will read about *adaptations*. An **adaptation** is a way that the structure of a cell, tissue or organ is suited (adapted) to its function.

What is hard to understand about this process is that, through mitosis, all the cells of the body have the *same* genes. For cells to function differently, they must produce different proteins, and different genes code for the production of these different proteins. How is it that some genes are 'switched on' and others are 'switched off' to produce different cells? The answer to this question is very complicated, and scientists are only just beginning to understand it.

CELLS, TISSUES AND ORGANS

Cells that have a similar function are grouped together as **tissues**. For example, the muscle of your arm contains millions of muscle cells, all specialised for one function – contracting to move the arm bones (see Chapter 7). This is muscle tissue or, more accurately, **voluntary muscle** tissue. The word 'voluntary' refers to the fact that the contraction of muscles like this is under the conscious control of the brain. The smooth muscle cell shown in Figure 1.17 makes up **involuntary muscle** tissue, since the muscles in the gut are not consciously controlled by the brain. Involuntary muscle is present in the walls of organs such as the intestine, bladder and blood vessels. There is a third type of muscle tissue called **cardiac muscle**, which makes up the muscular wall of the heart (see Chapter 6). It is interesting to note that all muscle (voluntary, involuntary and cardiac) contains the same special protein filaments that are able to bring about contraction. However, these filaments are arranged differently in each type of muscle, so the cells have a distinct structure that depends on the type of contraction carried out by the tissue (Table 1.2).

▼ Table 1.2 The three types of muscle cell.

Type	Structure	Function
Voluntary	Striped (striated) due to alignment of protein filaments in the cell. Many nuclei per cell. Not branched.	Rapid contraction to move bones (also called skeletal muscle). Under voluntary control by the brain.
Involuntary	Non-striated because protein filaments are not aligned in the cell (hence also called smooth muscle). One nucleus per cell. Not branched. Cell tapered at ends.	Slow, rhythmic contraction in walls of gut, blood vessels etc. Not under voluntary control by the brain.
Cardiac	Striated. Many nuclei per cell. Branched cells forming a strong mesh-like network.	Only present in the heart. Contracts rhythmically and constantly throughout life without tiring. Not under voluntary control.

Tissues that line organs are called **epithelia** (singular epithelium). Figure 1.17 shows a **ciliated epithelium** cell, which has tiny hair-like projections called cilia. Cilia are able to beat (wave backwards and forwards) to move liquids along. There are several other types of epithelia, such as the flattened cells lining the human cheek (Figure 1.3). This is called a **squamous epithelium**. You will read about several types of epithelia in this book.

Bone is a tissue made of cells that secrete a hard material made of calcium salts (see Chapter 7). Other tissues include **blood** (Chapter 6), which is made of various types of red and white blood cells in a liquid matrix called plasma, and **nervous tissue** (Chapter 9), which makes up the brain, spinal cord and nerves.

A collection of different tissues carrying out a particular function is called an **organ**. The main organs of the human body are shown in Figure 1.18.

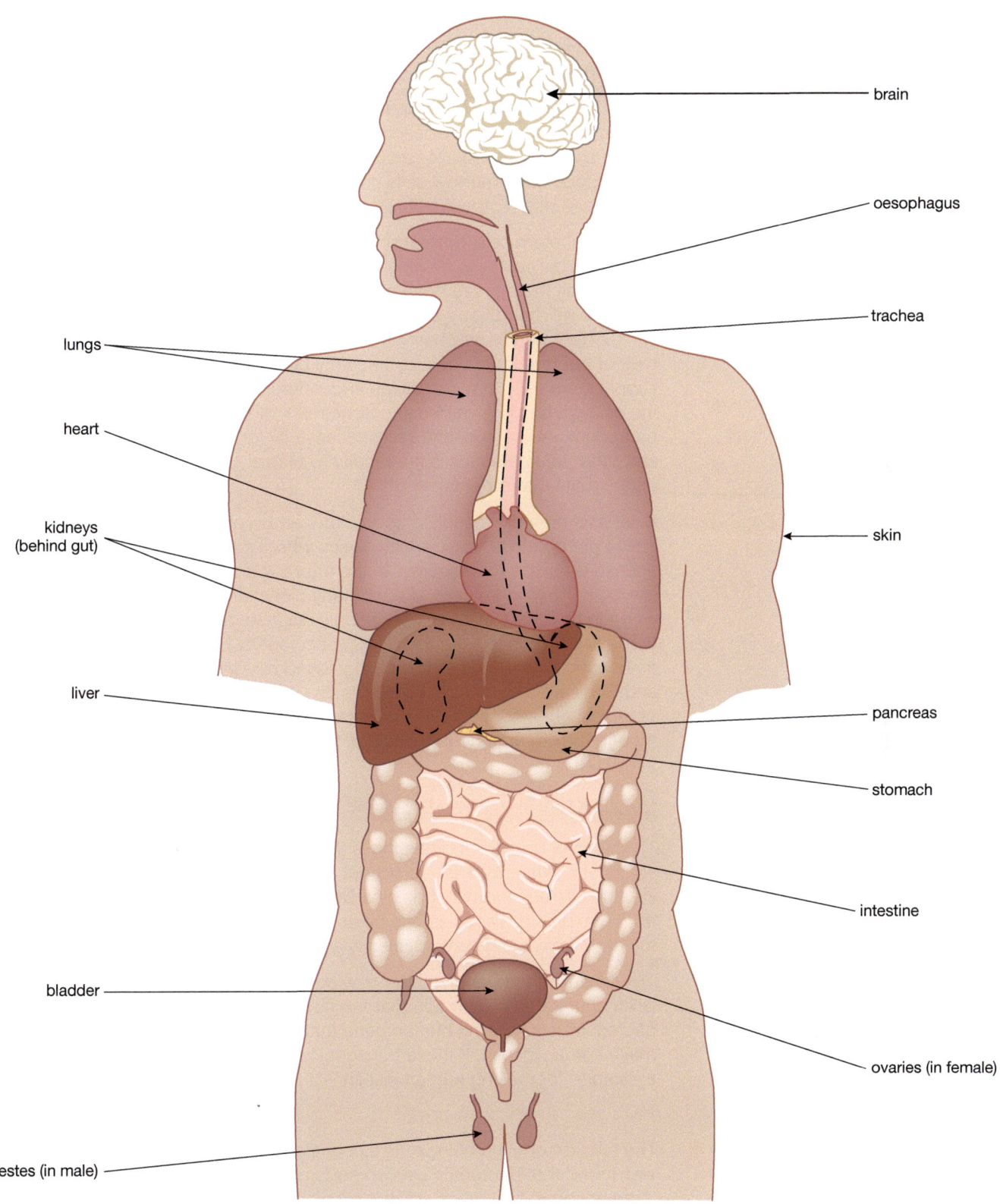

▲ Figure 1.18 Some of the main organs of the human body.

STEM CELLS

A **stem cell** is a cell that is able to divide many times by mitosis without undergoing differentiation. Later, it can differentiate into specialised cells such as muscle or nerves. In humans, there are two main types of stem cell:

- **Embryonic stem cells** are found in the early stage of development of the embryo. Embryonic stem cells can differentiate into any type of body cell.

- **Adult stem cells** are found in certain adult tissues, such as bone marrow, skin, and the lining of the intestine. They have lost the ability to differentiate into any type of cell but can form a number of specialised tissues. For example, bone marrow cells can divide many times but are only able to produce different types of red and white blood cell.

The use of stem cells to treat or prevent a disease, or to repair damaged tissues, is called **stem cell therapy**. At present, the only widely used types of stem cell therapy are bone marrow transplants. These are used to treat patients with conditions such as leukaemia (a type of blood cancer). Some cancer treatments use chemicals that kill cancer cells (chemotherapy) but this type of treatment also destroys healthy body cells. Bone marrow transplants supply stem cells that can divide and differentiate, replacing cells lost from the body during chemotherapy. Bone marrow transplants are now a routine procedure and have been used successfully for over 30 years. Bone marrow and other adult stem cells are readily available, but they have limited ability to differentiate into other types of cell.

Scientists are able to isolate and culture embryonic stem cells (Figure 1.19). These are obtained from fertility clinics where parents choose to donate their unused embryos for research. In the future, it is hoped that we will be able to use embryonic stem cells to treat many diseases such as diabetes, as well as brain disorders such as Parkinson's disease. Stem cells may also be able to repair nervous tissues damaged in accidents. So far, treatments using embryonic stem cells have not progressed beyond the experimental stage, and there are a number of problems. In particular, many people have moral or ethical objections to using cells from embryos for medical purposes, even though they might one day be used to cure many diseases.

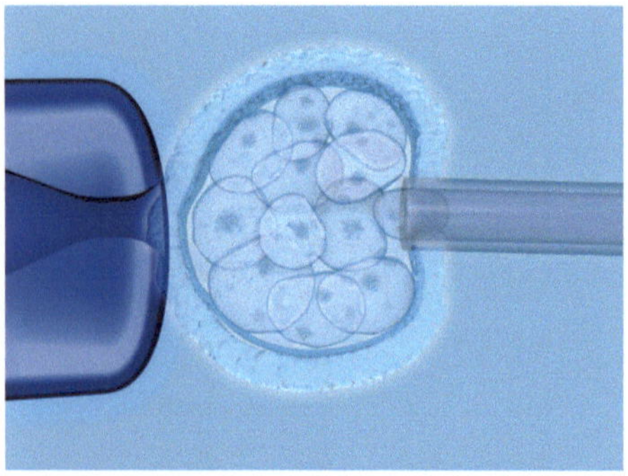

> **DID YOU KNOW?**
> Marrow is the soft tissue inside bones that produces blood cells. A bone marrow transplant is a medical operation that replaces unhealthy blood-forming cells with healthy ones.

▲ Figure 1.19 Extracting a stem cell from an embryo at an early stage of its development. The embryo consists of a ball of about 20 cells. A single cell is removed by drawing it into a fine glass capillary tube.

GENETIC ENGINEERING

The basis of genetic engineering is the production of **recombinant DNA**. A section of DNA – a gene – is cut out of the DNA of one species and inserted into the DNA of another. This new DNA is called 'recombinant' because DNA from two different species has been 'recombined'. The organism that receives the gene from a different species is a **transgenic** organism.

The organism that received the new gene now has an added capability. It will manufacture the protein that the new gene codes for. For example, if a bacterium receives the human gene that codes for insulin production, it will make human insulin. If these transgenic bacteria are cultured, they will become a 'factory' for making human insulin.

PRODUCING GENETICALLY MODIFIED (TRANSGENIC) BACTERIA

The breakthrough in being able to transfer DNA from cell to cell came when it was found that bacteria have two sorts of DNA – the DNA found in a bacterial 'chromosome' and much smaller circular pieces of DNA called **plasmids**.

UNIT 1   CELLS

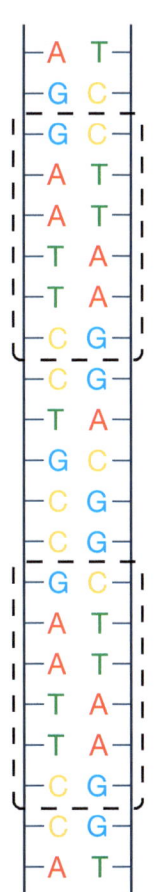

▲ Figure 1.20 Part of a DNA molecule containing the base sequence G-A-A-T-T-C. Notice that the sequence is present on both strands, but running in opposite directions.

> **DID YOU KNOW?**
> A bacterial chromosome is not like a human chromosome. It is a continuous loop of DNA rather than a strand. Also, the structure of a bacterial chromosome is simpler and does not contain the histone proteins present in eukaryotic chromosomes. The structure of a bacterial cell is described in Chapter 13.

Bacteria naturally 'swap' plasmids, and biologists found ways of transferring plasmids from one bacterium to another. The next stage was to find molecular 'scissors' and a molecular 'glue' that could cut out genes from one molecule of DNA and stick them back into another. Further research found the following enzymes that were able to do this:

- **Restriction endonucleases** (usually shortened to **restriction enzymes**) are enzymes that cut DNA molecules at specific points. Different restriction enzymes cut DNA at different places. They can be used to cut out specific genes from a molecule of DNA.
- **Ligases** (or DNA ligases) are enzymes that join the cut ends of DNA molecules.

Each restriction enzyme recognises a certain base sequence in a DNA strand. Wherever it encounters that sequence, it will cut the DNA molecule. Suppose a restriction enzyme recognises the base sequence G-A-A-T-T-C. It will only cut the DNA molecule if it can 'see' this base sequence on both strands. Figure 1.20 illustrates this.

Some restriction enzymes make a straight cut and the fragments of DNA they produce are said to have 'blunt ends' (Figure 1.21(a)). Other restriction enzymes make a staggered (step-shaped) cut. These produce fragments of DNA with overlapping ends with complementary bases (Figure 1.21(b)). These overlapping ends are called 'sticky ends' because fragments of DNA with exposed bases are more easily joined by ligase enzymes.

> **EXTENSION WORK**
> There is a lot more to producing recombinant DNA and transgenic bacteria than is described here. You could carry out some research to find out more about this subject.

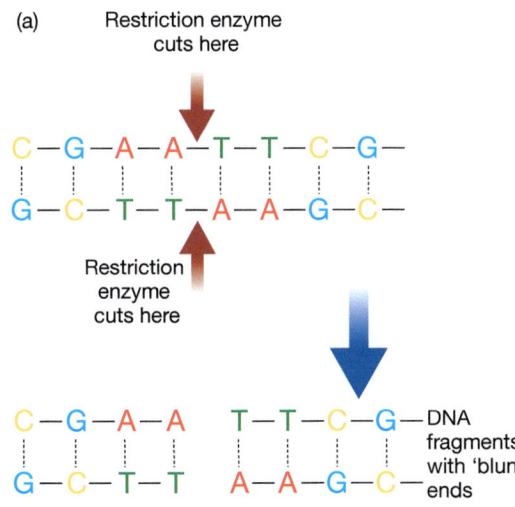

▲ Figure 1.21 How restriction enzymes cut DNA.

Biologists now had a method of transferring a gene from any cell into a bacterium. They could insert the gene into a plasmid and then transfer the plasmid into a bacterium. The plasmid is called a **vector** because it is the means of transferring the gene. The main processes involved in producing a transgenic bacterium are shown in Figure 1.22.

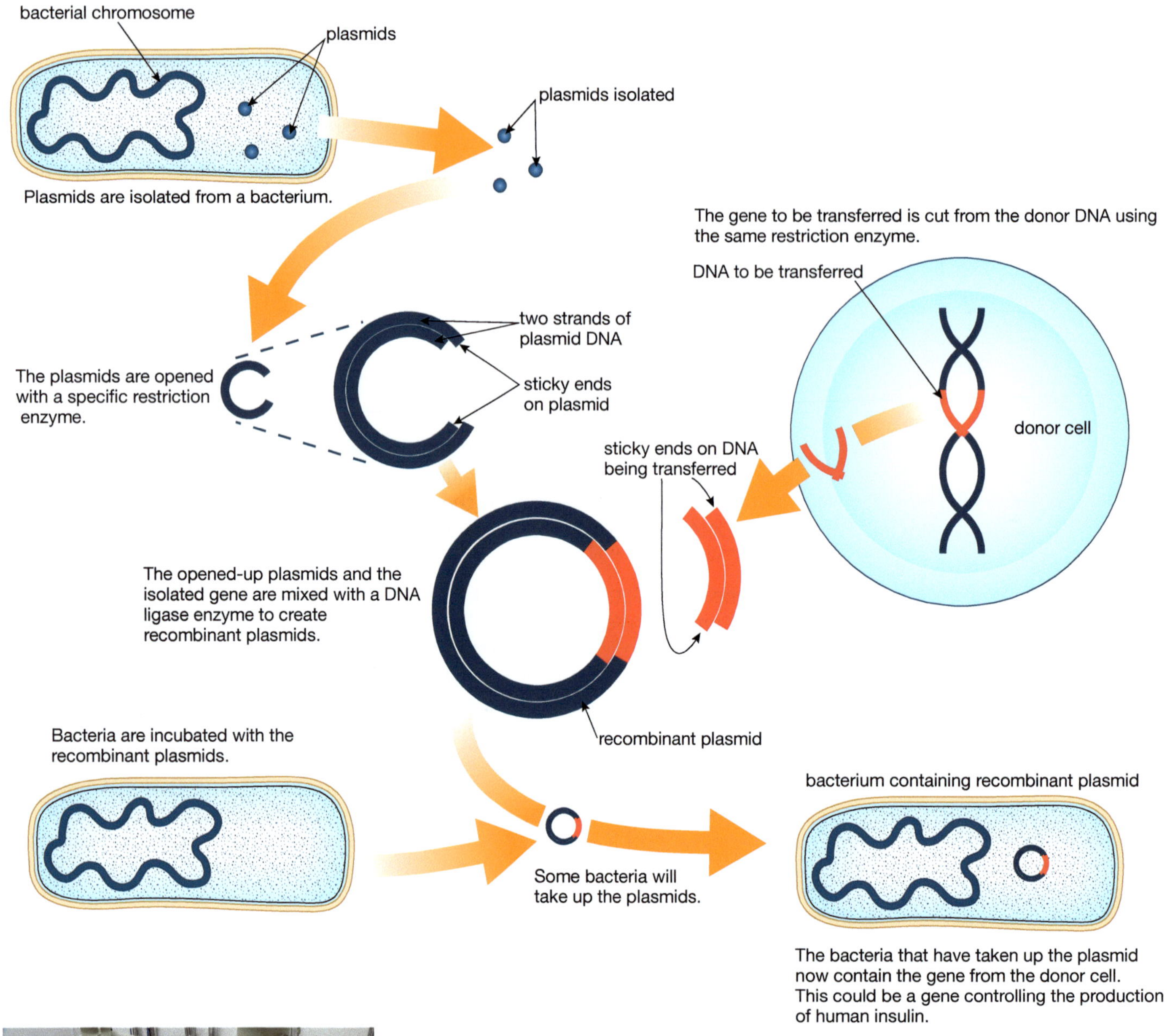

▲ Figure 1.22 The stages in producing a transgenic bacterium.

▲ Figure 1.23 An industrial fermenter holds hundreds of thousands of dm³ of a liquid culture.

Different bacteria have been genetically modified to manufacture a range of products. Once they have been genetically modified, they are grown or 'cultured' in tanks called **fermenters** to produce large amounts of the product (Figure 1.23).

SOME PRODUCTS OF GENETICALLY MODIFIED MICROORGANISMS

Since the basic techniques of transferring genes were developed, many simple single-celled organisms, such as bacteria and yeasts (unicellular fungi), have been genetically modified to produce useful products. Some examples of medical products made using genetically modified microorganisms are:

- **Human insulin:** People suffering from diabetes need a reliable source of the drug **insulin** to treat their condition (see Chapter 9). Before the use of genetic engineering, the only insulin available was extracted from the pancreases of other animals such as cattle or pigs. The chemical structure of insulin from these animals is not quite the same as that of human insulin and does not give the same degree of control of blood glucose levels. Now, however, the human gene for insulin production can be placed in bacteria, which then produce human insulin.

> **DID YOU KNOW?**
> More insulin is required every year because the number of diabetics worldwide increases each year, and also because diabetics now have longer life spans.

- **Growth hormone:** In some children, the pituitary gland does not produce enough growth hormone and their growth is restricted. Injections of human growth hormone from genetically modified bacteria can restore normal growth patterns.

> **DID YOU KNOW?**
> Before human growth hormone from genetically modified bacteria was available, the only source of the hormone was from human corpses (dead bodies). This was a rather unpleasant procedure and had health risks. A number of children treated in this way developed Creutzfeld–Jacob disease (the human form of 'mad cow' disease). When this became known, the treatment was withdrawn.

- **Hepatitis B vaccine:** Yeast cells can be genetically modified to produce the surface proteins (**antigens**) of the hepatitis B virus. These proteins are used to make a vaccine against hepatitis B. When the vaccine is injected into a patient, their body makes antibodies against the proteins, so the person becomes immune to the virus.

PRODUCING GENETICALLY MODIFIED PLANTS

The gene technology described so far can transfer DNA from one cell to another cell. In the case of bacteria, this is fine – a bacterium only has one cell. But plants have billions of cells and to genetically modify a plant, each cell must receive the new gene. Any procedure for genetically modifying plants has two main stages:

- introducing the new gene or genes into plant cells
- producing whole plants from just a few cells.

At first, biologists had problems inserting genes into plant cells. Then they discovered a soil bacterium called *Agrobacterium*, which regularly inserts plasmids into plant cells. Now that a vector had been found, the rest became possible. Figure 1.24 outlines one procedure that uses *Agrobacterium* as a vector.

Figure 1.24 Genetically modifying plants using *Agrobacterium*.

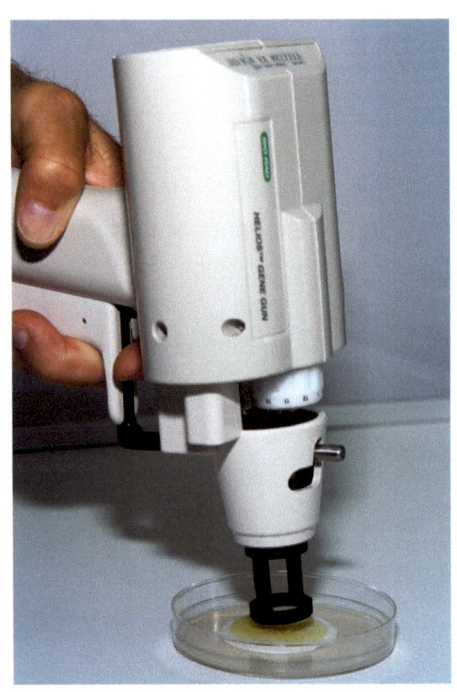

Figure 1.25 The gene gun.

This technique cannot be used on all plants. *Agrobacterium* will not infect cereals, so another technique was needed for these plants. The 'gene gun' was invented. This is a piece of laboratory equipment that fires tiny pellets made of gold (Figure 1.25). The pellets are coated with DNA that contains the required gene. These are fired directly into plant tissue. The gene gun has made it possible to genetically modify cereals and other crop plants.

Using *Agrobacterium* as a vector, biologists have produced genetically modified rice called 'golden rice'. This rice has three genes added to its normal DNA content. Two of these genes come from daffodils and one comes from a bacterium. Together, these genes allow the rice to make beta-carotene – the chemical that gives carrots their colour. It also colours the rice, which explains the name 'golden rice'. More importantly, the beta-carotene is converted to vitamin A when eaten. This could save the eyesight of millions of children in less economically developed countries, who go blind because they do not have enough vitamin A in their diet.

EXTENSION WORK

Golden rice sounds like a good idea but there have been several problems with it. Some people believe that there are ethical and environmental reasons why golden rice should not be grown and that it is better to provide other, natural crops containing enough beta-carotene. You could research the pros and cons of golden rice on the internet.

Genetically modified plants are also helping humans to fight infection. Biologists have succeeded in genetically modifying several species of plant in order to produce vaccines against different infectious diseases. For example, potatoes, bananas, lettuce, carrots and tobacco plants have all been engineered to produce proteins from the virus that causes hepatitis B. These proteins can be extracted from the plants and used to make a vaccine, which can be given to the patients by mouth or as an injection. At the time when this book was written, this vaccine had not been developed to the stage where it could replace vaccine from genetically modified yeast cells (described above). However, research in this area continues.

> **DID YOU KNOW?**
> What if you could receive a 'vaccination' every time you ate a banana? Instead of extracting the proteins and using them to make a vaccine, it might be possible to 'vaccinate' a person by getting them to eat a GM banana. The person's immune system would make antibodies against the virus proteins in the banana, and these antibodies would be able to destroy the virus without any need for an injection.

LOOKING AHEAD – MEMBRANES IN CELLS

If you continue to study biology beyond International GCSE, you will learn more about the structure and function of cells. You might like to look on the internet for some electron micrographs and do some further research about cell structure.

Electron microscopes allow us to see cells at a much greater magnification than by using a light microscope. They also reveal more detail. The image produced by a light microscope can only distinguish features that are about the size of a mitochondrion, but the electron microscope has a much greater *resolution*. Resolution is the ability to distinguish two points in an image as being separate. The maximum resolution of a light microscope is about 200 nanometres (nm) but with an electron microscope we can distinguish structures less than 1 nm in size. This is why ribosomes are only visible using an electron microscope – they are about 25 nm in diameter.

Electron microscopy (using an electron microscope) reveals that much of the cytoplasm is made up of membranes. As well as the cell surface membrane and the endoplasmic reticulum, there are membranes around organelles such as the nucleus and mitochondria, and sometimes there are membranes inside organelles as well.

All these membranes are needed because there are thousands of different chemical reactions happening in cells. A key function of membranes is to separate the different reactions into different compartments, so that they are not all happening in one big 'test tube'. For example, the reactions and enzymes of aerobic respiration (respiration that needs oxygen) are kept inside the mitochondria, separate from the rest of the cytoplasm (Figure 1.26).

> **KEY POINT**
> A nanometre (nm) is 10^{-9} m, or one millionth of a millimetre.

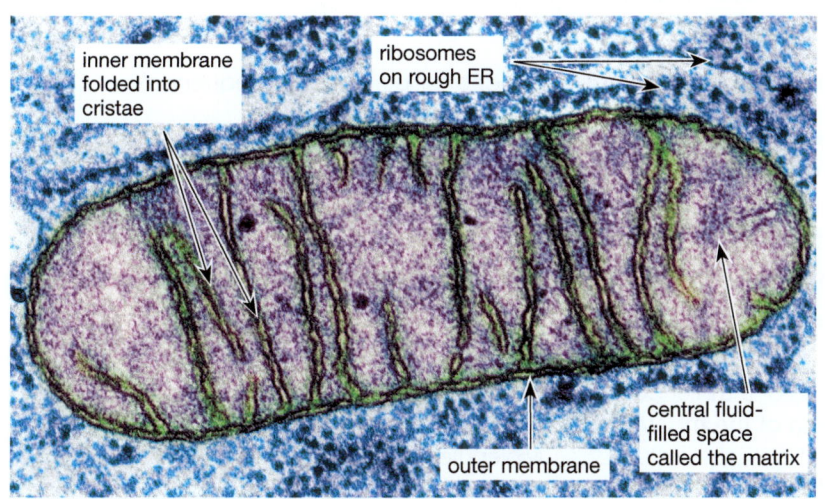

◀ Figure 1.26 An electron micrograph of a mitochondrion (magnification ×60 000). The mitochondrion has two membranes – an outer membrane separating its contents from the rest of the cytoplasm and an inner membrane forming folds called cristae. The reactions of aerobic respiration take place in the mitochondria of a cell. Different stages of the process happen in different parts of the mitochondrion.

CHAPTER QUESTIONS

SKILLS INTERPRETATION

1. Draw a diagram of a generalised animal cell as seen through a light microscope. Label all of the parts. Alongside each label, write the function of that part.

SKILLS REASONING

2. The diagram represents part of a molecule of DNA.

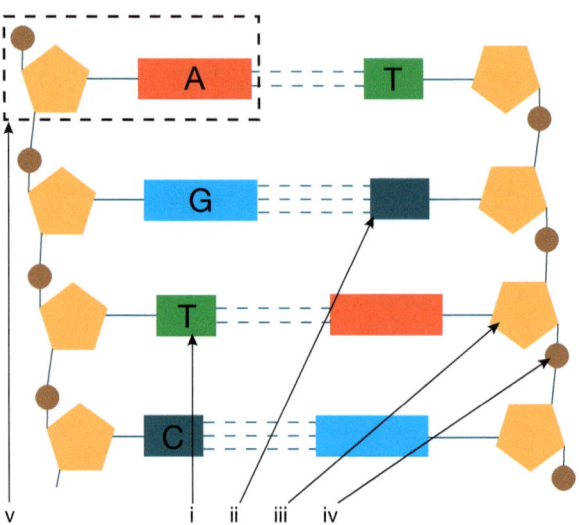

   a. Name the parts labelled i, ii, iii, iv and v.
   b. Use the diagram to explain the base-pairing rule.

SKILLS INTERPRETATION

3. DNA is the only molecule capable of replicating itself. Sometimes, mutations occur during replication.
   a. Draw a flow diagram to describe the process of DNA replication.

SKILLS REASONING

   b. Explain how a single gene mutation can lead to the formation of a protein in which:
      i. many of the amino acids are different from those coded for by the non-mutated gene
      ii. only one amino acid is different from those coded for by the non-mutated gene.

SKILLS CRITICAL THINKING

4. Below is a base sequence from part of the coding strand of a DNA molecule.

   TAC CTC GGT CAT CCC

   a. How many amino acids are coded for by this base sequence?
   b. The sequence of the coding strand was transcribed to form mRNA. Write down the base sequence of this mRNA.
   c. Write down the corresponding base sequence of the non-coding strand of the DNA.
   d. Copy and complete this description of the next stage in protein synthesis:

   The mRNA base sequence is converted into the amino acid sequence of a protein during a process called _____ . The mRNA sequence consists of a triplet code. Each triplet of bases is called a _____ . Reading of the mRNA base sequence begins at a _____ and ends at a _____ . Molecules of tRNA carrying an amino acid bind to the mRNA at an organelle called the _____ .

SKILLS ANALYSIS

5 In an investigation into mitosis, the distance between a chromosome and the pole (end) of a cell was measured. The graph shows the result of the investigation.

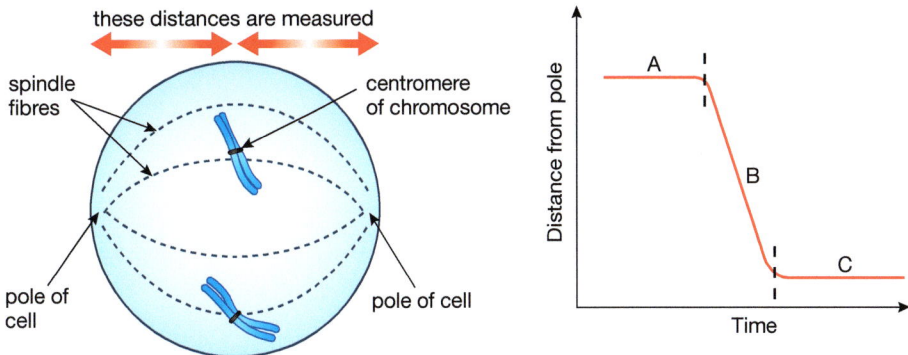

a Describe two events that occur during stage A.
b Explain what is happening during stage B.
c Describe two events that occur during stage C.

6 a What is a stem cell?

SKILLS REASONING

b State one difference between an embryonic stem cell and an adult stem cell.

c Suggest how stem cell therapy can be used in the treatment of leukaemia (blood cancer).

SKILLS ANALYSIS

7 The diagram shows the main stages in transferring the human insulin gene to a bacterium.

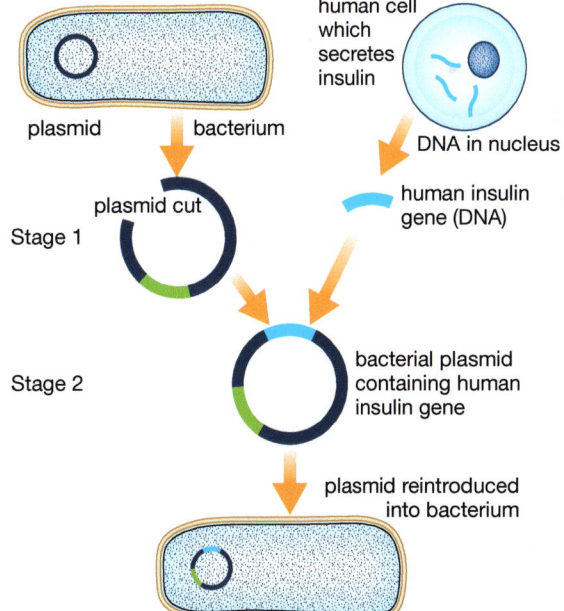

a Name the enzymes used at stages 1 and 2.

SKILLS CRITICAL THINKING

b What is the role of the plasmid in this procedure?
c How would the insulin-producing bacteria be used to produce significant amounts of insulin?

SKILLS REASONING

d Suggest why insulin produced in this way is preferred to insulin extracted from the pancreases of cows.

2 MOVEMENT OF SUBSTANCES INTO AND OUT OF CELLS

Cells need to take in certain substances (such as oxygen) from their surroundings, and get rid of other substances, such as carbon dioxide. The cell surface membrane is selective about which chemicals can pass through it. Molecules and ions can move through the membrane in three main ways: diffusion, active transport and osmosis.

LEARNING OBJECTIVES

- Know simple definitions of diffusion, osmosis and active transport
- Understand that movement of substances into and out of cells can be by diffusion, osmosis (understanding of water potential is required) and active transport
- Understand the factors that affect the rate of movement of substances into and out of cells, including surface area to volume ratio, temperature and concentration gradient

DIFFUSION

Many substances can pass through the membrane by **diffusion**. Diffusion happens when a substance is more concentrated in one place than another. For example, if a cell is producing carbon dioxide by respiration (see Chapter 5), the concentration of carbon dioxide will be higher inside the cell than outside. This difference in concentration is called a *concentration gradient*. The molecules of carbon dioxide are constantly moving about (they have kinetic energy). The cell membrane is permeable to carbon dioxide, so the molecules can move through it in either direction. Over time, more molecules will move out of the cell than into it, because there is a higher concentration of carbon dioxide molecules inside the cell than outside it. We say there is a *net* movement of molecules out of the cell (Figure 2.1).

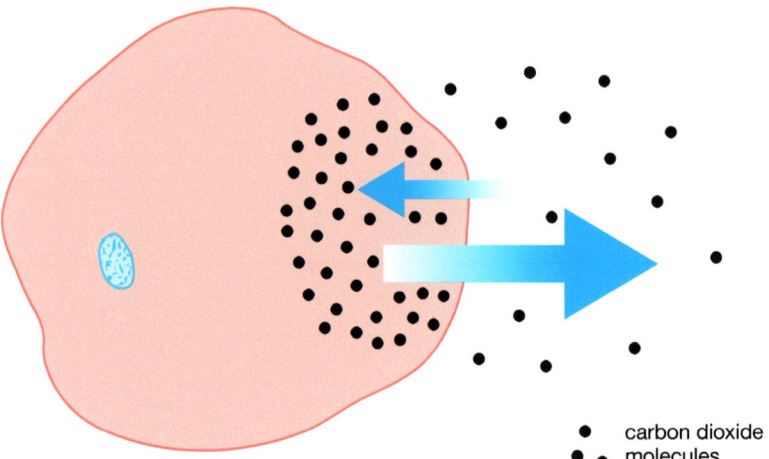

▲ Figure 2.1 Respiration produces carbon dioxide, so the concentration of carbon dioxide inside the cell increases. Although the carbon dioxide molecules diffuse in both directions across the cell membrane, the overall (net) movement is out of the cell, down the concentration gradient.

The opposite happens with oxygen. Respiration uses up oxygen, so there is a concentration gradient of oxygen from outside the cell to inside. There is therefore a net movement of oxygen into the cell by diffusion.

UNIT 1 — MOVEMENT OF SUBSTANCES INTO AND OUT OF CELLS

KEY TERM
Diffusion is the net movement of particles (molecules or ions) from a region of high concentration to a region of low concentration, i.e. down a concentration gradient.

Various factors affect the rate of diffusion.

- **The concentration gradient:** Diffusion happens more quickly when there is a steep concentration gradient (i.e. a big difference in concentration between two areas).

- **The surface area to volume ratio:** A larger surface area in proportion to the volume will increase the rate of diffusion.

- **The distance:** The greater the distance over which diffusion has to take place, the slower the rate of diffusion.

- **The temperature:** The rate of diffusion is greater at higher temperatures. This is because a high temperature provides the particles with more kinetic energy.

Safety Note: Wear eye protection and avoid skin contact with the permanganate and acid. When cutting the blocks of agar jelly, wear disposable gloves and use a sharp scalpel on a cutting board.

ACTIVITY 1

▼ PRACTICAL: DEMONSTRATING DIFFUSION IN A JELLY

Agar is a jelly that is used for growing cultures of bacteria. It has a consistency similar to the cytoplasm of a cell. Like cytoplasm, it has a high water content. Agar can be used to show how substances diffuse through a cell.

This demonstration uses the reaction between hydrochloric acid and potassium permanganate solution. When hydrochloric acid comes into contact with potassium permanganate, the purple colour of the permanganate disappears.

A Petri dish is prepared which contains a 2 cm deep layer of agar jelly, dyed purple with potassium permanganate. Three cubes of different sizes are cut out of the jelly, with side lengths 2 cm, 1 cm and 0.5 cm. The cubes have different volumes and total surface areas. They also have different surface area to volume ratios, as shown in the table below.

Length of side of cube / cm	Volume of cube / cm^3 (length × width × height)	Surface area of cube / cm^2 (length × width of one side) × 6	Ratio of surface area to volume of cube (surface area / volume)
2	(2 × 2 × 2) = 8	(2 × 2) × 6 = 24	24/8 = 3
1	(1 × 1 × 1) = 1	(1 × 1) × 6 = 6	6/1 = 6
0.5	(0.5 × 0.5 × 0.5) = 0.125	(0.5 × 0.5) × 6 = 1.5	1.5/0.125 = 12

Notice that the smallest cube has the largest surface area to volume ratio. The same is true of cells – a small cell has a larger surface area to volume ratio than a large cell.

The cubes are dropped, carefully and at the same time, into a beaker of dilute hydrochloric acid (Figure 2.2).

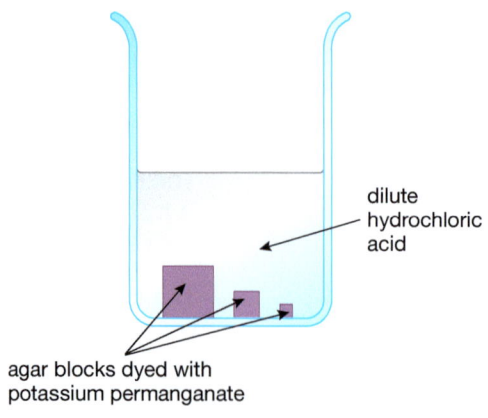

▲ Figure 2.2 Investigating diffusion in a jelly.

The time taken for each cube to turn colourless is noted.

Which cube would be the first to turn colourless and which the last? Explain the reasoning behind your prediction.

If the three cubes represented cells of different sizes, which cell would have the most difficulty in obtaining substances by diffusion?

It may be possible for you to try this experiment, using similar apparatus.

ACTIVE TRANSPORT

Diffusion happens because of the kinetic energy of the particles. It does not need any extra energy from respiration. Sometimes, however, a cell needs to take in a substance *against* the concentration gradient – when there is less of the substance outside the cell than inside it. It can do this by another process, called **active transport**.

During active transport, a cell uses energy from respiration to take up substances, rather like a pump uses energy to move a liquid from one place to another. In fact, biologists speak of the cell 'pumping' ions and molecules in or out. The pumps are large protein molecules located in the cell membrane. An example of a place where this happens is in the human small intestine, where epithelial cells lining the intestine absorb glucose by active transport.

Cells use active transport to control the uptake of many substances. A definition of active transport is given in the key term box.

KEY TERM
Active transport is the movement of particles against a concentration gradient, using energy from respiration.

OSMOSIS

Water moves across cell membranes by a special sort of diffusion, called **osmosis**. Osmosis happens when the total concentrations of all dissolved substances inside and outside the cell are different. Water will move across the membrane from the more dilute solution to the more concentrated one. Notice that this is still obeying the rules of diffusion – the water moves from where there is a higher concentration of water molecules to a lower concentration of

KEY POINT

The cell membrane is both selectively permeable and partially permeable. 'Selectively permeable' means that the membrane has control over which molecules it lets through (by active transport). 'Partially permeable' means that small molecules such as water and gases can pass through it, while larger molecules cannot.

water molecules. Osmosis can only happen if the membrane is permeable to water but not to some other solutes. We say that it is partially permeable.

One artificial partially permeable membrane is called **Visking tubing**. This is used in kidney dialysis machines (Chapter 10). Visking tubing has microscopic holes in it, which let small molecules such as water pass through (it is permeable to them). However, it is not permeable to some larger molecules, such as the sugar, sucrose. This is why it is called 'partially' permeable. You can show the effects of osmosis by filling a Visking tubing 'sausage' with concentrated sucrose solution, attaching it to a capillary tube and placing the Visking tubing in a beaker of water (Figure 2.3).

The level in the capillary tube rises as water moves from the beaker into the Visking tubing. This movement is due to osmosis. You can understand what is happening if you imagine a highly magnified view of the Visking tubing separating the two liquids (Figure 2.4).

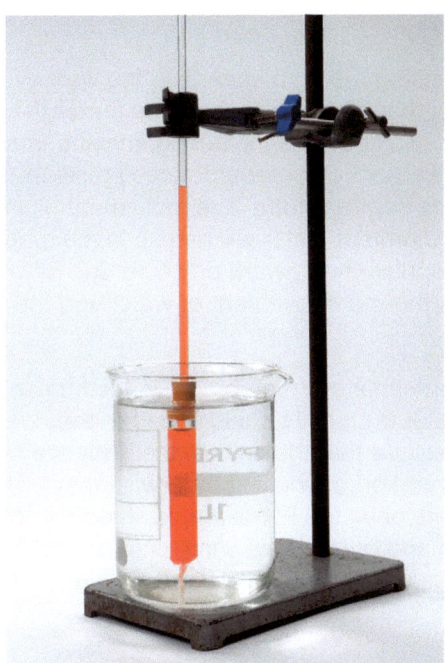

▲ Figure 2.3 Water enters the Visking tubing 'sausage' by osmosis. This causes the level of liquid in the capillary tube to rise. In the photograph, the contents of the Visking tubing have had a red dye added to them to make it easier to see the movement of the liquid.

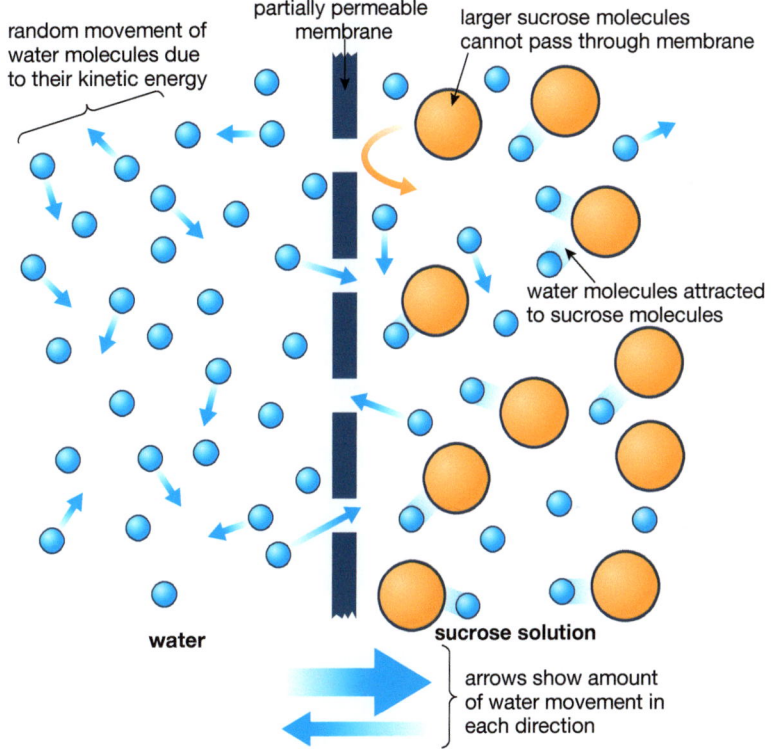

▲ Figure 2.4 In this model of osmosis, more water molecules diffuse from left to right than from right to left.

The sucrose molecules are too big to pass through the holes in the partially permeable membrane. The water molecules can pass through the membrane in either direction, but those on the right are attracted to the sugar molecules. This slows them down and means that they are less free to move – they have less kinetic energy. As a result of this, more water molecules diffuse from left to right than from right to left. In other words, there is a greater diffusion of water molecules from the more dilute solution (in this case pure water) to the more concentrated solution. This leads us to a simple definition of osmosis, as shown in the key term box.

KEY TERM

Osmosis is the net diffusion of water across a partially permeable membrane, from a dilute solution to a more concentrated solution.

> **KEY POINT**
>
> It is important to realise that neither of the two solutions has to be pure water. As long as there is a difference in their concentrations (and their water potentials), and they are separated by a partially permeable membrane, osmosis can still take place.

There is an alternative way of describing osmosis, which uses the idea of **water potential**. The water potential of a solution is a measure of how 'free' the water molecules are to move. The molecules in pure water can move most freely, so pure water has the highest water potential. The more concentrated a solution is, the lower its water potential. In the model in Figure 2.4, water moves from a higher to a lower water potential. This is a law that applies whenever water moves by osmosis. We can bring these ideas together in a 'water potential' definition of osmosis, as shown in the key term box.

> **KEY TERM**
>
> Osmosis is the net diffusion of water across a partially permeable membrane, from a solution with a higher water potential to one with a lower water potential.

In the demonstration of osmosis using Visking tubing (Figure 2.3), the water in the beaker has a higher water potential than the sucrose solution inside the Visking tubing. Osmosis takes place, and water enters the tubing, moving from a high to a lower water potential. The movement of water produces pressure, which pushes the coloured solution up the capillary tube. The movement does not go on forever – eventually the column of water will rise up in the tube to a point where its downward pressure equals the upward pressure due to osmosis. At this point, there will be no further net movement of water and the column will stop rising.

All cells are surrounded by a partially permeable cell membrane. In the human body, osmosis is important in moving water from cell to cell, and from the blood to the tissues (Chapter 6). It is important that the cells of the body are bathed in a solution with the right concentration of solutes; otherwise they could be damaged by osmotic movements of water. For example, if red blood cells are put into water, they will swell up and burst. If the same cells are put into a concentrated salt solution, they lose water by osmosis and shrink, producing cells with crinkly edges (Figure 2.5).

> **Safety Note:** Wear eye protection and avoid skin contact with the blood and solutions.

ACTIVITY 2

▼ PRACTICAL: DEMONSTRATING THE EFFECTS OF OSMOSIS ON RED BLOOD CELLS

Blood plasma has a concentration equivalent to a 0.85% salt solution. If fresh blood is placed into salt solutions with different concentrations, the blood cells will gain or lose water by osmosis. This can be demonstrated using sterile animal blood (available from suppliers of biological materials).

Three test tubes are set up, containing the following solutions:

A: 10 cm^3 of distilled water

B: 10 cm^3 of 0.85% salt solution

C: 10 cm^3 of 3% salt solution

1 cm^3 of blood is added to each tube and the tubes are shaken. A sample from each tube is examined under the microscope. The sample from tube A is found to contain no intact cells. Figure 2.5 shows cells from tubes B and C. The cells from tube B look normal, but those from tube C are shrunken, with crinkly edges.

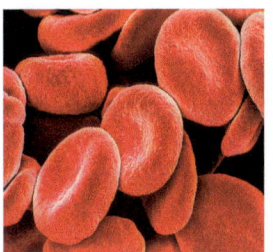 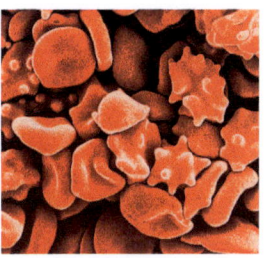

▲ Figure 2.5 Compare the blood cells on the right, which were placed in a 3% salt solution, with the normal blood cells on the left, from a 0.85% salt solution.

Using your knowledge of osmosis, can you explain what has happened to the red blood cells in each tube?

The three tubes are now placed in a centrifuge and spun around at high speed to separate any solid particles from solution. The results are shown in Figure 2.6.

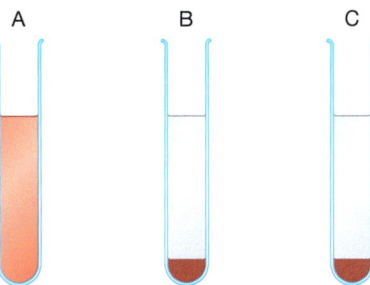

▲ Figure 2.6 The three tubes after centrifugation.

Tube A contains a clear red solution and no solid material at the bottom of the tube. Tubes B and C both contain a colourless liquid and a red precipitate at the bottom.

Can you explain these results?

Activity 2 shows how important it is that animal cells are surrounded by a solution containing the correct concentration of dissolved solutes. If the surrounding solution does not have the right concentration, cells can be damaged by the effects of osmosis. The red blood cells placed in pure water absorb the water by osmosis, swell up and burst, leaving a red solution of haemoglobin in the test tube. When placed in 3% salt solution, the red blood cells lose water by osmosis and shrink.

We will return to the idea that cells need a correct constant 'environment' in Chapter 10.

EXCHANGE SURFACES IN THE BODY

All cells exchange substances with their surroundings, but some parts of the body are specially adapted for the exchange of materials because they have a very large surface area in proportion to their volume. Two examples are the air sacs of the lungs (Chapter 5) and the lining of the small intestine (Chapter 4). Diffusion is a slow process, and organs that rely on diffusion need a large surface over which it can take place. The air sacs of the lungs have a very large surface area that allows oxygen and carbon dioxide to move between the air and the blood, during breathing. The lining of the small intestine is covered with tiny projections called villi, which provide a large surface area for the absorption of digested food. This takes place partly by diffusion and partly by active transport.

CHAPTER QUESTIONS

SKILLS REASONING

1 a Explain the differences between diffusion and active transport.

SKILLS CRITICAL THINKING

b State three factors that affect the rate of diffusion.

SKILLS PROBLEM SOLVING

c Calculate the ratio of surface area to volume ($\frac{\text{surface area}}{\text{volume}}$) of a cube with a side length of:
  i 2 cm
  ii 3 cm.

SKILLS REASONING

d From your answer to **c**, describe what happens to the surface area to volume ratio when the size of the cube is increased. What is the significance of this observation for the diffusion of substances into and out of cells?

SKILLS CRITICAL THINKING

2 The diagram shows a cell from the lining of a human kidney tubule. One role of this cell is to absorb glucose from the fluid passing along the tubule and pass it into the blood, as shown by the arrows on the diagram.

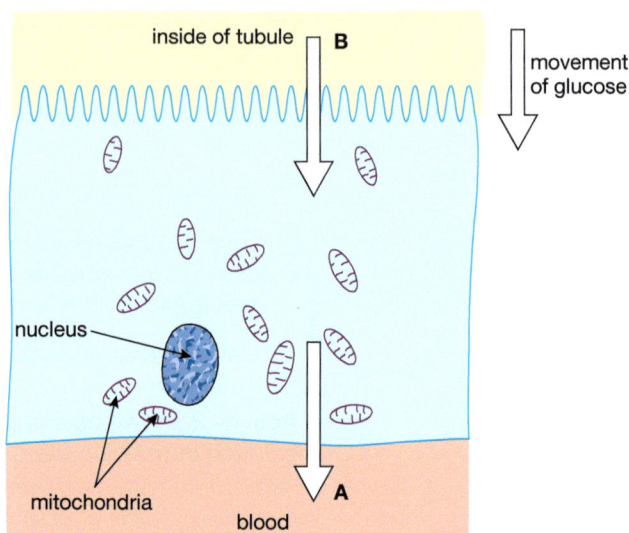

a What is the function of the mitochondria?

b The tubule cell contains a large number of mitochondria. They are needed for the cell to transport glucose across the cell membrane into the blood at 'A'. Suggest the method that the cell uses to do this and explain your answer.

c The mitochondria are not needed to transport glucose into the cell from the tubule at 'B'. Name the process by which the glucose molecules move across the membrane at 'B' and explain your answer.

SKILLS REASONING

d The surface membrane of the tubule cell at 'B' is greatly folded. Explain how this adaptation helps the cell to carry out its function.

SKILLS CRITICAL THINKING

3 An experiment was carried out to find the effects of osmosis on blood cells. Three test tubes were filled with different solutions. 10 cm³ of water was placed in tube A, 10 cm³ of 0.85% salt solution in tube B, and 10 cm³ of 3% salt solution in tube C. 1 cm³ of fresh blood was added to each tube. The tubes were shaken and a sample from each tube was observed under the microscope at a high magnification.

The tubes were then placed in a centrifuge and spun around at high speed to separate any solid particles from solution. The results are shown in the diagram below.

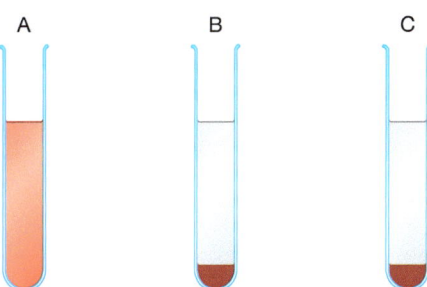

a Which solution had a salt concentration that was similar to that of blood?

b Describe what you would expect to see when viewing the samples from tubes A, B and C through the microscope.

c Explain the results shown in the diagram.

d When a patient has suffered severe burns, damage to the skin results in a loss of water from the body. This condition can be treated by giving the patient a saline drip. This is a 0.85% salt solution, which is fed into the patient's blood through a needle inserted into a vein. Explain why 0.85% salt solution is used, and not water.

3 BIOLOGICAL MOLECULES

Many of the molecules found in the body are made from simple 'building blocks'. For example, DNA is built up from smaller molecules called nucleotides. In the same way, carbohydrates, lipids and proteins are made of simpler molecular sub-units. In this chapter, you will read about the structure of these biological molecules, and the functions of an important group of proteins called enzymes.

LEARNING OBJECTIVES

- Know the chemical elements present in carbohydrates, proteins and lipids (fats and oils)
- Understand the structure of carbohydrates, proteins and lipids as large molecules made up of smaller basic units:
  - carbohydrates (starch and glycogen) from simple sugars
  - lipids from fatty acids and glycerol
  - protein from amino acids
- Describe chemical tests for starch, glucose (reducing sugar), lipid and protein
- Explain the role of enzymes as biological catalysts in metabolic reactions
- Explain the action of enzymes
- Explain how enzyme activity is affected by:
  - temperature
  - pH
  - substrate concentration
  - competitive and non-competitive inhibitors
- Describe how to investigate the effect of temperature on enzyme activity
- Describe how to investigate the effect of pH on enzyme activity
- Describe the advantages of using immobilised enzymes in:
  - the production of lactose-free milk
  - the conversion of sucrose into glucose and fructose
  - glucose testing strips for diabetics
- Describe how to investigate the action of immobilised enzymes, including the preparation of alginate beads

CARBOHYDRATES, LIPIDS AND PROTEINS

There are three main types of biological molecule in the human body – **carbohydrates**, **lipids** and **proteins**. All three are composed of just a few chemical elements. Carbohydrates and lipids are entirely made of the elements carbon (C), hydrogen (H) and oxygen (O). Proteins also contain these elements, along with nitrogen (N) and small amounts of sulfur (S). However, the structures of these three types of molecule are very different.

THE STRUCTURE OF CARBOHYDRATES

The basic units that make up carbohydrates are simple sugars such as **glucose**. Glucose is an example of a **monosaccharide**, which is a 'single' sugar unit. This means that it cannot be broken down into a simpler sugar. Another monosaccharide is **fructose**, a sugar found in fruits. The chemical formula for glucose is $C_6H_{12}O_6$ – you can see from this formula that glucose contains only carbon, hydrogen and oxygen.

DID YOU KNOW?
In the word 'carbohydrates', the 'carbo' part refers to carbon and the 'hydrate' part refers to the fact that the hydrogen and oxygen atoms in a carbohydrate are often in the ratio 2 : 1, the same as in water (H_2O).

When two monosaccharides are joined together, they form a 'double sugar' or **disaccharide**. For example, ordinary table sugar (**sucrose**) is a disaccharide of glucose and fructose. The sugar found in milk (**lactose**) is a disaccharide made of glucose and galactose.

Monosaccharides can also join together in long chains to produce a **polysaccharide**. One important polysaccharide is **starch**. The 'staple diets' of people from around the world are foods containing starch, such as rice, potatoes, bread and pasta. Starch is made up of long chains of hundreds of glucose molecules joined together. In chemistry terms, it is a *polymer* of glucose (Figure 3.1).

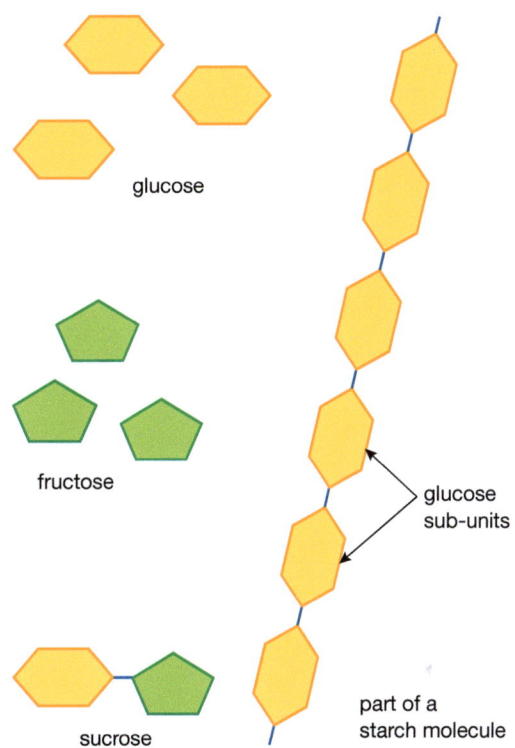

▲ Figure 3.1 Glucose and fructose are 'single sugar' molecules or monosaccharides. A molecule of glucose joined to a molecule of fructose forms the 'double sugar' or disaccharide called sucrose. Starch is a polysaccharide of many glucose sugars.

Starch is only found in plant tissues, but some animal cells contain a very similar carbohydrate called **glycogen**. This is also a polysaccharide of glucose, and is found in tissues such as liver and muscle, where it acts as an energy store.

THE STRUCTURE OF LIPIDS

Lipids are fats and oils. The difference between the two is that a fat is solid at room temperature, whereas an oil is liquid. Fats are more common in animal tissues, while oils are mainly found in plants.

Lipids contain the same three elements as carbohydrates – carbon, hydrogen and oxygen – but the proportion of oxygen in a lipid is much lower than in a carbohydrate. For example, meat contains a fat called tristearin, which has the formula $C_{51}H_{98}O_6$.

The chemical 'building blocks' of lipids are two types of molecule called **glycerol** and **fatty acids**. Glycerol is an oily liquid, also known as glycerine. It is used in making some cosmetics. In lipids, a molecule of glycerol is joined to three fatty acid molecules. There are many different fatty acid molecules, which result in the wide variety of lipids found in food (Figure 3.2).

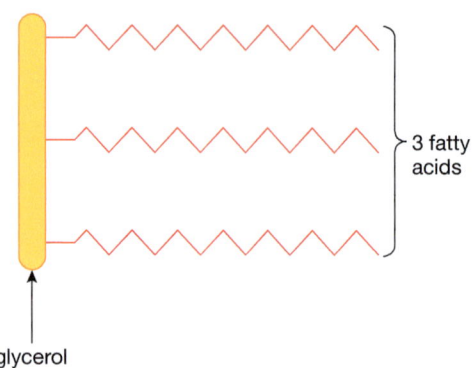

▲ Figure 3.2 Lipids are made up of a molecule of glycerol joined to three fatty acids. The many different fatty acids form the variable part of the molecule.

THE STRUCTURE OF PROTEINS

Whereas starch is a polymer made from a single 'building block' (glucose), proteins are made from 20 different sub-units called **amino acids**. All amino acids contain four chemical elements: carbon, hydrogen, oxygen and nitrogen. Two amino acids also contain sulfur. The amino acids are linked together in long chains, which are usually folded up or twisted into spirals. Cross-links hold the chains together (Figure 3.3).

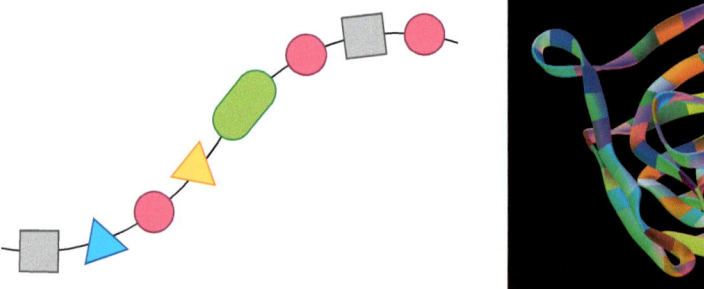

▲ Figure 3.3 (a) A chain of amino acids forming part of a protein molecule. Each shape represents a different amino acid. (b) A computer model of a protein in the blood that is involved in forming a blood clot. The coloured bands represent different amino acids in a chain.

The shape of a protein is very important in allowing it to perform its function. The shape depends on the order of amino acids in the protein. Because there are 20 different amino acids, and they can be arranged in any order, the number of different protein structures that can be made is enormous. There are thousands of different kinds of protein in organisms, from structural proteins such as collagen and keratin in skin and nails, to proteins with more specific functions, such as enzymes and haemoglobin.

TESTS FOR BIOLOGICAL SUBSTANCES

You can carry out simple chemical tests for starch, glucose, protein or lipid. Activity 3 uses pure substances for the tests, but it is possible to do them on normal foods too. Unless the food is a liquid like milk, it needs to be cut up into small pieces and ground with a pestle and mortar, then shaken with some water in a test tube. This is done to extract the components of the food and dissolve any soluble substances, such as sugars.

UNIT 1 BIOLOGICAL MOLECULES

Safety Note: Wear eye protection and avoid skin contact with all the reactants. Always use a water bath for the glucose test – never use a Bunsen burner to heat the test tube directly.

(a)
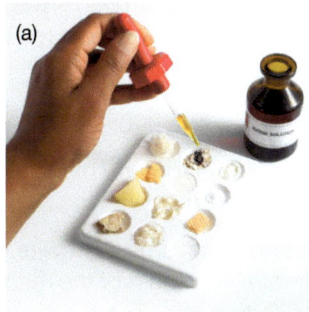

(b)
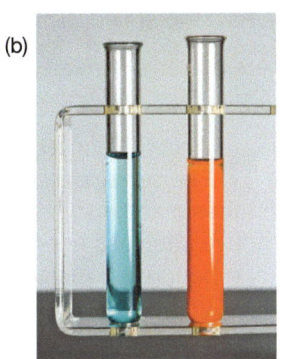

(c)

(d)

▲ Figure 3.4 (a) Testing for starch using iodine; (b) testing for glucose using Benedict's solution; (c) testing for protein using biuret; and (d) testing for lipid.

ACTIVITY 3

▼ PRACTICAL: TESTING FOR SOME BIOLOGICAL SUBSTANCES

Test for starch
A little starch is placed on a spotting tile. A drop of yellow-brown iodine solution is added to the starch. The iodine reacts with the starch, forming a very dark blue, or 'blue-black', colour (Figure 3.4(a)). Starch is insoluble but this test will work on a solid sample of food, such as potato, or a suspension of starch in water

Test for glucose
Glucose is called a **reducing sugar**. This is because the test for glucose involves reducing an alkaline solution of copper (II) sulfate to copper (I) oxide.

A small spatula measure of glucose is placed in a test tube and a little distilled water is added (about 2 cm deep). The tube is shaken to dissolve the glucose. Several drops of *Benedict's solution* are added to the tube, enough to colour the mixture blue.

A water bath is prepared by half-filling a beaker with water and heating it on a tripod and gauze. The test tube is placed in the beaker and the water is allowed to boil (using a water bath is safer than heating the tube directly with the Bunsen burner). After a few seconds, the clear blue solution will gradually change colour, forming a cloudy orange or 'brick red' precipitate of copper (I) oxide (Figure 3.4(b)).

All other single sugars (monosaccharides), such as fructose, are reducing sugars, and so are some double sugars (disaccharides), such as the milk sugar, lactose. However, ordinary table sugar (sucrose) is not. If sucrose is boiled with Benedict's solution it will stay a clear blue colour.

Test for protein
The test for protein is sometimes called the 'biuret' test, after the coloured compound that is formed.

A little protein, such as powdered egg white (albumen), is placed in a test tube and about 2 cm depth of water is added. The tube is shaken to mix the powder with the water. An equal volume of dilute (5%) potassium hydroxide solution is added and the tube is shaken again. Finally, two drops of 1% copper sulfate solution are added. A pale purple colour will develop (Figure 3.4(c)). (Sometimes these two solutions are supplied already mixed together as 'biuret solution'.)

Test for lipid
Fats and oils are insoluble in water but will dissolve in ethanol (alcohol). The test for lipid uses this fact.

A pipette is used to place one drop of olive oil in the bottom of a test tube. About 2 cm depth of ethanol is added and the tube is shaken to dissolve the oil. The solution is poured into a test tube that is about three-quarters full with cold water. A white cloudy layer will form on the top of the water (Figure 3.4(d)). The white layer forms as the ethanol dissolves in the water and leaves the lipid behind as a suspension of tiny droplets, called an emulsion.

> **KEY POINT**
>
> The chemical reactions taking place in a cell are known as metabolic reactions. The sum of all the metabolic reactions is known as the **metabolism** of the cell. The function of enzymes is to catalyse metabolic reactions.

ENZYMES: CONTROLLING REACTIONS IN THE CELL

The chemical reactions that take place in a cell are controlled by a group of proteins called enzymes. Enzymes are biological **catalysts**. A catalyst is a chemical that speeds up a reaction without being used up itself. It takes part in the reaction but afterwards it is unchanged and free to catalyse more reactions. Cells contain hundreds of different enzymes, each catalysing a different reaction. This is how the activities of a cell are controlled – the nucleus contains the genes, the genes control the production of enzymes, and the enzymes catalyse reactions in the cytoplasm:

genes → proteins (enzymes) → catalyse reactions

Everything a cell does depends on which enzymes it can make. This in turn depends on which of the genes in its nucleus are working.

Enzymes are needed because the temperatures inside organisms are low (e.g. the human body temperature is about 37 °C). Without catalysts, most of the reactions that happen in cells would be far too slow to allow life to go on. The reactions can only take place quickly enough when enzymes are present to speed them up.

There are thousands of different sorts of enzyme because enzymes are proteins, and protein molecules have an enormous range of structures and shapes.

The molecule that an enzyme acts on is called its **substrate**. Each enzyme has a small area on its surface called the **active site**. The substrate attaches to the active site of the enzyme. The reaction then takes place and products are formed. When the substrate joins up with the active site, less energy is needed for the reaction to start and the products can be formed more easily.

Enzymes also catalyse reactions where large molecules are built up from smaller ones. In this case, several substrate molecules attach to the active site, the reaction takes place and the larger product molecule is formed. The product then leaves the active site.

The substrate fits into the active site of the enzyme rather like a key fitting into a lock. Just as a key will only fit one lock, a substrate will only fit into the active site of a particular enzyme. This is known as the **lock and key model** of enzyme action (Figure 3.5). It is the reason why enzymes are *specific*, i.e. an enzyme will only catalyse one reaction.

> **DID YOU KNOW?**
>
> You have probably heard of enzymes being involved in the digestion of food. In the intestine, enzymes are secreted onto the food to break it down. These are called *extracellular* enzymes, which means they function outside cells. However, most enzymes are *intracellular* – they stay inside cells and carry out their function there. You will find out about digestive enzymes in Chapter 4.

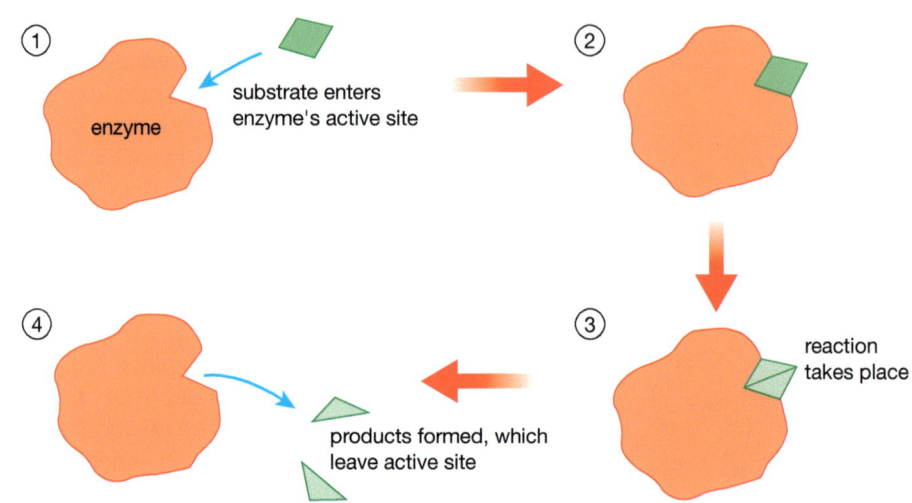

▲ Figure 3.5 Enzymes catalyse reactions at their active site. The active site acts like a 'lock' to the substrate 'key'. The substrate fits into the active site and products are formed. This happens more easily than without the enzyme – enzymes act as catalysts.

After an enzyme molecule has catalysed a reaction, the product is released from the active site and the enzyme is free to act on more substrate molecules.

FACTORS AFFECTING ENZYMES

A number of factors affect the activity of enzymes, including:

- temperature
- pH
- substrate concentration
- the presence of **inhibitors**.

TEMPERATURE

The effect of temperature on the action of an enzyme is easiest to see as a graph, where we plot the rate of reaction against temperature (Figure 3.6).

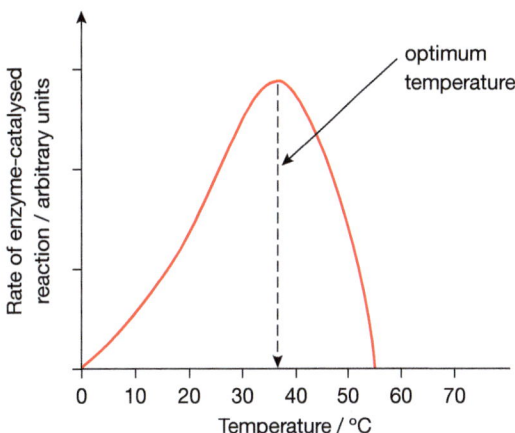

▲ Figure 3.6 Effect of temperature on the action of an enzyme.

Enzymes in the human body have evolved (changed over time) to work best at body temperature (37 °C). The graph in Figure 3.6 shows a peak on the curve at this temperature, which is called the *optimum temperature* for the enzyme.

As the enzyme is heated up to the optimum temperature, the rate of reaction increases. This is because higher temperatures give the molecules of the enzyme and the substrate more kinetic energy, so they collide more often. More collisions mean that the reaction will take place more frequently.

However, if the enzyme is heated too much, the rate of reaction decreases again. This is because enzymes are made of protein, and proteins are broken down by heat. At temperatures above 40 °C, the heat destroys the enzyme – we say that the enzyme is **denatured**. You can see the effect of denaturing when you boil an egg. The egg white is made of protein and turns from a clear runny liquid into a white solid as the heat denatures the protein. Denaturing changes the shape of the active site, so the substrate will no longer fit into it. Denaturing is permanent – the enzyme molecules will no longer catalyse the reaction.

pH

The pH around the enzyme is also important. The pH inside cells is usually around 7 (neutral) and most enzymes have evolved to work best at this pH. At extremes of pH, the enzyme activity decreases, as shown in Figure 3.7. The pH at which the enzyme works best is called its *optimum pH*. Either side of the

> **KEY TERM**
> The 'optimum' temperature is the 'best' temperature, in other words the temperature at which the reaction takes place most rapidly.

> **KEY TERM**
> Kinetic energy is the energy an object has because of its movement. The molecules of enzyme and substrate are moving faster, so they have more kinetic energy.

> **DID YOU KNOW?**
> Not all enzymes have an optimum temperature near 37 °C, only those of animals such as mammals and birds, which have body temperatures close to this value. Enzymes have evolved to work best at the normal body temperature of the organism. Bacteria that always live at an average temperature of 10 °C will probably have enzymes with an optimum temperature near 10 °C.

optimum, the pH affects the structure of the enzyme molecule and changes the shape of its active site, so the substrate will not fit into it so well.

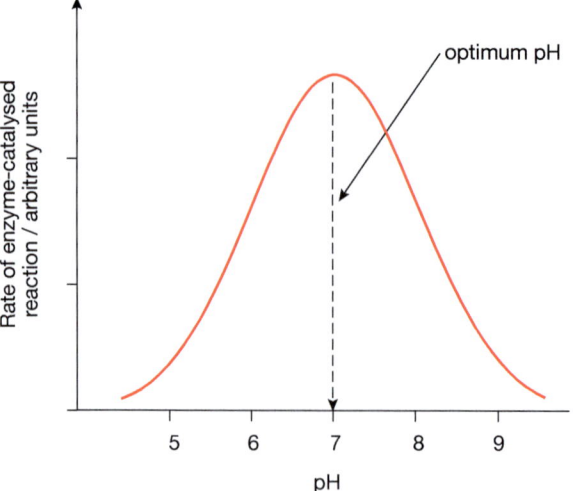

▲ Figure 3.7 Most enzymes work best at a neutral pH.

KEY POINT

Although most enzymes work best at a neutral pH, a few have an optimum below or above pH 7. The stomach produces hydrochloric acid, which makes its contents very acidic (see Chapter 4). Most enzymes stop working at a low pH, but the stomach makes an enzyme called pepsin which has an optimum pH of about 2. Pepsin is adapted to work well in these unusually acidic surroundings.

SUBSTRATE CONCENTRATION

When there is a low concentration of substrate, the active sites of some enzyme molecules will be empty, so the rate of reaction will be low. If the concentration of substrate is increased, there will be more collisions between enzyme and substrate molecules and more active sites will be filled. As a result, the rate of reaction will increase. This can be shown by a graph (Figure 3.8).

Notice that the curve becomes flatter or 'levels off' at high concentrations of substrate. This is because all the active sites become filled with substrate (we say they are *saturated*), so the rate of reaction reaches a maximum. It is as if each substrate molecule has to 'wait' for an empty active site to become available. The enzyme concentration is limiting the rate at which the reaction can take place – enzyme concentration is now the limiting factor.

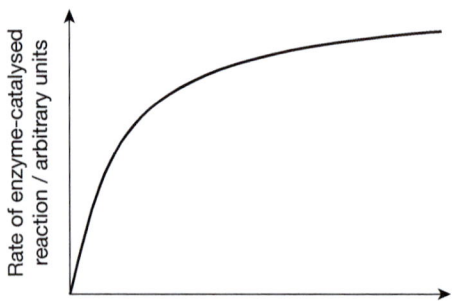

▲ Figure 3.8 The effect of substrate concentration on the action of an enzyme.

THE PRESENCE OF INHIBITORS

Inhibitors are substances that reduce the rate of an enzyme-catalysed reaction. Two types are called competitive and non-competitive inhibitors (Figure 3.9).

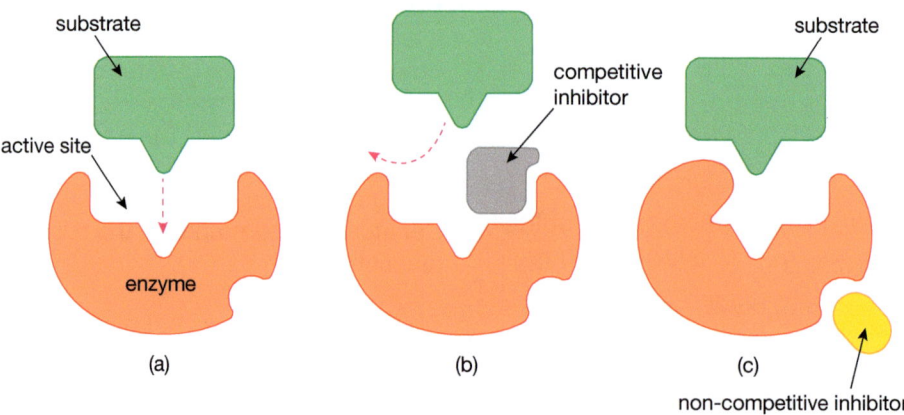

▲ Figure 3.9 The actions of enzyme inhibitors. (a) Normal enzyme–substrate reaction. (b) The effect of a competitive inhibitor. (c) The effect of a non-competitive inhibitor.

> **DID YOU KNOW?**
> Methanol (methyl alcohol) is a highly poisonous (toxic) chemical. If a person drinks methanol, an enzyme in the body oxidises it to formaldehyde and formic acid. These chemicals attack the optic nerve leading from the eye to the brain, causing blindness. Doctors give ethanol (ethyl alcohol) as an antidote for methanol poisoning, because ethanol competitively inhibits the oxidation of methanol. The enzyme oxidises ethanol in preference to methanol, so the toxic by-products do not build up and cause nerve damage.

Competitive inhibitors are molecules with a similar shape to the substrate (Figure 3.9(b)). They fit into the active site, stopping the substrate from entering. However, this is *temporary* – the inhibitor can leave and the substrate can take its place. This is why these inhibitors are called competitive – the inhibitor and the substrate are *competing* for the active site.

The presence of the inhibitor slows the rate of reaction. The greater the concentration of inhibitor (relative to substrate) the more effect it will have on the rate of reaction. Figure 3.10 shows how a competitive inhibitor affects enzyme activity at different concentrations of substrate. Notice that at high concentrations of substrate, the inhibitor has no effect on the activity of the enzyme.

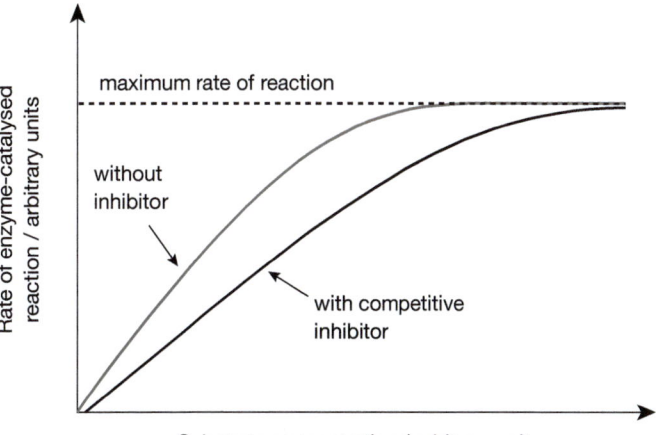

▲ Figure 3.10 The effect of a competitive inhibitor at different concentrations of substrate.

Non-competitive inhibitors do not have a shape like that of the substrate (Figure 3.9(c)). They do not attach to the active site, but to other parts of the enzyme. When they attach, they change the shape of the whole enzyme molecule, including the active site. The active site can no longer receive the substrate, so the reaction slows down.

This time, the enzyme will not be able to catalyse the reaction even if the concentration of substrate is increased, because the substrate cannot fit into the active site. The relative concentrations of inhibitor and substrate do not affect the rate of reaction.

> **DID YOU KNOW?**
> The ions of 'heavy metals' such as silver (Ag^+), mercury (Hg^{2+}) and lead (Pb^{2+}) are highly poisonous (toxic). This is because they bind irreversibly to proteins and act as non-competitive inhibitors of enzymes.

ACTIVITY 4

▼ PRACTICAL: INVESTIGATING THE EFFECT OF TEMPERATURE ON THE ACTIVITY OF AMYLASE

The digestive enzyme amylase breaks down starch into the sugar maltose. The activity of the amylase can be measured by recording the speed at which the starch disappears.

Figure 3.11 shows apparatus which can be used to record how quickly the starch is used up.

Safety note: Wear eye protection and avoid skin contact with the reactants.

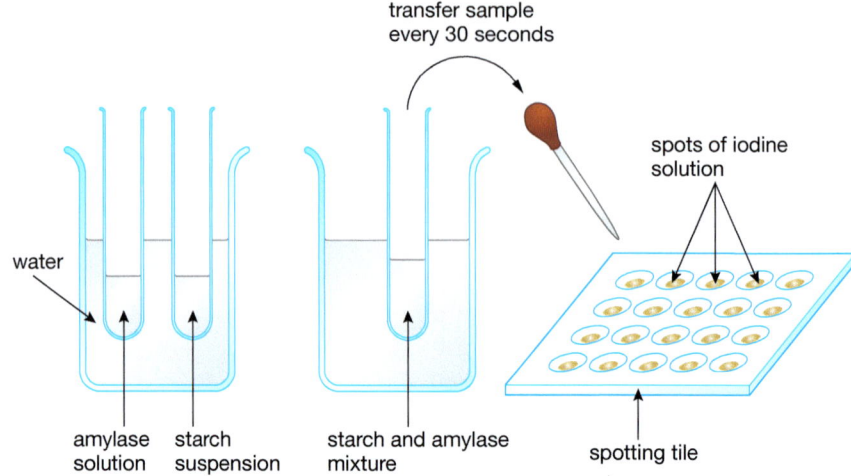

▲ Figure 3.11 Apparatus used to find the effect of temperature on the activity of amylase.

Spots of iodine solution are placed in the dips on the spotting tile. A syringe is used to place 5 cm^3 of starch suspension in one boiling tube. A different syringe is used to place 5 cm^3 of amylase solution in another tube. The beaker is filled with water at 20 °C. Both boiling tubes are placed in the beaker of water for 5 minutes, and the temperature of the water bath is recorded.

The amylase solution is then poured into the starch suspension, and the tube containing the mixture is left in the water bath. Immediately, a pipette is used to remove a small sample of the mixture from the tube and add it to the first drop of iodine solution on the spotting tile. The colour of the iodine solution is recorded.

A sample of the mixture is taken every 30 seconds for 10 minutes and tested for starch as above, until the iodine solution remains yellow, showing that all the starch has been used up.

The experiment is repeated, with the water bath at different temperatures between 20 °C and 60 °C. A set of results showing the colour changes at each temperature is given in the table opposite.

Time / min	Temperature / °C				
	20	30	40	50	60
0.0	blue-black	blue-black	blue-black	blue-black	blue-black
0.5	blue-black	blue-black	brown	blue-black	blue-black
1.0	blue-black	blue-black	yellow	blue-black	blue-black
1.5	blue-black	blue-black	yellow	blue-black	blue-black
2.0	blue-black	blue-black	yellow	brown	blue-black
2.5	blue-black	blue-black	yellow	brown	blue-black
3.0	blue-black	blue-black	yellow	brown	blue-black
3.5	blue-black	blue-black	yellow	yellow	blue-black
4.0	blue-black	blue-black	yellow	yellow	blue-black
4.5	blue-black	blue-black	yellow	yellow	blue-black
5.0	blue-black	blue-black	yellow	yellow	blue-black
5.5	blue-black	blue-black	yellow	yellow	blue-black
6.0	blue-black	brown	yellow	yellow	blue-black
6.5	blue-black	brown	yellow	yellow	blue-black
7.0	blue-black	yellow	yellow	yellow	blue-black
7.5	blue-black	yellow	yellow	yellow	brown
8.0	blue-black	yellow	yellow	yellow	brown
8.5	brown	yellow	yellow	yellow	yellow
9.0	brown	yellow	yellow	yellow	yellow
9.5	yellow	yellow	yellow	yellow	yellow
10.0	yellow	yellow	yellow	yellow	yellow

The rate of reaction can be calculated from the time taken for the starch to be broken down fully, as shown by the colour change from blue-black to yellow. For example, at 50 °C the starch had all been digested after 3.5 minutes. The rate is found by dividing the volume of starch (5 cm³) by the time:

$$\text{Rate} = \frac{5.0 \text{ cm}^3}{3.5 \text{ min}} = 1.4 \text{ cm}^3 \text{ per min}$$

Plotting a graph of rate against temperature should produce a curve similar to the one shown in Figure 3.6. Try this, using the results in the table. Better still, you may be able to do this experiment and provide your own results.

If your curve is not exactly the same as the one in Figure 3.6, can you suggest reasons for this? How could you improve the experiment to get more reliable results?

Safety note: Wear eye protection and avoid skin contact with the reactants. Wash hands after chopping the potato. Take care with the blender – the blades are very sharp.

ACTIVITY 5

▼ PRACTICAL: INVESTIGATING THE EFFECT OF pH ON THE ACTIVITY OF CATALASE

Buffer solutions are solutions of salts that resist changes in pH. Different buffer solutions can be prepared to maintain different values of pH. Buffer solutions are useful for finding the effect of pH on enzyme activity.

Hydrogen peroxide (H_2O_2) is a product of metabolism. Hydrogen peroxide is toxic (poisonous), so it must not be allowed to build up in cells. The enzyme catalase protects cells by breaking down hydrogen peroxide into the harmless products water and oxygen:

$$2H_2O_2 \rightarrow 2H_2O + O_2$$

Potato cells contain a high concentration of catalase. A large potato is chopped into small pieces and placed in a blender with an equal volume of distilled water. The blender is switched on to mince up the potato tissue and release the catalase from the cells. The remains of the potato tissue are allowed to settle to the bottom and the liquid extract above the debris is removed.

The extract is tested for catalase activity at different values of pH. A graduated syringe is used to place 5 cm³ of extract in a boiling tube. Another syringe is used to add 5 cm³ of pH 7 buffer solution to the same boiling tube. The tube is shaken gently to mix the buffer with the potato extract. The mixture is left for 5 minutes, then 5 cm³ of 5% hydrogen peroxide solution is added to the tube from a third syringe. A bung and delivery tube are quickly inserted into the boiling tube and the end of the delivery tube is placed in a beaker of water (Figure 3.12).

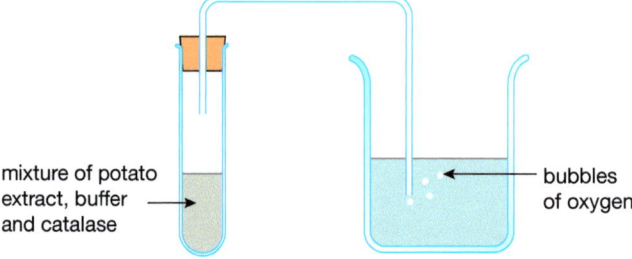

▲ Figure 3.12 Apparatus to investigate the effect of pH on catalase activity.

The bubbles of oxygen gas produced in the first minute after the hydrogen peroxide is added are counted. The number of bubbles per minute is a measure of the initial reaction rate.

The experiment is repeated, using different buffers. Some results are shown in the table below.

pH	Rate of reaction / bubbles per minute
5	6
6	39
7	47
8	14

Plot a bar chart of the results.

- Which pH gives the fastest rate of reaction?
- How could you modify this experiment to find a more accurate value for the optimum pH for the enzyme catalase? (Think – are four values of pH enough?)
- Is there a better way to measure the rate of reaction than counting bubbles of gas?
- How could you control the temperature during the experiment?
- How could you modify the experiment to make the results more reliable? (Think – is one set of results enough?)

IMMOBILISED ENZYMES

Enzymes can be extracted from cells or tissues and used for industrial or medical purposes. For example, the amylase used in Activity 4 is usually obtained from a fungus. Preparations of enzymes have a wide range of uses, for example:

- stain removers in biological detergents
- conversion of starch to glucose in the food industry
- preservatives in food and drinks
- clarification (clearing) of fruit juices
- making cheese
- manufacture of paper
- detection of blood glucose levels in people with diabetes
- measuring the level of the waste product called urea in the blood.

Some of these examples use enzymes dissolved in water ('in solution'). However, some enzymes are more useful if they are not in solution, but are instead *immobilised* by being attached to, or trapped within, an insoluble material. Using an **immobilised enzyme** has several advantages over the use of a dissolved enzyme:

- The enzymes are much more stable at high temperatures and are less likely to denature.
- The enzymes are more resistant to changes in pH.
- The enzymes are less likely to be broken down by organic solvents.
- The products are uncontaminated by enzyme and can be collected more easily.
- The enzyme can be kept and re-used.
- An industrial process can use columns of immobilised enzyme, allowing large-scale production.

Enzymes can be immobilised in a number of different ways. Some methods involve attaching the enzyme molecules to the surface of a material such as porous glass or cellulose. Other methods involve trapping the enzyme molecules in a permeable membrane, such as nylon, or in a polymer such as **alginate** (see Activity 6).

Three uses of immobilised enzymes are described below.

LACTOSE-FREE MILK

Milk contains a disaccharide sugar called lactose (see the section on carbohydrates earlier in this chapter). Babies get their nutrition from milk, and they produce an enzyme called **lactase**, which breaks down lactose into glucose and galactose.

Many adults also drink milk and eat milk products such as butter and cheese. However, many adults are unable to produce lactase, so they cannot digest lactose. This is called 'lactose intolerance'. If a lactose-intolerant person drinks milk, the lactose is instead broken down by bacteria in their gut, producing waste gases. The person suffers from painful stomach cramps, nausea and diarrhoea.

Immobilised lactase can be used to produce lactose-free milk and milk products for lactose-intolerant people (see Activity 6).

The lactase is immobilised in porous beads and held in a column. Milk is passed through the column, where the lactase breaks down the lactose. Lactase is an expensive enzyme, and this method means it can be kept and re-used, lowering the cost.

> **DID YOU KNOW?**
> It has been estimated that three-quarters of the adult population of the world is lactose-intolerant to some degree.

> **DID YOU KNOW?**
> Despite their liking for milk and cream, many cats are lactose-intolerant. Dairy products can give cats problems similar to those in lactose-intolerant humans.

BREAKDOWN OF SUCROSE

Sucrose is a disaccharide of glucose and fructose (see Figure 3.1). The enzyme **invertase** is produced by yeast cells, and breaks down sucrose into its component monosaccharides.

Invertase is used commercially to break down sucrose to produce a mixture of glucose and fructose, called 'invert syrup'. Invert syrup is widely used for sweetening products in the food and drinks industry.

Commercial invertase is extracted from yeast and trapped in alginate beads (like the lactase in Activity 6). It is then used in this immobilised form.

TESTING FOR GLUCOSE IN BLOOD OR URINE

The disease called **diabetes** can result in higher than normal levels of glucose in the blood and the appearance of glucose in the urine (see Chapter 9). People with diabetes have to test their blood and urine for glucose at regular intervals. This can be done in two ways – using test strips or with biosensors.

Test strips use two immobilised enzymes. An enzyme called glucose oxidase catalyses the oxidation of glucose to gluconic acid and hydrogen peroxide:

$$\text{glucose} + O_2 \rightarrow \text{gluconic acid} + H_2O_2 \qquad \text{(Reaction 1)}$$

The hydrogen peroxide produced in Reaction 1 oxidises a colourless organic substance (XH_2) to its coloured form, X, by removing hydrogen:

$$\underset{\text{(colourless)}}{XH_2} + H_2O_2 \rightarrow 2H_2O + \underset{\text{(coloured)}}{X} \qquad \text{(Reaction 2)}$$

('X' represents the organic substance, which is colourless when reduced and coloured when oxidised.)

Reaction 2 is catalysed by another enzyme called peroxidase. The amount of the coloured substance X produced is a direct measure of the concentration of glucose present in the sample of blood or urine. This method is very specific for glucose – the reactions will not be affected by other sugars in the sample. The glucose oxidase, peroxidase and XH_2 are attached to a cellulose pad on a test strip, which is dipped in the blood or urine sample. The colour change of the test strip gives an approximate measure of the concentration of glucose (Figure 3.13).

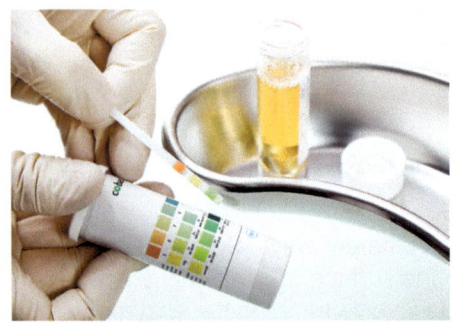

▲ Figure 3.13 Coloured test strips are used to detect glucose in urine.

> **Safety note:** Wear eye protection and avoid skin contact with the lactase, alginate solution and gel. Do not taste the whole milk or the lactose-free milk.

ACTIVITY 6

▼ PRACTICAL: USING IMMOBILISED LACTASE TO PRODUCE LACTOSE-FREE MILK

Lactase catalyses the breakdown of lactose into glucose and galactose. This is used to make lactose-free milk for people who are lactose-intolerant. The lactase is immobilised in beads of calcium alginate gel.

A small syringe is used to add 2 cm^3 of lactase solution to a beaker containing 8 cm^3 of 2% sodium alginate solution. The two are mixed thoroughly using a stirring rod. Using the same syringe, some of the mixture is removed from the beaker and added drop by drop to another beaker, containing 100 cm^3 of a solution of 1.5% calcium chloride.

As the mixture drops into the solution, the sodium alginate is converted into insoluble calcium alginate, forming gelatinous (jelly-like) beads. The beads contain the enzyme immobilised in the calcium alginate gel. (At this stage, it is important that the tip of the syringe does not touch the calcium chloride solution, or it will block with gel.)

The beads are left to set for five minutes. Then the mixture is poured through a sieve to separate the beads from the solution. The beads are washed with distilled water.

The beads are packed into the barrel of a 10 cm^3 syringe, which has its outlet covered by a small piece of nylon gauze. The syringe is clamped upright, with the outlet tube closed by an adjustable clip. Fresh milk (not UHT) is poured into the open end of the syringe barrel (Figure 3.14).

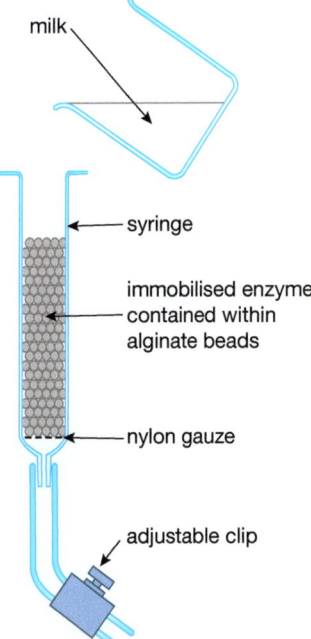

▲ Figure 3.14 Using lactase immobilised in alginate beads to produce lactose-free milk.

The milk leaving the syringe is tested using glucose test strips. After a while, glucose will be present in the milk that has passed through the column of beads. As a Control, the milk is tested before treatment – it will not contain any glucose.

LOOKING AHEAD – BIOSENSORS

Another way of measuring blood glucose is to use a **biosensor**. This is an instrument that quickly and easily measures the concentration of glucose, and gives a more accurate reading than using test strips. A drop of the person's blood is placed onto the probe of the biosensor (Figure 3.15).

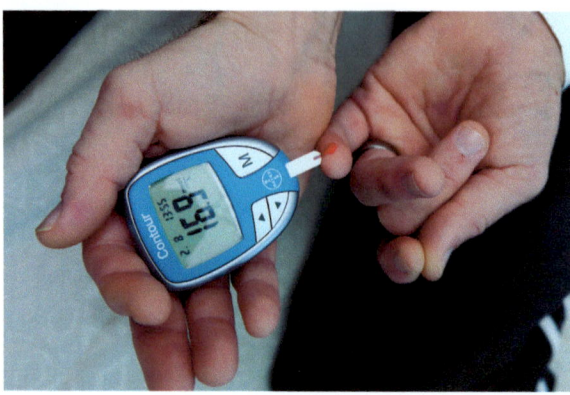

▲ Figure 3.15 A biosensor for measuring blood glucose levels. A drop of blood has been placed onto the probe of the biosensor.

The biosensor uses the enzyme glucose oxidase described above. At the end of the probe, the enzyme is immobilised in a gel attached to an instrument called an oxygen electrode. This measures the concentration of oxygen dissolved in a solution. When the drop of blood is placed on the probe, glucose in the blood diffuses into the gel. The enzyme breaks down this glucose (see Reaction 1 on page 44), using up oxygen. The electrode measures the oxygen level and changes it into an electrical signal, which is converted into a reading of blood glucose concentration on a meter.

Biosensors using immobilised enzymes are used in medicine to measure a number of other substances in the blood, including alcohol and other drugs.

CHAPTER QUESTIONS

SKILLS INTERPRETATION

1 Copy and complete the following table.

Biological molecule	'Building blocks' of the molecule	Chemical elements in the molecule
carbohydrate	simple sugars (monosaccharides)	
lipid		carbon, hydrogen, oxygen
protein		

SKILLS EXECUTIVE FUNCTION

2 Describe how to carry out a chemical test for the following substances. State the result if the test is positive.

a starch

b glucose

SKILLS CRITICAL THINKING

3 a What is an enzyme?

SKILLS INTERPRETATION

b Draw a labelled diagram to explain the 'lock and key' model of enzyme action.

SKILLS REASONING

c *'The lock and key model helps us to understand the effect of a competitive inhibitor on an enzyme-controlled reaction'.* Explain the meaning of this statement.

d From your knowledge of inhibitors, suggest how you could find out if an enzyme-catalysed reaction was being inhibited by a competitive or non-competitive inhibitor.

SKILLS EXECUTIVE FUNCTION

4 The enzyme lactase catalyses the breakdown of lactose into glucose and galactose:

lactose —lactase→ glucose + galactose

One of the products of the reaction (galactose) acts as a competitive inhibitor of lactase.

The apparatus below uses immobilised lactase to produce lactose-free milk.

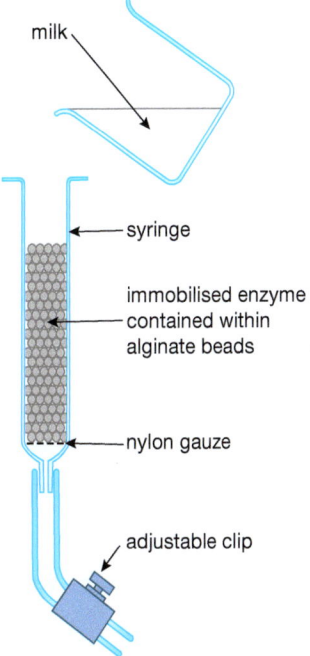

The speed of flow of the milk through the column of alginate beads affects the rate of the enzyme-catalysed reaction. If the flow is too fast, there is not enough time for the reaction to take place. If the flow is too slow, galactose builds up in the syringe and inhibits the reaction.

a Using this apparatus, how could you investigate the effect of the speed of flow of the milk on the rate of breakdown of lactose?

SKILLS CRITICAL THINKING

b Name the independent variable and the dependent variable in this investigation.

c State three advantages of using immobilised enzymes to make useful products.

UNIT QUESTIONS

SKILLS CRITICAL THINKING

 1 Which of the following is the best definition of 'differentiation'?

  A The organisation of the body into cells, tissues and organs

  B A type of cell division resulting in the growth of an embryo

  C The adaptation of a cell for its function

  D The process by which the structure of a cell becomes specialised for its function

(Total 1 mark)

 2 Which of the following are components of DNA?

  A Deoxyribose, uracil and phosphate

  B Ribose, adenine and guanine

  C Deoxyribose, phosphate and adenine

  D Ribose, thymine and cytosine

(Total 1 mark)

 3 Which of the following is the function of transfer RNA (tRNA)?

  A Transporting amino acids to a ribosome

  B Coding for the order of amino acids

  C Transcription of the DNA

  D Translation of the DNA

(Total 1 mark)

 4 The base sequence for the same length of DNA before and after a gene mutation was as follows:

  Before mutation: ATT TCC GTT ATC CGG

  After mutation: ATT CCG TTA TCC GGA

Which type of mutation took place?

  A duplication

  B deletion

  C substitution

  D inversion

(Total 1 mark)

 5 The statements below show some of the stages in the production of human insulin from genetically modified bacteria.

  1. DNA for insulin inserted into plasmids

  2. Bacteria cloned

  3. Plasmids inserted into bacteria

  4. DNA for insulin cut out using restriction enzyme

Which of the following shows the correct sequence of steps in the process?

A 2 → 1 → 4 → 3
B 4 → 2 → 3 → 1
C 4 → 1 → 3 → 2
D 2 → 3 → 4 → 1

(Total 1 mark)

SKILLS CRITICAL THINKING

6 Which of the following enzymes is used to join together pieces of DNA?

A ligase
B DNA polymerase
C protease
D restriction enzyme

(Total 1 mark)

7 Which of the following cell processes needs a source of metabolic energy?

A diffusion
B osmosis
C respiration
D active transport

(Total 1 mark)

SKILLS REASONING

8 A bag made of Visking tubing contains a 5% solution of sucrose. It is placed in a beaker containing a 10% solution of sucrose. Which of the following will take place?

A More water will enter the Visking tubing than leaves it and the volume of the tubing will increase
B More water will leave the Visking tubing than enters it and the volume of the tubing will decrease
C There will be no net movement of water and the volume will not change
D At first the volume of the Visking tubing will increase, but it will then decrease as pressure builds up in the bag

(Total 1 mark)

SKILLS CRITICAL THINKING

9 Biuret solution is used to test for which of the following substances?

A starch
B glucose
C protein
D lipid

(Total 1 mark)

SKILLS REASONING

10 How does denaturing by heat prevent an enzyme from working?

A By providing kinetic energy to the substrate molecules
B By preventing the substrate from binding with the active site
C By preventing the products from leaving the active site
D By increasing the optimum temperature for the enzyme

(Total 1 mark)

SKILLS CRITICAL THINKING

11 Enzymes increase the rate of a reaction. Consider the following statements about the 'lock and key' model of enzyme action:

1 The reaction can only take place in the active site.

2 The active site is the 'key' to the substrate 'lock'.

3 The enzyme is unable to take part in further reactions after the substrate has left the active site.

4 When the substrate enters the active site, the energy needed to start the reaction is reduced.

Which of the statements is/are true?

A 1 and 2

B 3 and 4

C 2 and 3

D 4 only

(Total 1 mark)

 12 Which of the following is *not* an advantage of using an immobilised enzyme?

A The enzyme can be collected with the products.

B The enzyme is less affected by changes in pH.

C The enzyme is more stable at high temperatures.

D The enzyme can be re-used.

(Total 1 mark)

13 Copy and complete the following passage about genes:

A gene is a section of a molecule called _____.
The molecule is found within the _____ of a cell, within thread-like structures called _____.
The strands of the molecule form a double helix joined by paired bases. The base adenine is always paired with its complementary base _____, and the base cytosine is paired with _____. During the process of transcription, the order of bases in one strand of the molecule is used to form _____, which carries the code for making proteins out to the cytoplasm.

(Total 6 marks)

SKILLS PROBLEM SOLVING

14 In a section of double-stranded DNA there are 100 bases, of which 30 are cytosine (C).

Calculate the total number of each of the following in this section of DNA:

a complementary base pairs (1)

b guanine (G) bases (1)

c thymine (T) bases (1)

d adenine (A) bases (1)

e deoxyribose sugar groups. (1)

(Total 5 marks)

SKILLS INTERPRETATION **15** Different particles (molecules or ions) move across cell membranes using different processes. The table below shows some ways in which active transport, osmosis and diffusion are similar and some ways in which they are different. Copy and complete the table, using a tick (✓) if the feature applies to a process or a cross (✗) if it does not apply.

Feature	Active transport	Osmosis	Diffusion
particles must have kinetic energy			
requires energy from respiration			
particles move down a concentration gradient			

(Total 3 marks)

SKILLS ANALYSIS **16** The graph shows the effect of temperature on an enzyme. The enzyme was extracted from a microorganism that lives in hot mineral springs near a volcano.

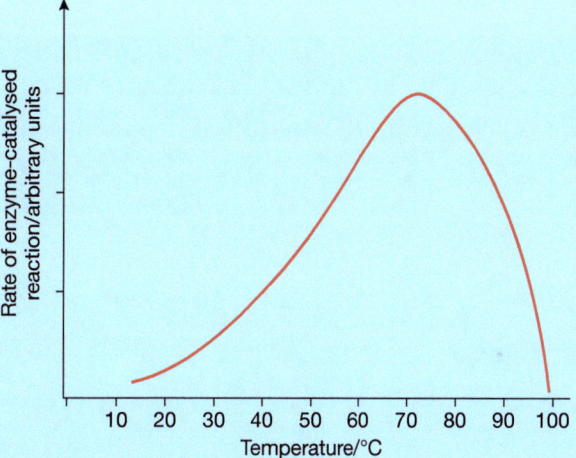

SKILLS REASONING

a What is the optimum temperature of this enzyme? (1)

 b Explain why the activity of the enzyme is greater at 60 °C than at 30 °C. (3)

 c The optimum temperature of enzymes in the human body is about 37 °C. Explain why this enzyme is different. (2)

 d Describe what happens to the enzyme at 90 °C. (2)

(Total 8 marks)

| NUTRITION AND ENERGY 53 | RESPIRATION AND GAS EXCHANGE 74 | INTERNAL TRANSPORT 92 |

UNIT 2

The human body is a very complex machine. It needs food for growth and repair of tissues, and as a supply of energy. In Chapter 4, you will learn how the digestive system obtains the nutrients we need from the food we eat.

In Chapter 5, you will learn how we obtain energy from our food, through the process of respiration. Aerobic respiration needs a supply of oxygen from the air, and produces carbon dioxide as a waste product. These gases are exchanged in the lungs, which are also described in Chapter 5.

In Chapter 6, you will find out about the structure and function of the blood and circulatory system.

4 NUTRITION AND ENERGY

Food is essential for life – the nutrients obtained from it are used in many different ways by the body. In this chapter, you will look at the different classes of food substance and how they are broken down by the digestive system and absorbed. You will also find out about some of the problems that happen when we do not eat a balanced diet.

LEARNING OBJECTIVES

- Explain the importance of a balanced diet and the recommended dietary intake of carbohydrates, lipids, proteins, vitamins A and C, calcium, iron, and fibre

- Know the sources and functions of carbohydrates, lipids, proteins, vitamins A, C and D, and the mineral ions calcium and iron

- Describe the causes and symptoms of deficiency diseases – kwashiorkor (due to lack of protein), anaemia (lack of iron), scurvy (lack of vitamin C) and blindness (lack of vitamin A)

- Describe how to investigate the vitamin C content of a sample of food

- Understand differences in dietary needs related to age, pregnancy, climate and occupation

- Describe how to investigate the energy content of a sample of food

- Know the structure of the human alimentary canal

- Know the types, structure and functions of teeth, the factors affecting their growth, and how to care for teeth and gums

- Explain how food is moved through the gut by peristalsis and the role of dietary fibre in the process

- Describe the functions of the mouth, oesophagus, stomach, small intestine, large intestine and pancreas in digestion

- Understand the role of digestive enzymes, including:
  - their site of production and action
  - the digestion of starch to glucose by amylase and maltase
  - the digestion of proteins to amino acids by proteases (pepsin, trypsin)
  - the digestion of lipids to fatty acids and glycerol by lipases

- Know that bile is produced by the liver and stored in the gall bladder, and understand the role of bile in neutralising stomach acid and emulsifying lipids

- Understand how the structure of the villi helps absorption of the products of digestion in the small intestine

- Understand the meaning of Body Mass Index (BMI) and how to calculate it, the role of obesity as a risk factor in early onset of diabetes and the significance of high cholesterol levels in atherosclerosis*

- Explain the importance of hygienic methods of food preparation, cooking, storage and preservation

*These topics are discussed in more detail in Chapter 6 and Chapter 9.

A BALANCED DIET

We need food for three main reasons:

- to supply us with a 'fuel' for energy
- to provide materials for growth and repair of tissues
- to help prevent disease and keep our bodies healthy.

The food that we eat is called our diet. No matter what you like to eat, if your body is to work properly and stay healthy, your diet must include carbohydrates, lipids, proteins, **minerals** and **vitamins**, as well as **dietary fibre** and water. Food should provide you with all of these things, but it is important to ensure your diet includes the *right amount* of each substance. A diet that provides enough of these substances and in the correct proportions to keep you healthy is called a **balanced diet** (Figure 4.1). We will look at each type of food in turn, to find out about its chemistry and the role it plays in the body.

▲ Figure 4.1 A balanced diet contains all the types of food the body needs, in just the right amounts.

CARBOHYDRATES

Carbohydrates only make up about 1% of the mass of the human body, but they have a very important role. They are the body's main 'fuel' for supplying cells with energy. Cells release this energy by breaking down the sugar glucose, in the process called cell respiration (see Chapter 5).

Glucose is found naturally in many sweet-tasting foods, such as fruits and vegetables. Other foods contain different sugars, such as the fruit sugar called fructose, and the milk sugar, lactose. Ordinary table sugar, the sort some people put in their tea or coffee, is called sucrose. Sucrose is the main sugar that is transported through plant stems. This is why we can extract it from sugar cane, which is the stem of a large grass-like plant. Sugars have two physical properties that you will probably know: they all taste sweet, and they are all soluble in water.

We can get all the sugar we need from natural foods such as fruits and vegetables, and from the digestion of starch. Many processed foods contain large amounts of *added* sugar. For example, a typical can of cola can contain up to seven teaspoons (27 g) of sugar! There is hidden sugar in many other foods. A tin of baked beans contains about 10 g of added sugar. This is on top of all the food that we eat with a more obvious sugar content, such as cakes, biscuits and sweets.

In fact, we get most of the carbohydrate in our diet not from sugars, but from starch, which is a polysaccharide of glucose (see Chapter 3). Starch is a large, *insoluble* molecule. Because it does not dissolve, it is found as a storage carbohydrate in many plants, such as potato, rice, wheat and millet, and in foods made from these plants, such as bread and pasta.

Another polysaccharide of glucose is glycogen, which is found in animal tissues such as liver and muscle. Polysaccharides are broken down into simple sugars during digestion, so that the sugars they contain can be absorbed into the blood.

Another carbohydrate that is a polymer of glucose is **cellulose**, a material that makes up plant cell walls. Humans are not able to digest cellulose, because the human gut does not make the enzyme needed to break down the cellulose molecule. This means that we are not able to use cellulose as a source of energy. However, it still has a vitally important function in our diet. It forms dietary fibre or 'roughage', which gives the muscles of the gut something to push against as food is moved through the intestine. This keeps the gut contents moving, avoiding constipation and helping to prevent serious diseases of the intestine, such as colitis and bowel cancer.

How much carbohydrate is there in a balanced diet? It is difficult to give one value for this, because different people need different amounts of food for energy. However, the recommended total carbohydrate in a healthy diet should supply about 50% of the body's daily energy needs, with less than 5% coming from sugars. Total daily carbohydrate for an adult should include about 30 g of dietary fibre.

LIPIDS (FATS AND OILS)

Lipids are fats and oils. Fats are solid at room temperature and oils are liquid. They are made of fatty acids and glycerol (see Chapter 3). Meat, butter, cheese, milk, eggs and oily fish are all rich in animal fats; so are foods fried in animal fat. Vegetable oils include many types used for cooking, such as olive oil, corn oil and rapeseed oil, as well as products made from oils, such as margarine (Figure 4.2).

▲ Figure 4.2 These foods are all rich in lipids.

Lipids make up about 10% of our body's mass. They form an essential part of the structure of all cells, and fat is also deposited in certain parts of the body as a long-term store of energy (for example, under the skin and around the heart and kidneys). The fat layer under the skin acts as insulation, reducing heat loss through the surface of the body. Fat around organs such as the kidneys helps to protect them from mechanical damage.

Although lipids are an essential part of our diet, too much lipid is unhealthy. In particular, you must make sure your diet does not contain too much **cholesterol** (a lipid compound) or saturated fat. Cholesterol is a substance that the body gets from food such as eggs and meat, but we also make cholesterol in our liver. It is an essential part of all cells, but too much cholesterol is linked to heart disease (see Chapter 6).

The recommended total lipid in a healthy diet should supply less than 35% of the body's daily energy needs, with less than a third of this coming from saturated fat.

> **DID YOU KNOW?**
> *Saturated* lipids (saturated fats) are more common in food from animal sources, such as meat and dairy products. 'Saturated' is a word used in chemistry, which means that the fatty acids of the lipids contain no double bonds. Other lipids are *unsaturated*, which means that their fatty acids contain double bonds. These are more common in plant oils. There is evidence that unsaturated lipids are healthier for us than saturated ones.

PROTEINS

Proteins are large molecules made of long chains of amino acids (see Chapter 3). They make up about 18% of the mass of the body. This is the second largest fraction after water. All cells contain protein, so we need it for growth and repair of tissues. Many compounds in the body are made from protein, including enzymes.

Most foods contain some protein, but certain foods such as meat, fish, cheese and eggs are particularly rich in it. You will notice that these foods are animal products. Plant material generally contains less protein, but some foods, especially beans, peas and nuts, are richer in protein than others.

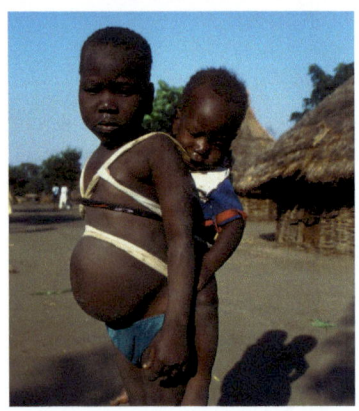

▲ Figure 4.3 This child is suffering from a lack of protein in his diet, a disease called kwashiorkor. His swollen belly is not due to a full stomach, but is caused by fluid collecting in the tissues. Other symptoms include loss of weight, poor muscle growth, general weakness and flaky skin.

> **DID YOU KNOW?**
> Humans need 20 different amino acids. They can make 10 of them, but the other 10 have to be taken in as part of the diet. These 10 are called essential amino acids. There are higher amounts of essential amino acids in meat, fish, eggs and dairy products. If you are a vegetarian, you can still get all the essential amino acids you need, as long as you eat a varied diet that includes a range of different plant materials.

> **DID YOU KNOW?**
> A disease that is caused by a lack of a particular nutrient is called a *nutrient deficiency disease*. Later in this chapter you will learn about other deficiency diseases caused by a lack of certain vitamins and minerals.

However, we do not need much protein in our diet to stay healthy. Nutrition experts recommend a daily intake of 0.75 grams of protein per kilogram bodyweight per day for healthy adults. For example:

- if an adult weighs 60 kg, they need: (60 × 0.75) = 45 g of protein per day
- if an adult weighs 75 kg, they need: (75 × 0.75) = 56 g of protein per day

In more economically developed countries, people often eat far more protein than they need. In many poorer countries, a disease called **kwashiorkor** – caused by a lack of protein – is common (Figure 4.3).

MINERALS

All the foods you have read about so far are made from just five chemical elements: carbon, hydrogen, oxygen, nitrogen and sulfur. Our bodies contain many other elements which we get from our food as minerals or mineral ions. Some are present in large amounts in the body, for example calcium, which is used for making teeth and bones. Others are present in much smaller amounts, but still have essential jobs to do. For instance, our bodies contain about 3 g of iron. This may not sound like much, but without it our blood would not be able to carry oxygen. Table 4.1 shows just a few of these minerals and the reasons they are needed.

Table 4.1 Some examples of minerals needed by the body.

Mineral	Approximate mass in an adult body / g	Location or role in body	Examples of foods rich in minerals
calcium	1000	making teeth and bones	dairy products, fish, bread, vegetables
phosphorus	650	making teeth and bones; part of many chemicals, e.g. DNA	most foods
sodium	100	in body fluids, e.g. blood	common salt, most foods
chlorine	100	in body fluids, e.g. blood	common salt, most foods
magnesium	30	making bones; found inside cells	green vegetables
iron	3	part of haemoglobin in red blood cells, helps carry oxygen	red meat, liver, eggs, some vegetables, e.g. spinach

If a person does not get enough of a particular mineral from their diet, they will show the symptoms of a mineral deficiency disease. For example, a one-year-old child needs to consume about 0.6 g (600 mg) of calcium every day, to make the bones grow properly and harden. Anything less than this over a prolonged period could result in poor bone development. The bones become deformed, a disease called rickets (Figure 4.4). Rickets can also be caused by a lack of vitamin D in the diet (see below).

Similarly, 16-year-olds need about 12 mg of iron in their daily food intake. If they do not get this amount, they cannot make enough haemoglobin for their red blood cells (see Chapter 6). This causes a condition called anaemia. People of any age can become anaemic; a person suffering from anaemia becomes tired and lacks energy, because their blood does not carry enough oxygen.

Figure 4.4 The legs of this child show the symptoms of rickets. This is due to lack of calcium or a lack of vitamin D in the diet, leading to poor bone growth. The bones stay soft and cannot support the weight of the body, so they become deformed.

VITAMINS

During the early part of the twentieth century, experiments were carried out that identified another class of food substances. When young laboratory rats were fed a diet of pure carbohydrate, lipid and protein, they all became ill and died. If they were fed on the same pure foods with a little added milk, they grew normally. The milk contained chemicals that the rats needed in small amounts to stay healthy. These chemicals are called vitamins. The results of one of these experiments are shown in Figure 4.5.

At first, the chemical nature of vitamins was not known, and they were given letters to distinguish between them, such as vitamin A, vitamin B and so on. Each vitamin was identified by the effect on the body of a lack of that vitamin (vitamin deficiency). For example, vitamin D is needed for growing bones to take up calcium salts. A deficiency of this vitamin can result in rickets (Figure 4.4), just as a lack of calcium can.

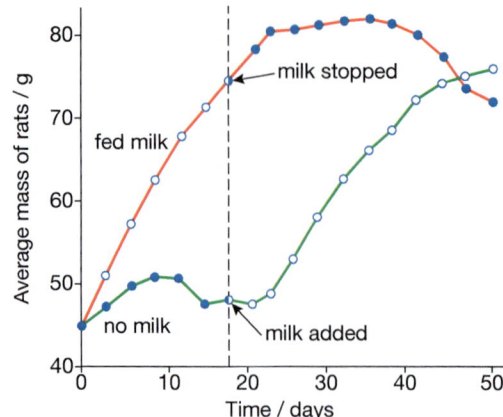

▲ Figure 4.5 Rats were fed on a diet of pure carbohydrate, lipid and protein, with and without added milk. Vitamins in the milk had a dramatic effect on their growth.

We now know the chemical structure of the vitamins and the exact ways in which they work in the body. Vitamin A is needed to make a light-sensitive chemical in the retina of the eye (see Chapter 8). One of the first signs of vitamin A deficiency is night blindness, where the person finds it difficult to see in dim light. Long-term deficiency of vitamin A results in complete blindness. This is especially common in children in poorer countries. It is estimated that between 250 000 and 500 000 badly nourished children in developing countries go blind each year as a result of vitamin A deficiency. About half of these children die within a year of becoming blind. This is because lack of vitamin A has other effects in the body, so the children are more likely to develop a severe illness from childhood infections such as measles.

Vitamin C is needed to make fibres of a material called connective tissue. This acts as a 'glue', binding cells together in a tissue and supporting internal organs. Connective tissue is found in the walls of blood vessels, in the skin and in many other tissues such as cartilage, tendons and ligaments. Vitamin C deficiency leads to a disease called scurvy, where wounds fail to heal, and bleeding occurs in various places in the body. This is especially noticeable in the gums (Figure 4.6).

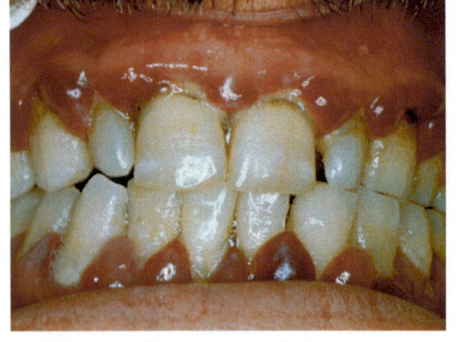

▲ Figure 4.6 Vitamin C helps lining cells, such as those in the mouth and gums, stick to each other. Lack of vitamin C causes scurvy, where the mouth and gums become damaged and bleed.

DID YOU KNOW?
The cure for scurvy was discovered as long ago as 1753. Sailors on long voyages often got scurvy because they ate very little fresh fruit and vegetables (the main source of vitamin C). A ship's doctor called James Lind wrote an account of how the disease could quickly be cured by eating fresh oranges and lemons. A British explorer called Captain Cook, on his world voyages in 1772 and 1775, kept his sailors healthy by making sure that they ate fresh fruit. By 1804, all British sailors were made to drink lime juice to prevent scurvy.

Vitamin B is not a single substance, but a collection of many different substances called the vitamin B group. It includes vitamins B1 (thiamine), B2 (riboflavin) and B3 (niacin). These compounds are involved in the process of cell respiration. Different deficiency diseases result if any of them are missing from the diet. For example, lack of vitamin B1 results in the weakening of the muscles and paralysis, a disease called beri-beri.

The main vitamins, their roles in the body and some foods which are good sources of them, are summarised in Table 4.2. Notice that we only need very small amounts of vitamins but we cannot stay healthy without them.

Table 4.2 Summary of the main vitamins. Note that you only need to remember the details of vitamins A, C and D.

Vitamin	Recommended daily amount in diet[1]	Use in the body	Effect of deficiency	Some foods that are a good source of the vitamin
A	0.8 mg	making a chemical in the retina; also protects the surface of the eye	night blindness, leading to total blindness and death, especially in children; damage to the cornea of the eye	fish liver oils, liver, butter, margarine, carrots
B1	1.1 mg	helps with cell respiration	beri-beri	yeast extract, cereals
B2	1.4 mg	helps with cell respiration	poor growth, dry skin	green vegetables, eggs, fish
B3	16 mg	helps with cell respiration	pellagra (dry red skin, poor growth, and digestive disorders)	liver, meat, fish
C	80 mg	sticks together cells lining surfaces such as the mouth	scurvy	fresh fruit and vegetables
D	5 µg	helps bones absorb calcium and phosphate	rickets, poor teeth	fish liver oils; also made in skin in sunlight

[1]Figures are the European Union's recommended daily intake for an adult (2012). 'mg' stands for milligram (a thousandth of a gram) and 'µg' for microgram (a millionth of a gram).

ACTIVITY 1

▼ PRACTICAL: INVESTIGATING THE VITAMIN C CONTENT OF LEMON JUICE

Vitamin C will remove the colour from a blue dye called DCPIP*. This colour change can be used to measure the amount of vitamin C present in a food. One method is to add DCPIP drop by drop to a solution containing vitamin C until the dye no longer changes colour.

A graduated pipette or syringe is used to place 2 cm^3 of a 1% solution of vitamin C in a test tube.

Using a 5 cm^3 graduated pipette or syringe, a 1% solution of DCPIP is added drop by drop to the vitamin C solution; the mixture is shaken gently after each drop. The DCPIP is added until the blue colour of the final drop does not disappear. The exact volume of DCPIP used is noted.

The procedure is then repeated, starting with 2 cm^3 of lemon juice instead of vitamin C solution. If more than 5 cm^3 of DCPIP is completely decolourised, the juice is diluted and the test repeated. The volume of DCPIP that was decolourised by the lemon juice is compared with the volume decolourised by the standard vitamin C solution. For example:

- The volume of DCPIP decolourised by 2 cm^3 of a 1% solution of vitamin C = 4.3 cm^3
- The volume of DCPIP decolourised by 2 cm^3 of lemon juice = 1.9 cm^3
- A 1% solution of vitamin C contains 1 g per 100 cm^3, or 0.01 g per cm^3
- Therefore, you can calculate the amount of vitamin C in the lemon juice:

    $\frac{1.9}{4.3} \times 0.01$ g per cm^3 = 0.0044 g per cm^3 (or 4.4 mg per cm^3)

> **Some points to note**
>
> - The procedure should be repeated several times and average values for the standard and sample calculated. This will check the reliability of the results.
> - The precision of the volume measurements is better if you use a graduated pipette rather than a syringe.
> - With strongly acidic juices such as lemon juice, the DCPIP does not decolourise completely – the end-point is a pale pink.
> - To measure vitamin C content in a solid food, such as an apple, you need to grind up the apple with some distilled water to extract the vitamin.
>
> If you have the opportunity to try this method, you could use it to test a number of hypotheses, such as:
>
> - Fresh fruit juice contains more vitamin C than old fruit juice.
> - Lemons contain more vitamin C than oranges.
> - Exposure to high temperatures (as in cooking) destroys vitamin C.
>
> *DCPIP stands for dichlorophenolindophenol.

Safety Note: Wear eye protection. Avoid skin contact with the DCPIP. Do not taste any of the juices or fruit.

ENERGY FROM FOOD

Some foods contain more energy than others. The energy content depends on the proportions of carbohydrate, lipid and protein that a food contains. Energy in a food is measured in kilojoules (kJ). If a gram of carbohydrate is fully oxidised, it produces about 17 kJ. A gram of lipid yields over twice as much as this (39 kJ). Protein can produce about 18 kJ per gram. Food labels usually show the energy content of the food, and list the amounts of different nutrients that it contains (Figure 4.7).

▲ Figure 4.7 Food packaging is labelled to show the proportions of different food types that the food contains, along with its energy content. The energy in units called kilocalories (kcal) is also shown, but scientists no longer use this old-fashioned unit.

Foods with a high percentage of lipid, such as butter or nuts, contain a large amount of energy. Others, such as fruits and vegetables, are mainly composed of water and have a much lower energy content (Table 4.3).

Table 4.3 Energy content of some common foods.

Food	kJ per 100g
margarine	3200
butter	3120
peanuts	2400
samosa	2400
chocolate	2300
Cheddar cheese	1700
table sugar	1650
cornflakes	1530
rice	1500
spaghetti	1450
fried beefburger	1100

Food	kJ per 100g
white bread	1060
chips	990
grilled beef steak	930
fried cod	850
roast chicken	770
boiled potatoes	340
milk	270
baked beans	270
yoghurt	200
boiled cabbage	60
lettuce	40

EXTENSION WORK

Food scientists measure the amount of energy in a sample of food by burning it in a calorimeter (Figure 4.8). The calorimeter is filled with oxygen, to make sure the food will burn easily. A heating filament carrying an electrical current ignites the food. The energy given out by the burning food is measured by using it to heat up water flowing through a coil in the calorimeter.

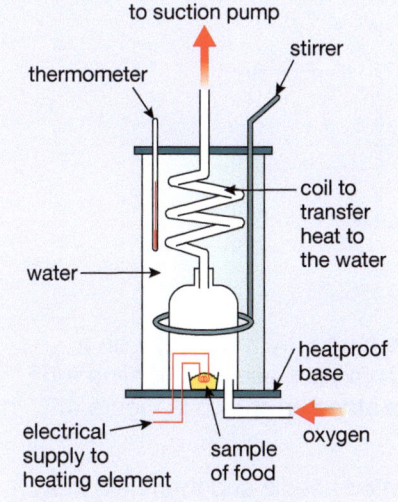

▲ Figure 4.8 A food calorimeter.

If you have samples of food that will easily burn in air, you can measure the energy in them by a similar method, using the heat from the burning food to warm up water in a test tube (see Activity 2).

Even while you are asleep you need a supply of energy – in order to keep warm, for your heart to keep beating, to allow messages to be sent through your nerves, and for other body functions. However, the energy you need at other times depends on the physical work you do. The total amount of energy that a person needs to keep healthy depends on their age and body size, and on the amount of activity they do. It also depends on the climate they live in: people in cold countries need more energy to maintain their body temperature. Table 4.4 shows some examples of how much energy is needed each day by people of different age, sex and occupation.

▼ Table 4.4 The daily energy needs of different types of people.

Age/sex/occupation of person	Energy needed per day / kJ
newborn baby	2000
child aged 2	5000
child aged 6	7500
girl aged 12–14	9000
boy aged 12–14	11 000
girl aged 15–17	9000
boy aged 15–17	12 000
female office worker	9500
male office worker	10 500
heavy manual worker	15 000
pregnant woman	10 000
breast-feeding woman	11 300

Remember that these are approximate figures and averages. Generally, the greater a person's weight, the more energy that person needs. This is why men, with a greater average body mass, need more energy than women. The energy needs of a pregnant woman are increased, mainly because of the extra weight she has to carry. A heavy manual worker, such as a labourer, needs extra energy for increased muscle activity. A woman who is breast-feeding needs extra energy to produce milk for her baby.

ACTIVITY 2

▼ PRACTICAL: MEASURING THE ENERGY CONTENT OF A FOOD

If a sample of food will burn well in air, its energy content can be measured using a simplified version of the food calorimeter (Figure 4.9). Suitable foods are dry pasta, crispbread or biscuits. It is not advisable to use nuts, since some people are allergic to them.

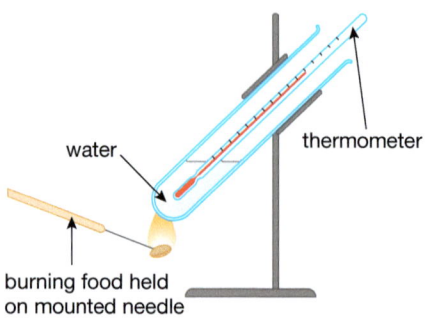

▲ Figure 4.9 Measuring the energy content of a food sample.

Safety Note: Do not taste any of the foods. Wear eye protection when burning the sample. When spearing each food sample with the needle, always spear downwards onto a cutting board to avoid injury. Wash your hands after the experiment. Do not use nuts.

First of all, the mass of the food sample is found, by weighing it on a balance. A measured volume of water (20 cm³) is placed in a boiling tube, and the tube is supported in a clamp on a stand as shown in Figure 4.9. The temperature of the water is recorded.

The food is speared on the end of a mounted needle and then held in a Bunsen burner flame until it catches fire (this may take 30 seconds or so). When the food is alight, the mounted needle is used to hold the burning food underneath the boiling tube of water so that the flame heats up the water. This is continued, relighting the food if it goes out, until the food will no longer burn.

The thermometer is used to stir the water gently, to make sure the heat is evenly distributed, and the final temperature of the water is measured.

Two facts are needed to calculate the energy content of the food:

- The energy needed to raise the temperature of 1 g of water by 1 °C is 4.2 joules.
- A volume of 1 cm³ of water has a mass of 1 g.

Multiplying the rise in temperature of the water by the mass of the water and then by 4.2 gives the number of joules of energy that were transferred to the water. Dividing this by the mass of the food gives the energy per gram:

Energy in joules per gram (J per g)

$$= \frac{(\text{final temperature} - \text{temperature at start}) \times 20 \text{ (g)} \times 4.2 \text{ (J per g per °C)}}{\text{mass of food (g)}}$$

For example, imagine you had a piece of pasta weighing 0.55 g. The starting temperature of the 20 g of water was 21 °C. After the burning pasta was used to heat up the water, the temperature of the water was 43 °C.

The energy content of the pasta is:

$$\frac{(43 - 21) \times 20 \times 4.2}{0.55} = 3360 \text{ J per g (3400 J per g to 2 significant figures)}$$

> **Comparison of the energy content of different foods**
>
> You may be able to use a similar method to find the energy content of suitable foods that will burn easily. Suggest a hypothesis about the energy content of the foods, and design an experiment to test your hypothesis. Explain how you will make sure your results are reliable.

It is not only the recommended energy requirements that vary with age, sex and pregnancy; the *content* of the diet will also vary (Table 4.5). For instance, younger women need extra iron in their diet because they lose blood during menstruation (periods). Women who experience 'heavy' periods (i.e. large volumes of blood loss) may need still more iron to avoid developing anaemia. Women who are breast-feeding require larger amounts of calcium, for milk production.

▼ Table 4.5 The daily mineral needs of different types of people.

Age and sex of person	Calcium / mg	Iron / mg
male 15–18 years	1000	11
female 15–18 years	1000	15 (may be higher in girls who experience heavy periods)
male 19–50 years	700	9
female 19–50 years	700	15 (may be higher in women who experience heavy periods)
pregnant woman	700*	27
breast-feeding woman	1250	9

*During pregnancy, a woman's body adapts by absorbing more calcium from her food and decreasing calcium excretion by the kidneys. This means that a pregnant woman needs the same amount of calcium as a woman who is not pregnant.

TEETH

In humans, only the lower jaw is movable, while the upper jaw is fused to the skull. Chewing is brought about by the cheek muscles, which move the lower jaw up and down, and allow some side-to-side movement.

Like most other mammals, humans have four types of teeth, called incisors, canines, premolars and molars (Figure 4.10). The arrangement of an animal's teeth is called its *dentition*.

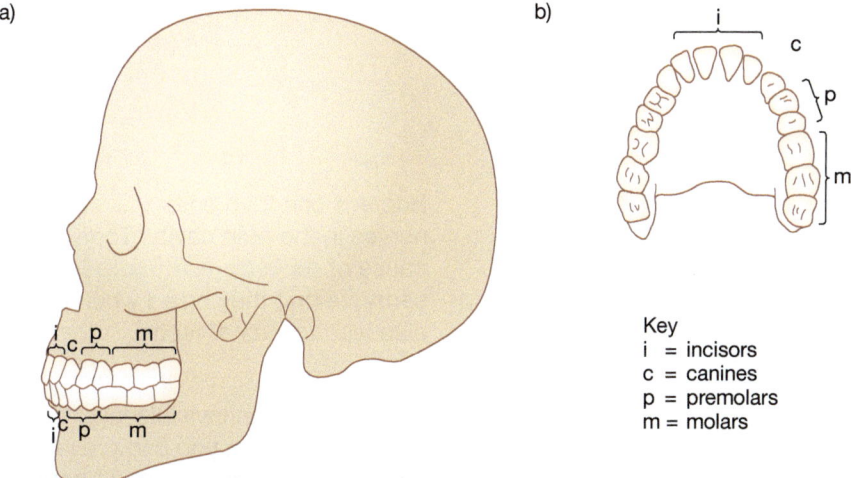

Key
i = incisors
c = canines
p = premolars
m = molars

▲ Figure 4.10 (a) Side view of human skull and teeth; (b) upper teeth shown from below.

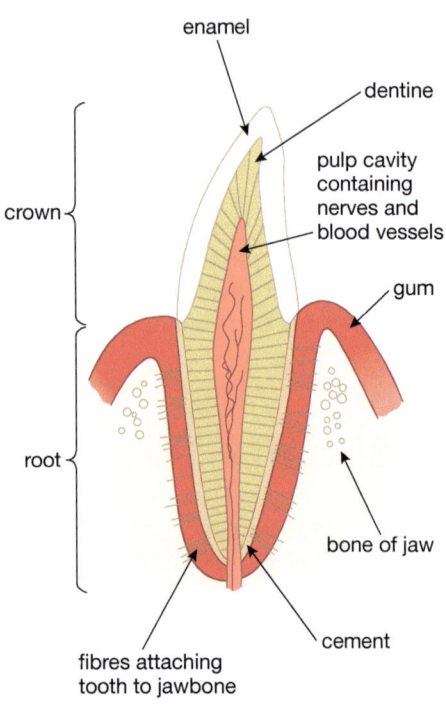

▲ Figure 4.11 The internal structure of an incisor tooth.

At the front of the mouth are the **incisors**. They are relatively sharp, chisel-shaped teeth, used for biting off pieces of food. Behind the incisors in each jaw is a pair of **canines** (one on each side of the mouth). In humans, the canines are similar in shape to the incisors and have the same biting function. Behind the canines are the 'cheek teeth' (**molars** and **premolars**).

The cheek teeth have a flatter top surface (crown), and the top and bottom sets of cheek teeth meet crown to crown, so that they can be used for chewing or crushing food. Human teeth are adapted to deal with a wide range of food types, from meat and fish to plant roots, stems and leaves.

The crown of a tooth is covered with a non-living material called **enamel** (Figure 4.11), which is the hardest substance in the body. Underneath the enamel is a softer material (but still about as hard as bone) called **dentine**. The middle of the tooth is called the **pulp cavity**. It contains blood vessels and nerves. There are fine channels running through the dentine, filled with cytoplasm. These cytoplasmic strands are kept alive by nutrients and oxygen from the blood vessels in the pulp cavity. The root of the tooth is covered with cement, containing fibres. This material anchors the tooth in the jawbone but allows a slight degree of movement when the person is chewing.

During their lifetime, mammals have two sets of teeth. The first set of teeth is called the milk teeth. In humans, these start to grow through the gum when a child is a few months old. By the age of about 21 or 22 months, a child will have 20 teeth, mainly milk teeth. From around the age of 7 years, the milk teeth are pushed out by permanent teeth growing underneath them. The molars at the back are present only as permanent teeth. Eventually a full set of 32 adult teeth is formed (Figure 4.10).

TOOTH DECAY

Tooth decay or **dental caries** is one of the most common diseases in the world. It is caused by bacteria in the mouth feeding on sugar. The bacteria break down the sugar, forming acids which dissolve the tooth enamel. Once the enamel is penetrated, the acid breaks down the softer dentine underneath. Eventually a cavity is formed in the tooth (Figure 4.12).

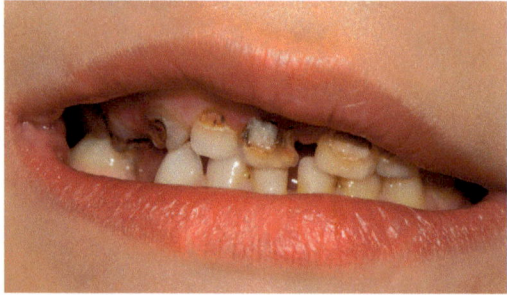

▲ Figure 4.12 A bad case of tooth decay. One of the causes was too much sugar in the person's diet.

Bacteria can then enter this cavity and enlarge it until the decay reaches the nerves in the pulp cavity. Then you feel the pain! These bacteria are also the cause of **periodontal disease**, where the gums become inflamed and so sensitive that they bleed when the teeth are brushed. Periodontal disease can also lead to loss of teeth.

DENTAL HYGIENE

Bacteria form an invisible layer on the surface of teeth, called **plaque**. One of the obvious ways you can prevent tooth decay is by regularly brushing your teeth. Plaque takes about 24 hours to form again, so teeth (and gums) should be brushed twice a day. Dentists advise the use of a toothbrush with a small head that will allow the bristles to reach into the crevices between the teeth or, better still, an electric toothbrush. Modern electric toothbrushes have a timed cycle, so that you know you have brushed your teeth for the right length of time. You should also use dental floss to clean between the teeth, where it is difficult to brush. If plaque is left on the teeth, it soon forms a hard deposit called **tartar**. This has to be removed by a dentist. You should have a dental check-up every 6 months, even when you have no obvious problems – the dentist may spot a developing tooth or gum problem before it gets worse.

Fluoride has been shown to reduce tooth decay by strengthening the enamel, especially when teeth are growing. In many places, fluoride is added to drinking water, and this has made a big difference to the incidence of tooth decay. If fluoride is not added to your drinking water, you can take fluoride tablets, or use fluoride toothpaste.

Finally, a good balanced diet is essential for teeth to grow healthily. Avoiding sweets and sugary drinks will reduce the supply of nutrients for the bacteria.

DIGESTION

> **KEY TERM**
>
> Digestion is the chemical and mechanical breakdown of food. It converts large insoluble molecules into small soluble molecules, which can be absorbed into the blood.

Food, such as a piece of bread, contains carbohydrates, lipids and proteins, but they are not the same carbohydrates, lipids and proteins as in our tissues. The components of the bread must be broken down into their 'building blocks' before they can be absorbed through the wall of the gut. This process is called **digestion**. The digested molecules – sugars, fatty acids, glycerol and amino acids – along with minerals, vitamins and water, can then be carried around the body in the blood. When they reach the tissues, they are reassembled into the molecules that make up our cells.

Digestion is speeded up by enzymes, which are biological catalysts (see Chapter 3). Digestive enzymes are made by the tissues and glands in the gut. Although most enzymes stay inside cells, digestive enzymes pass out of cells into the gut contents, where they act on the food. This *chemical* digestion is helped by *mechanical* digestion. Mechanical digestion is the physical breakdown of food. The most obvious place where this happens is in the mouth, where the teeth bite and chew the food, cutting it into smaller pieces that have a larger surface area. This means that enzymes can act on the food more quickly. Other parts of the gut also help with mechanical digestion. For example, muscles in the wall of the stomach contract to churn up the food while it is being chemically digested.

PERISTALSIS

> **KEY POINT**
>
> Remember that dietary fibre has an important role in peristalsis: fibre gives the muscles of the gut wall something to push against. Fibre in our diet is essential for avoiding a number of diseases of the intestine (see earlier in this chapter).

Muscles are also responsible for moving the food along the gut. The walls of the intestine contain two layers of muscle. One layer has fibres arranged in rings around the gut. This is the circular muscle layer. The other layer has fibres running along the length of the gut, and is called the longitudinal muscle layer. These two layers work together to push food along. When the circular muscles contract and the longitudinal muscles relax, the gut is made narrower. When the opposite happens – i.e. the longitudinal muscles contract and the circular muscles relax – the gut becomes wider. Waves of muscle contraction like this pass along the gut, pushing the food along, rather like squeezing toothpaste from a tube. This is called **peristalsis** (Figure 4.13). It means that movement of food in the gut does not depend on gravity – we can still digest food while standing on our heads!

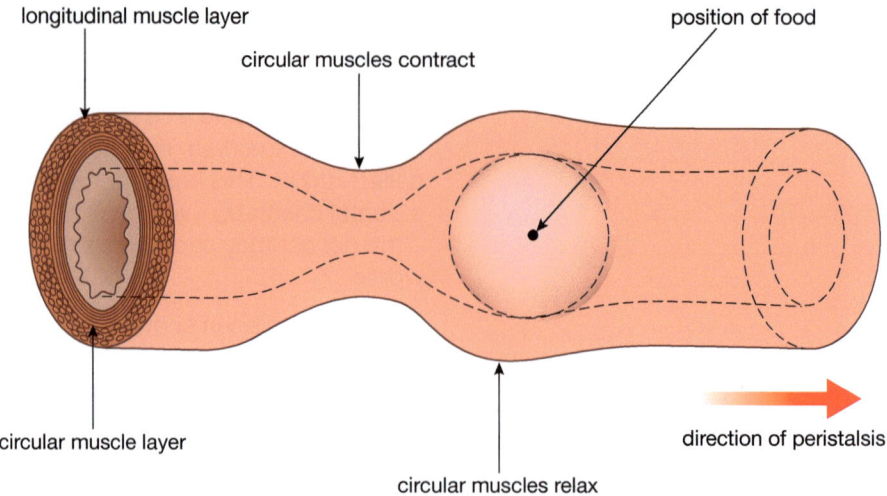

▲ Figure 4.13 Peristalsis: contraction of circular muscles behind the food narrows the gut, pushing the food along. When the circular muscles are contracted, the longitudinal ones are relaxed, and vice versa.

THE DIGESTIVE SYSTEM

Figure 4.14 shows a simplified diagram of the human digestive system. It is simplified so that you can see the order of the organs along the gut. The real gut is much longer than this, and coiled up so that it fills the whole space of the abdomen. Overall, its length in an adult is about 8 m. This gives plenty of time for the food to be broken down and absorbed as it passes through the gut.

Food is broken down in the mouth, in the stomach and in the first part of the small intestine (called the **duodenum**). The food is broken down using enzymes, which are made in the wall of the gut or by glands such as the **salivary glands** and the **pancreas**. Digestion continues in the last part of the small intestine (the **ileum**) and it is here that the digested food is absorbed. The last part of the gut, the large intestine, is mainly concerned with absorbing water out of the remains, and storing the waste products (**faeces**) before they are removed from the body.

The three main classes of food are broken down by three classes of enzyme. Carbohydrates are digested by enzymes called **carbohydrases**. Proteins are acted upon by **proteases**, and enzymes called **lipases** break down lipids. Some of the places in the gut where these enzymes are made are shown in Table 4.6.

▼ Table 4.6 Some of the enzymes that digest food in the human gut. The substances shown in bold are the end products of digestion that can be absorbed from the gut into the blood.

Class of enzyme	Examples	Digestive action	Source of enzyme	Where it acts in the gut
carbohydrases	amylase amylase maltase	starch → maltose[1] starch → maltose maltose → **glucose**	salivary glands pancreas wall of small intestine	mouth small intestine small intestine
proteases	pepsin trypsin peptidases	proteins → peptides[2] proteins → peptides peptides → **amino acids**	stomach wall pancreas wall of small intestine	stomach small intestine small intestine
lipases	lipase	lipids → **glycerol and fatty acids**	pancreas	small intestine

[1] Maltose is a disaccharide made of two glucose molecules joined together.
[2] Peptides are short chains of amino acids.

UNIT 2　　NUTRITION AND ENERGY

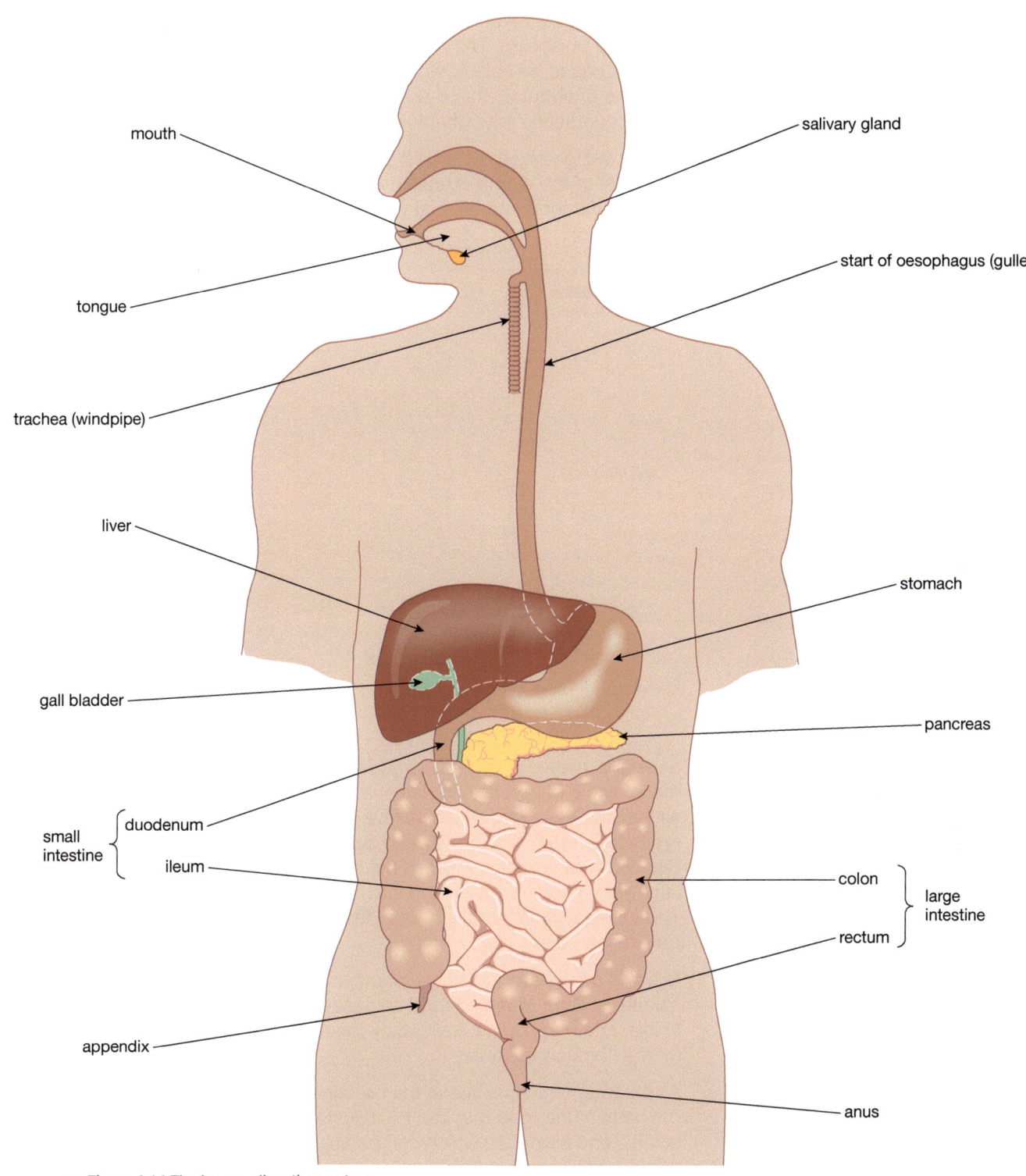

▲ Figure 4.14 The human digestive system.

Digestion begins in the mouth. **Saliva** helps moisten the food and contains the enzyme **amylase**, which starts the breakdown of starch. The chewed lump of food, mixed with saliva, then passes along the **oesophagus** (gullet) to the stomach.

The food is held in the stomach for several hours, while initial digestion of protein takes place. The stomach wall secretes hydrochloric acid, so the stomach contents are strongly acidic. This has a very important function. It kills

> **REMINDER**
>
> Amylase digests starch into maltose. Amylase is the enzyme, starch is the substrate and maltose is the product.

bacteria that are taken into the gut along with the food, helping to protect us from food poisoning. The protease enzyme that is made in the stomach, called **pepsin**, has to be able to work in these acidic conditions; it has an optimum pH value of about 2. This is unusually low – most enzymes work best at near neutral conditions (see Chapter 3).

The semi-digested food is held back in the stomach by a ring of muscle at the outlet of the stomach, called a **sphincter muscle**. When this relaxes, it releases the food into the first part of the small intestine, called the duodenum (Figure 4.15).

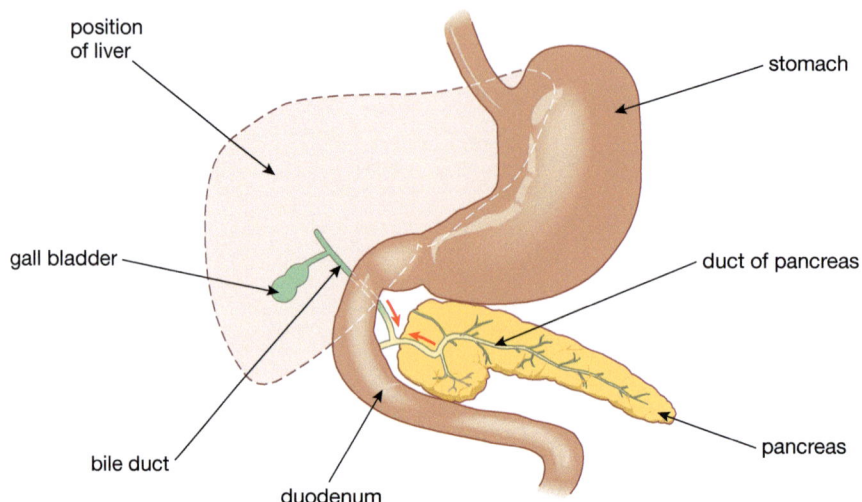

▲ Figure 4.15 The first part of the small intestine, the duodenum, receives digestive juices from the liver and pancreas through tubes called ducts.

Several digestive enzymes are added to the food in the duodenum. These are made by the pancreas and digest starch, proteins and lipids (Table 4.6). As well as this, the **liver** makes a digestive juice called **bile**. Bile is a green liquid that is stored in the **gall bladder** and passes down the **bile duct** on to the food. Bile does not contain enzymes but has another important function. It turns any large lipid globules in the food into an emulsion of tiny droplets (Figure 4.16). This increases the surface area of the lipid, so that lipase enzymes can break it down more easily.

Bile and pancreatic juice have another function. They are both alkaline. The mixture of semi-digested food and enzymes coming from the stomach is acidic, and must be neutralised by the addition of alkali before it continues on its way through the gut.

As the food continues along the intestine, more enzymes are added, until the parts of the food that can be digested have been fully broken down into soluble end products, which can be absorbed. This is the role of the last part of the small intestine, the ileum.

▲ Figure 4.16 Bile turns fats into an emulsion of tiny droplets for easier digestion.

ABSORPTION IN THE ILEUM

The ileum is highly adapted to absorb the digested food. The lining of the ileum has a very large surface area, which means that it can quickly and efficiently absorb the soluble products of digestion into the blood. The length of the intestine helps to provide a large surface area, and this is aided by folds in its lining. However, the greatest increase in area is due to tiny projections from the lining, called **villi** (Figure 4.17).

(a)

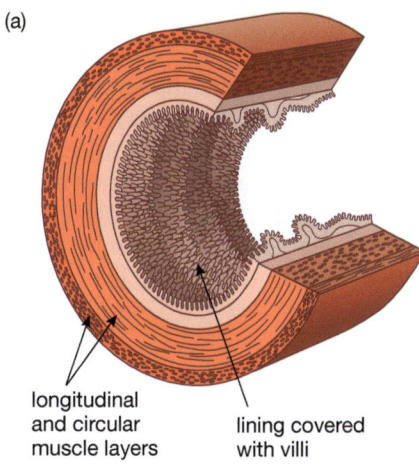

longitudinal and circular muscle layers

lining covered with villi

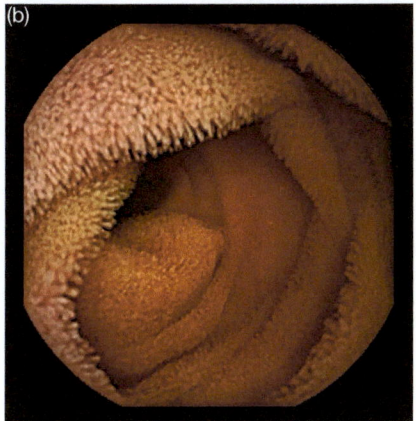

▲ Figure 4.17 (a) The inside lining of the ileum is adapted to absorb digested food by the presence of millions of tiny villi; (b) Photo of the inside of a patient's ileum, taken using a camera attached to an endoscope. You can see thousands of tiny villi covering the lining.

The singular of villi is 'villus'. Each villus is only about 1–2 mm long but there are millions of them, so the total area of the lining is thought to be around 300 m^2 – about the size of a tennis court. This provides a massive area in contact with the digested food. High-powered microscopy has revealed that the surface cells of each individual villus also have hundreds of minute projections, called **microvilli**. These increase the surface area for absorption even more (Figure 4.18).

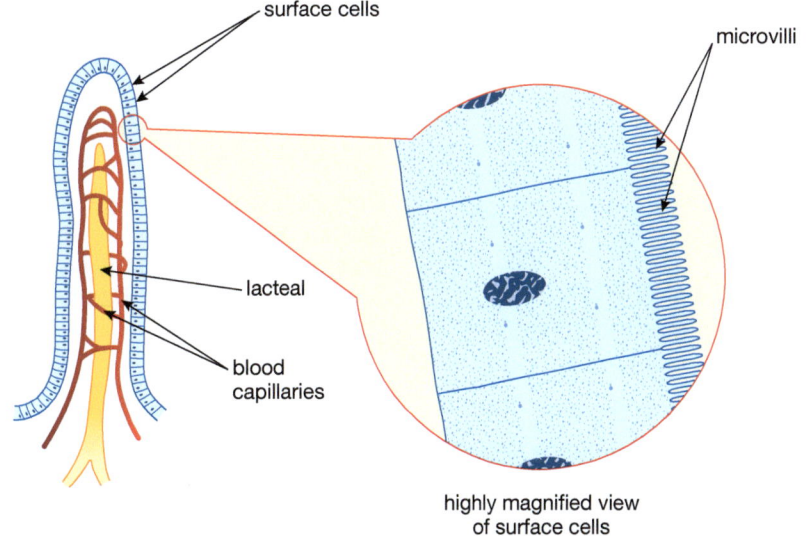

highly magnified view of surface cells

▲ Figure 4.18 Each villus contains blood vessels and a lacteal, which absorb the products of digestion. The surface cells of the villus are covered with microvilli, which further increase the surface area for absorption.

Each villus contains a network of blood capillaries. Most of the digested food enters these blood vessels but the products of fat digestion, as well as tiny fat droplets, enter a tube in the middle of the villus, called a **lacteal**. The lacteals form part of the body's **lymphatic system**, which transports a liquid called **lymph**. This lymph eventually drains into the blood system too. The lymphatic system is described in Chapter 6.

The surface of a villus is made of a single layer of cells called an epithelium. This means that there is only a short distance between the digested food in the ileum and the blood capillaries, making it easier for the products of digestion to diffuse through and enter the blood. The epithelium cells contain many mitochondria, which supply the energy needed for active transport of some substances.

In addition, each villus contains muscle fibres which contract to move the villus. The villi are in constant motion, keeping them in contact with the contents of the ileum and maintaining a steep concentration gradient for diffusion of the products of digestion.

The blood vessels from the ileum join up to form a large blood vessel called the **hepatic portal vein**, which leads to the liver (see Chapter 6). The liver acts rather like a food processing factory, breaking some molecules down, and building up and storing others. For example, glucose from carbohydrate digestion is converted into glycogen and stored in the liver. Later, the glycogen can be converted back into glucose when the body needs it (see Chapter 9).

The digested food molecules are distributed around the body by the blood system (see Chapter 6). The soluble food molecules are absorbed from the blood into cells of tissues, where they are used to build new parts of cells. This is called **assimilation**.

THE LARGE INTESTINE – ELIMINATION OF WASTE

> **DID YOU KNOW?**
> Removal of faeces by the body is sometimes incorrectly called excretion. The word 'excretion' only applies to materials that are the waste products of cells of the body, such as carbon dioxide. Faeces are not products of cell metabolism – they consist of waste that has passed through the gut and left the body via the anus, without entering the cells. The correct name for this process is **egestion**.

By the time the contents of the gut have reached the end of the small intestine, most of the digested food, as well as most of the water, has been absorbed. The waste material consists mainly of cellulose (fibre) and other indigestible remains, water, dead and living bacteria, and cells lost from the lining of the gut. The function of the first part of the large intestine, called the **colon**, is to absorb most of the remaining water from the contents, leaving a semi-solid waste material called faeces. This is stored in the **rectum**, until it is expelled out of the body through the **anus**.

THE BODY MASS INDEX

Being severely overweight is known as **obesity**. An obese person has a greater risk of developing several serious diseases when compared with a person of a healthy body weight. Two of these diseases are **coronary heart disease** (**CHD**, see Chapter 6) and a type of diabetes (see Chapter 9).

Whether or not a person is a healthy weight depends on both their body mass and their height. For example, a tall man and a short man could both weigh 80 kg. However, if the tall man is 190 cm in height, he is a healthy weight, while if the short man is only 160 cm in height, he is obese.

The tall man has a lower **body mass index**, or **BMI**. To calculate your BMI, take your mass in kilograms and divide by your height in metres, squared:

$$\text{BMI} = \frac{\text{mass in kg}}{(\text{height in metres})^2}$$

Ideally, you should have a BMI between 18.5 and 25. Outside this range, a person is underweight, overweight, or obese (Figure 4.19).

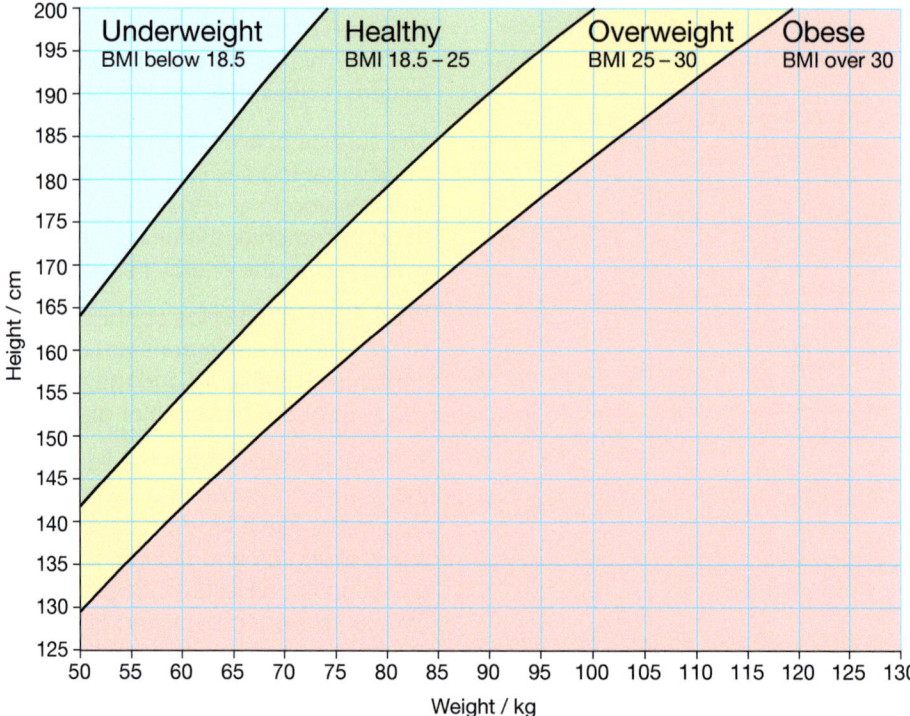

▲ Figure 4.19 The weight status of people based on their body mass index (BMI).

FOOD HYGIENE

Some bacteria that cause disease are transmitted in food, for example:

- *Salmonella enterica* – causes salmonellosis (a type of food poisoning)
- *Salmonella typhi* – causes typhoid fever
- *Listeria monocytogenes* – causes listeriosis (a type of food poisoning)
- *Clostridium botulinum* – causes botulism, a rare and sometimes fatal illness caused by a **toxin** (poison) made by the bacteria.

Most microorganisms transmitted in food or drink are bacteria, although some viruses are transmitted in this way. In addition, other microorganisms (bacteria and fungi) are the cause of food eventually going bad or 'spoiling'. Food hygiene is concerned with preventing harmful microorganisms from being transmitted in food, as well as preserving food and reducing its rate of spoilage.

PREVENTING TRANSMISSION OF MICROORGANISMS IN FOOD

We can prevent transmission of food-borne microorganisms at various stages:

- We can prevent microorganisms from getting into the food in the first place.
- We can treat the food to slow down the rate at which microorganisms multiply in the food (food preservation).
- We can cook food properly to kill any microorganisms present.

When food is sold, it must be stored in a way that minimises transmission of microorganisms from humans. Food packaging helps to do this, but unpackaged foods (e.g. loose fruit and vegetables) are at risk. Packaged foods show 'display until' and 'use by' dates, to tell us when a food may become unsafe to eat because of contamination by microorganisms or their toxins. After the 'use by' date, the number of microorganisms may have increased to dangerous levels (Figure 4.20).

In the home, food should also be stored in a way that minimises contamination. For example:

- Cooked and raw foods should not be stored together. This is because bacteria in the uncooked food may be transmitted to the cooked food.
- Foods that have been previously frozen and then thawed should be cooked and eaten straight away, and not re-frozen. This especially applies to foods such as meat (particularly chicken), fish and ice cream. Bacteria grow more quickly in previously frozen and thawed foods than in fresh foods.
- Food should not be left in the open air or on a work surface. Bacteria in the air could land on the food, and insects such as flies could carry bacteria to the food.

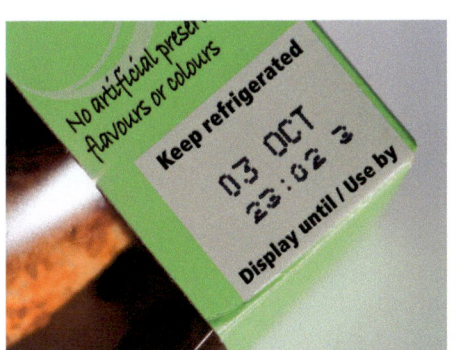

▲ Figure 4.20 Dates on packaged foods help to prevent transmission of microorganisms.

FOOD PRESERVATION

Some methods of food preservation have been used for hundreds of years, for example, salting, or pickling food in vinegar. Other methods are much more modern, for example, irradiating food to kill bacteria. Table 4.7 describes some methods of food preservation and how they work.

Table 4.7 Common methods of food preservation.

Method	Technique and principles involved	Examples of foods preserved by this method
salting	Some foods are covered in salt; others are soaked in salt water. High salt concentrations stop bacteria from multiplying. Bacterial cells lose water by osmosis and die.	some meats, fish, vegetables
pickling	Foods are bottled in vinegar, a weak solution of ethanoic (acetic) acid. The low pH inactivates most microorganisms by affecting enzyme activity.	fish, onions, cabbage, mayonnaise
pasteurisation	Food is heated to about 63 °C for 30 minutes. Alternatively, it is heated to 71.5 °C for 15 seconds. It is then quickly cooled to 10 °C. Heating kills bacteria (but not their reproductive spores). The spores are prevented from growing by the rapid cooling.	milk, cream, ice cream, fruit juices, beer
ultra-heat-treatment (UHT)	Superheated steam at up to 160 °C is blown through the food for 2 seconds. This kills all bacteria and their spores.	milk
canning	Food is packed in cans, heated to high temperatures, sealed and then re-heated to temperatures of 105–160 °C. The high temperature kills bacteria and their spores. Cans prevent more bacteria reaching the food.	meat, fish, vegetables, soup
drying	Food is dried by blowing hot air through it. Drying removes water from the food so bacteria cannot digest and absorb it.	cereals, some fruits
freezing	Foods are cooled rapidly and stored at −12 °C to −18 °C. Low temperatures prevent bacteria from multiplying. Rapid freezing prevents the formation of ice crystals, which could alter the texture and flavour of the food.	vegetables, meat, fish, prepared meals
irradiation	High-energy gamma radiation is passed through the food. All bacteria and spores are killed, although any toxins already produced by the bacteria remain.	vegetables, fruits, shellfish (irradiation is not allowed in some countries)

CHAPTER QUESTIONS

SKILLS EXECUTIVE FUNCTION

1 The diagram shows an experiment that was set up as a model to show why food needs to be digested.

The Visking tubing acts as a model of the small intestine because it has tiny holes in it that some molecules can pass through. The tubing was left in the boiling tube for an hour, then the water in the tube was tested for starch and glucose.

a Describe how you would test the water for starch and glucose. What would the results be for a 'positive' test in each case?

SKILLS REASONING

b The tests showed that glucose was present in the water but starch was not. Explain why.

SKILLS CRITICAL THINKING

c If the tubing takes the place of the intestine, what part of the body does the water in the boiling tube represent?

d What does 'digested' mean?

SKILLS DECISION MAKING

2 A student carried out an experiment to find out the best conditions for the enzyme pepsin to digest protein. For the protein, she used egg white powder, which forms a cloudy white suspension in water. The table below shows how the four tubes were set up.

Tube	Contents
A	5 cm³ egg white suspension, 2 cm³ pepsin, 3 drops of dilute acid. Tube kept at 37 °C
B	5 cm³ egg white suspension, 2 cm³ distilled water, 3 drops of dilute acid. Tube kept at 37 °C
C	5 cm³ egg white suspension, 2 cm³ pepsin, 3 drops of dilute acid. Tube kept at 20 °C
D	5 cm³ egg white suspension, 2 cm³ pepsin, 3 drops of dilute alkali. Tube kept at 37 °C

The tubes were left for 2 hours and the results were then observed. Tubes B, C and D were still cloudy. Tube A had gone clear.

a Three tubes were kept at 37 °C. Why was this temperature chosen?

b Explain what had happened to the protein in tube A.

c Why did tube D stay cloudy?

d Explain why tube B remained cloudy.

e Tube C was left for another 3 hours. Gradually it started to clear. Explain why digestion of the protein happened more slowly in this tube.

f The lining of the stomach secretes hydrochloric acid. Explain the function of this.

g When the stomach contents pass into the duodenum, they are still acidic. How are they neutralised?

3 Copy and complete the following table of digestive enzymes.

Enzyme	Food on which it acts	Products
amylase		
trypsin		
		fatty acids and glycerol

4 Describe four adaptations of the small intestine (ileum) that allow it to absorb digested food efficiently.

5 Bread is made mainly of starch, protein and lipid. Imagine a piece of bread about to start its journey through the human gut. Describe what happens to the bread as it passes through the mouth, stomach, duodenum, ileum and colon. Explain how the bread is moved along the gut. Illustrate your description using two or three simplified diagrams.

6 a The table lists four deficiency diseases. Copy and complete the table (one row has been done for you).

Deficiency disease	Nutrient that is lacking in the diet
kwashiorkor	
scurvy	vitamin C
anaemia	
night blindness	

b Describe a method that can be used to measure the amount of vitamin C in fruit juice.

c Oranges contain vitamin C. It is suggested that the amount of vitamin C in an orange decreases with time. Describe an investigation to test this hypothesis.

7 List three ways to preserve fish. In each case, explain how the method works.

5 RESPIRATION AND GAS EXCHANGE

Cells get their energy by oxidising foods such as glucose. This process is called respiration, and needs a continuous supply of oxygen from the blood. In addition, the waste product carbon dioxide needs to be removed from the body. These gases are exchanged between the blood and the air in the lungs. In this chapter, you will learn about the processes of respiration and gas exchange. You will also find out about ways that smoking can damage the lungs and stop these vital organs from working properly.

LEARNING OBJECTIVES

- Know that the process of respiration releases energy in living organisms
- Know the word equation and balanced chemical symbol equation for aerobic respiration in living organisms
- Understand the role of ATP in energy transfer – addition and removal of a phosphate group and associated energy requirement and release
- Explain the differences between aerobic and anaerobic respiration
- Know the word equation for anaerobic respiration
- Understand how an oxygen debt arises and how it is repaid after exercise
- Know the structure of the thorax, including the ribs, intercostal muscles, diaphragm, trachea, bronchi, bronchioles, alveoli and pleural membranes
- Explain the role of the intercostal muscles and diaphragm in ventilation
- Explain how the alveoli of the lungs are adapted for gas exchange by diffusion
- Describe how to investigate the difference in carbon dioxide concentration between inhaled (inspired) and exhaled (expired) air
- Describe how to investigate the effect of exercise on the rate of breathing
- Interpret spirometer traces showing breathing movements and understand the terms lung capacity, vital capacity and tidal volume
- Describe how to investigate lung capacity
- Describe the regulation of carbon dioxide content in the blood, including the role of chemoreceptors in the aorta and carotid arteries
- Understand the damage that smoking causes to the respiratory and cardiovascular systems*

* There is further information about the effect of smoking on the heart and circulation in Chapter 6.

RESPIRATION RELEASES ENERGY

A cell needs a source of energy in order to be able to carry out all the processes needed for life. It gets this energy by breaking down food molecules to release the stored chemical energy that they contain. This process is called respiration. Many people think that respiration means the same as 'breathing' but, although there are links between the two processes, the biological meaning of respiration is very different.

AEROBIC RESPIRATION

KEY POINT

Respiration is called an *oxidation* reaction, because oxygen is used to break down food molecules.

Respiration happens in all the cells of our body. Oxygen is used to oxidise food, and carbon dioxide (and water) are released as waste products. Because it uses oxygen, this is known as **aerobic respiration**. The main food oxidised is the sugar glucose. Glucose contains stored chemical energy that can be converted into other forms of energy that the cell can use. It is rather like burning a fuel to get the energy out of it, except that burning releases most of the energy as heat. Respiration releases some heat energy, but most of the energy is used to make a substance called **adenosine triphosphate** (ATP, see below). The energy stored in the ATP molecules can then be used for a variety of purposes, such as:

- contraction of muscle cells, producing movement
- active transport of molecules and ions
- building large molecules, such as proteins
- cell division.

The energy released as heat is also used to maintain a constant body temperature (see Chapter 10).

$$\text{glucose} + \text{oxygen} \rightarrow \text{carbon dioxide} + \text{water} (+ \text{energy})$$
$$C_6H_{12}O_6 + 6O_2 \rightarrow 6CO_2 + 6H_2O (+ \text{energy})$$

Aerobic respiration happens in the cells of humans, animals, plants and many other organisms. It is important to realise that the equation above is only a *summary* of the process. It actually takes place gradually, as a sequence of small steps, which release the energy of the glucose in small amounts. Each step in the process is catalysed by a different enzyme. The later steps in the process are the aerobic ones, and these release the most energy. They happen in the mitochondria of the cell.

> **DID YOU KNOW?**
> Carbon from the glucose passes out into the atmosphere as carbon dioxide. The carbon can be traced through this pathway using a radioactive form of carbon called carbon-14.

ATP: THE ENERGY 'CURRENCY' OF THE CELL

Respiration releases energy, while other cell processes use it up. Cells have a way of passing the energy from respiration to the other processes that need it. They do this using a chemical called adenosine triphosphate or ATP. ATP is present in all living cells.

ATP is composed of an organic molecule called adenosine attached to three phosphate groups. In a cell, ATP can be broken down, losing one phosphate group and forming adenosine diphosphate or ADP (Figure 5.1(a)). When this reaction takes place, chemical energy is released. This energy can be used to drive metabolic processes.

(a) When energy is needed ATP is broken down into ADP and phosphate (P):

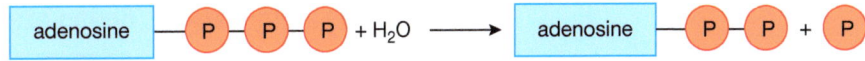

(b) During respiration ATP is made from ADP and phosphate:

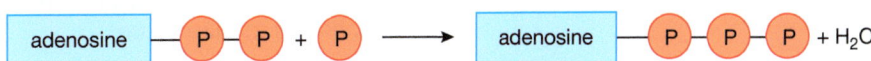

▲ Figure 5.1 ATP is the energy 'currency' of the cell.

During respiration, the opposite happens. Energy from the oxidation of glucose is used to drive the reverse reaction and a phosphate is added onto ADP (Figure 5.1(b)) to form ATP.

ATP is often described as the energy 'currency' of the cell. It transfers energy between the process that releases it (respiration) and the processes in a cell that use it up.

ANAEROBIC RESPIRATION

In some situations, cells can respire *without* using oxygen. This is called **anaerobic respiration**. In anaerobic respiration, glucose is broken down but not into carbon dioxide and water. Anaerobic respiration produces less energy but has the advantage that it can occur in situations where oxygen is in short supply. In humans, this happens in contracting muscle cells.

Muscle cells respire anaerobically when they are short of oxygen. If muscles are overworked, the blood cannot reach them fast enough to deliver enough oxygen for aerobic respiration. This happens when a person does a sudden 'burst' of activity, such as a sprint, or quickly lifting a heavy weight. In this situation, the glucose is broken down into a substance called **lactate**:

$$\text{glucose} \rightarrow \text{lactate (+ some energy)}$$

Anaerobic respiration provides enough energy to keep the overworked muscles going for a short period. Imagine you run a 100-metre sprint. During the race, lactate builds up in your muscles and bloodstream. When the race is over, the lactate must be removed from the body. It is broken down into carbon dioxide and water. Oxygen is needed for this, so you breathe quickly and deeply for some time after the race has finished.

The volume of oxygen needed to completely oxidise the lactate that builds up in the body during anaerobic respiration is called the **oxygen debt**. The maximum oxygen debt a person can have is about 17 dm^3. It can take up to an hour to fully 'repay' this debt and remove the lactate.

In a long-distance race (e.g. 10 000 metres or a marathon), lactate builds up at the start but your body soon adjusts and the lactate is broken down while you are running. Some sports scientists think that this is the cause of the 'second wind' – when an athlete who is too out of breath and tired to continue suddenly finds the strength to carry on. This probably happens because the athlete's body has adjusted and the cells now have the proper balance of oxygen to deal with the build-up of lactate.

> **DID YOU KNOW?**
> Many microorganisms such as bacteria and fungi can respire anaerobically. Different species produce different products.

> **DID YOU KNOW?**
> Lactate is sometimes called lactic acid. However, at the pH of the blood, most is present as the lactate ion rather than the acid, so it is better to call it 'lactate'.

> **DID YOU KNOW?**
> The muscles can work for a short time without oxygen and some sprinters hold their breath while running 100 metres. In a long-distance race, most of the athlete's energy comes from aerobic respiration.

> **DID YOU KNOW?**
> We used to think that lactate was toxic and caused muscle fatigue. We now know that this is *not* true. In fact, physiologists have shown that lactate *delays* muscle fatigue. Fatigue is caused by other changes that happen in the muscles during exercise.

Some people think that respiration and breathing mean the same thing, but you need to understand the difference between the two. As you have seen, respiration is the reaction that releases energy from foods such as glucose. Breathing is the mechanism that moves air into and out of the lungs, allowing gas exchange to take place.

THE STRUCTURE OF THE GAS EXCHANGE SYSTEM

The lungs are enclosed in the chest or **thorax** by the ribcage and a muscular sheet of tissue called the **diaphragm** (Figure 5.2). As you will see, the actions of these two structures bring about the movement of air into and out of the lungs. Each rib is joined to the next by two sets of muscles called **intercostal muscles** ('costals' are rib bones). The diaphragm separates the contents of the thorax from the abdomen. It is a shallow dome shape, with a fibrous middle part forming the 'roof' of the dome and muscular edges forming the walls.

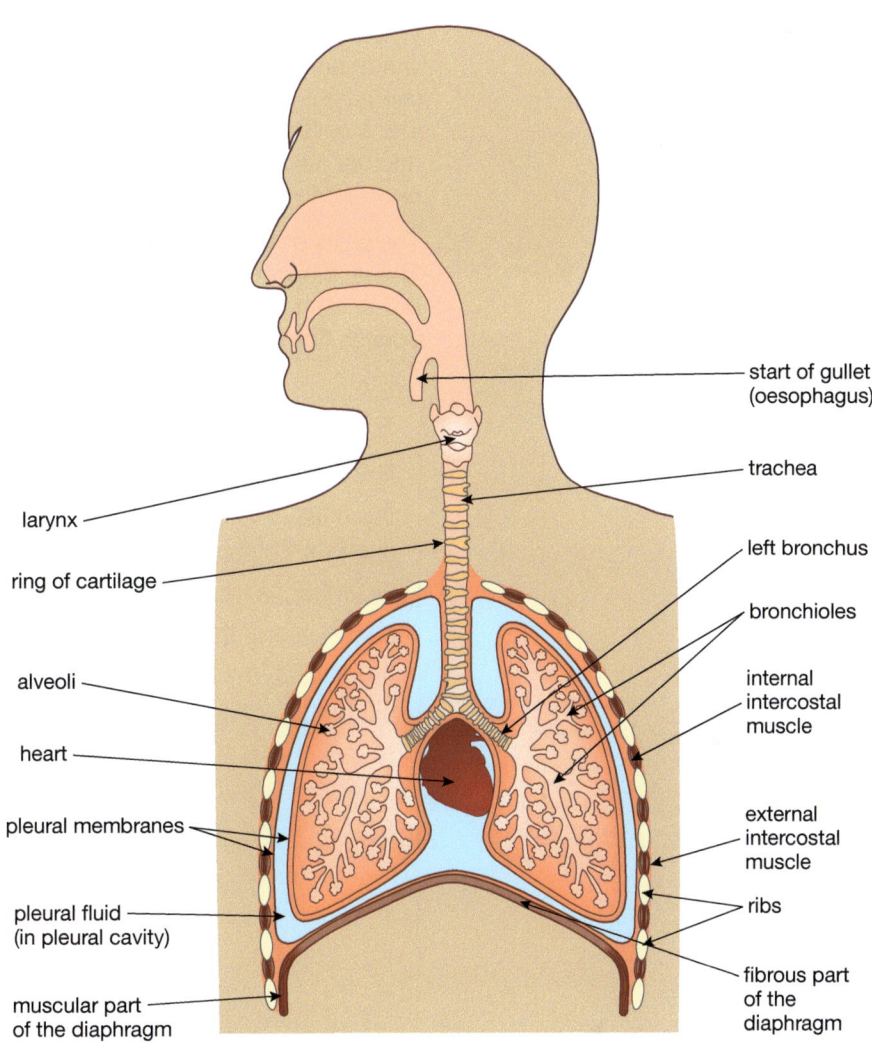

▲ Figure 5.2 The human gas exchange system.

The air passages of the lungs form a highly branching network (Figure 5.3) which is sometimes called the **bronchial tree**.

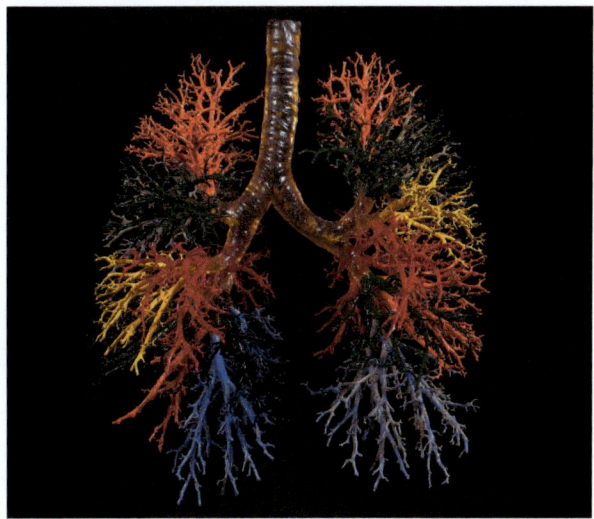

▲ Figure 5.3 This cast of the human lungs was made by injecting a pair of lungs with a liquid plastic. The plastic was allowed to set, then the lung tissue was dissolved away with acid.

When we breathe in, air enters our nose or mouth and passes down the windpipe or **trachea**. The trachea splits into two tubes called the **bronchi** (singular **bronchus**), one leading to each lung. Each bronchus divides into smaller and smaller tubes called **bronchioles**, eventually ending at microscopic air sacs called **alveoli** (singular **alveolus**). It is here that gas exchange with the blood takes place.

The walls of the trachea and the bronchi contain rings of gristle or **cartilage**. These support the airways and keep them open when we breathe in. They are rather like the rings in a vacuum cleaner hose – without them, the hose would squash flat when the cleaner sucks in air.

> **DID YOU KNOW?**
> In the bronchi, the cartilage forms complete, circular rings. In the trachea, the rings are incomplete, and shaped like a letter 'C' (Figure 5.4). The open part of each ring is at the back of the trachea, next to where the oesophagus (gullet) lies as it passes through the thorax. When food passes along the oesophagus by peristalsis (see Chapter 4) the gaps in the rings allow the lumps of food to pass through more easily, without the peristaltic wave 'catching' on the rings.
>
>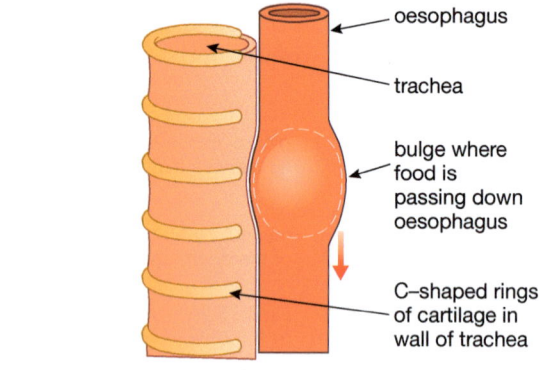
>
> ▲ Figure 5.4 C-shaped rings of cartilage in the trachea.

The inside of the thorax is separated from the lungs by two thin, moist membranes called the **pleural membranes**. They make up a continuous envelope around the lungs, forming an airtight seal. Between the two membranes is a space called the **pleural cavity**, filled with a thin layer of liquid called **pleural fluid**. This acts as lubrication, so that the surfaces of the lungs do not stick to the inside of the chest wall when we breathe.

KEEPING THE AIRWAYS CLEAN

The trachea and larger airways are lined with a layer of cells that have an important role in keeping the airways clean. Some cells in this lining secrete a sticky liquid called **mucus**, which traps particles of dirt or bacteria that are breathed in. Other cells are covered with tiny hair-like structures called **cilia** (Figure 5.5). The cilia beat backwards and forwards, sweeping the mucus and trapped particles out towards the mouth. In this way, dirt and bacteria are prevented from entering the lungs, where they might cause an infection. As you will see, one of the effects of smoking is that it destroys the cilia and stops this protection mechanism from working properly.

▲ Figure 5.5 This electron microscope picture shows cilia from the lining of the trachea.

VENTILATION OF THE LUNGS

Ventilation means moving air in and out of the lungs. This requires a difference in air pressure – the air moves from a place where the pressure is high to one where it is low. Ventilation depends on the fact that the thorax is an airtight cavity. When we breathe, we change the volume of our thorax, which alters the pressure inside it. This causes air to move in or out of the lungs.

There are two movements that bring about ventilation, those of the ribs and the diaphragm. If you put your hands on your chest and breathe in deeply, you can feel your ribs move upwards and outwards. They are moved by the intercostal muscles (Figure 5.6). The outer (external) intercostals contract, pulling the ribs up. At the same time, the muscles of the diaphragm contract, pulling the diaphragm down into a more flattened shape (Figure 5.7(a)). Both these movements increase the volume of the chest cavity and cause a slight drop in pressure inside the thorax compared with the air pressure outside. Air then enters the lungs (inhalation or inspiration).

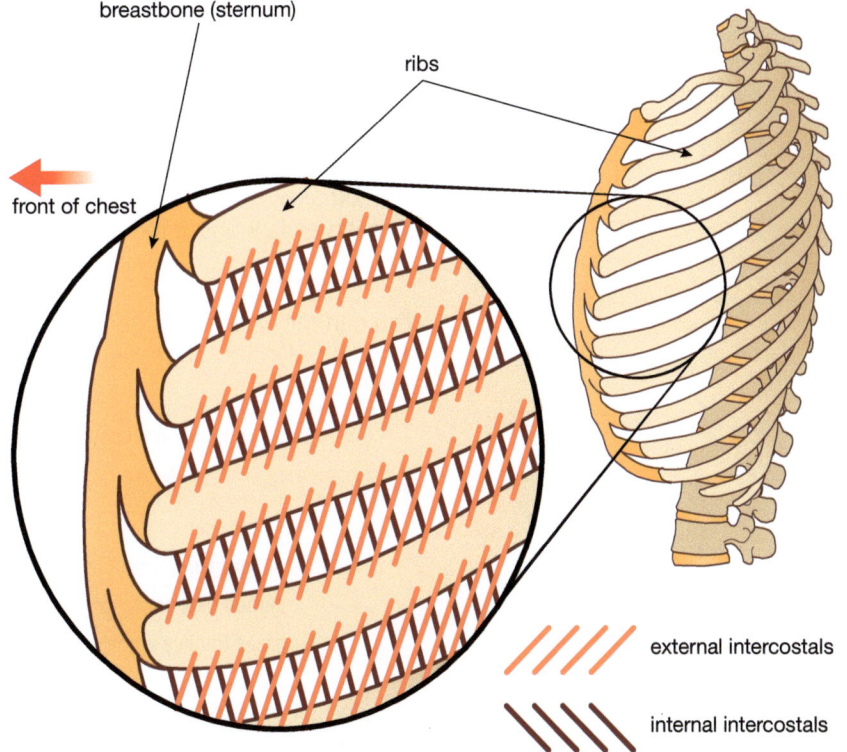

▲ Figure 5.6 Side view of the chest wall, showing the ribs. The diagram shows how the two sets of intercostal muscles run between the ribs. When the external intercostals contract, they move the ribs upwards. When the internal intercostals contract, the ribs are moved downwards.

The opposite happens when you breathe out deeply. The external intercostals relax and the internal intercostals contract, pulling the ribs down and in. At the same time, the diaphragm muscles relax and the diaphragm goes back to its normal dome shape. The volume of the thorax decreases and the pressure in the thorax is raised slightly above atmospheric pressure. This time, the difference in pressure forces air out of the lungs (Figure 5.7(b)). Exhalation (expiration) is helped by the fact that the lungs are elastic, so they have a tendency to collapse and empty like a balloon.

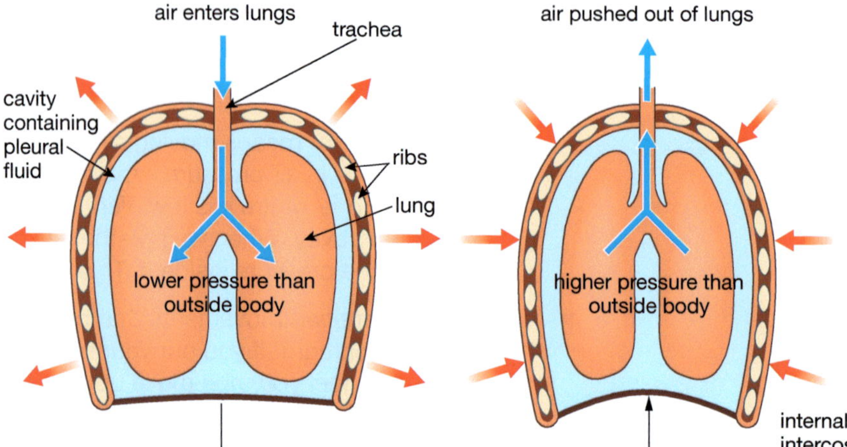

> **DID YOU KNOW?**
> During normal (shallow) breathing, the elasticity of the lungs and the weight of the ribs acting downwards are enough to cause exhalation. The internal intercostals are only really used for deep (forced) breathing out, for instance, when we are exercising.

▲ Figure 5.7 Changes in the position of the ribs and diaphragm during breathing. (a) Breathing in (inhalation). (b) Breathing out (exhalation).

It is important that you remember the changes in volume and pressure during ventilation. If you have trouble understanding them, think of what happens when you use a bicycle pump. If you push the pump handle, the air in the pump is squashed, its pressure rises and it is forced out of the pump. If you pull on the handle, the air pressure inside the pump falls a little and air is drawn in from outside. This is similar to what happens in the lungs. In exams, students sometimes talk about the lungs *forcing* the air in and out – this is not accurate!

GAS EXCHANGE IN THE ALVEOLI

You can tell what is happening during gas exchange if you compare the amounts of different gases in atmospheric air with the air breathed out (Table 5.1).

▼ Table 5.1 Approximate percentage volume of gases in atmospheric (inhaled) and exhaled air.

Gas	Atmospheric air	Exhaled air
nitrogen	78	79
oxygen	21	16
carbon dioxide	0.04	4
other gases (mainly argon)	1	1

Exhaled air is also warmer than atmospheric air, and is saturated with water vapour.

Clearly, the lungs are transferring oxygen into the blood and removing carbon dioxide from it. This happens in the alveoli. To do this efficiently, the alveoli

DID YOU KNOW?
Be careful when interpreting percentages! The *percentage* of a gas in a mixture can vary, even if the actual *amount* of the gas stays the same. This is easiest to understand from an example. Imagine you have a bottle containing a mixture of 20% oxygen and 80% nitrogen. If you use a chemical to absorb all the oxygen in the bottle, the nitrogen left will now be 100% of the gas in the bottle, despite the fact that the *amount* of nitrogen has not changed. This is why the percentage of nitrogen in inhaled and exhaled air is slightly different, even though no nitrogen is made by the body.

KEY POINT
Be careful. Students sometimes write 'The alveolus has *cell walls*'. This statement is not correct – a cell wall is part of a plant cell! The correct way to describe the structure is: 'The alveolus has *a wall made of cells*'.

must have a structure which brings the air and blood very close together, over a very large surface area. There are enormous numbers of alveoli. It has been calculated that the two lungs contain about 700 000 000 of these tiny air sacs, with a total surface area of 60 m² – bigger than the floor area of an average classroom! Viewed through a high-powered microscope, the alveoli look rather like bunches of grapes, and are covered with tiny blood capillaries (Figure 5.8).

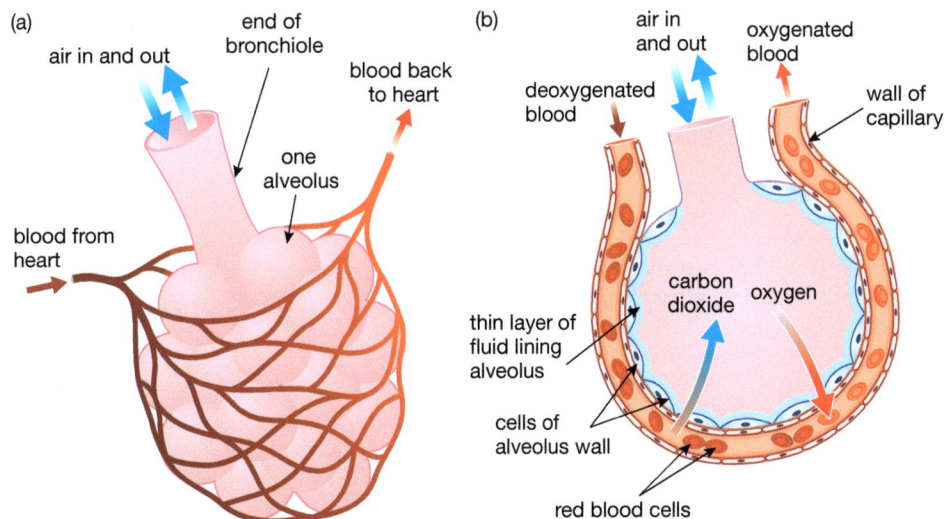

▲ Figure 5.8 (a) Alveoli and the surrounding capillary network. (b) Diffusion of oxygen and carbon dioxide takes place between the air in the alveolus and the blood in the capillaries.

Deoxygenated blood is pumped from the heart to the lungs and passes through the capillaries surrounding the alveoli. The blood has come from the respiring tissues of the body, where it has given up some of its oxygen to the cells, and gained carbon dioxide. Around the lungs, the blood is separated from the air inside each alveolus by only two cell layers: the cells making up the wall of the alveolus, and the capillary wall itself. This is a distance of less than a thousandth of a millimetre.

DID YOU KNOW?
The thin layer of fluid lining the inside of the alveoli comes from the blood. The capillaries and cells of the alveolar wall are 'leaky' and the blood pressure pushes fluid out from the blood plasma into the alveolus. Oxygen dissolves in this moist surface before it passes through the alveolar wall into the blood.

Because the air in the alveolus has a higher concentration of oxygen than the blood entering the capillary network, oxygen diffuses from the air, across the wall of the alveolus and into the blood. At the same time, there is more carbon dioxide in the blood than there is in the air in the lungs. This means that there is a diffusion gradient for carbon dioxide in the other direction, so carbon dioxide diffuses the other way – out of the blood and into the alveolus. The result is that the blood which leaves the capillaries and flows back to the heart has gained oxygen and lost carbon dioxide. The heart then pumps the oxygenated blood around the body again, to supply the respiring cells (see Chapter 6).

ACTIVITY 3

▼ PRACTICAL: COMPARING THE CARBON DIOXIDE CONTENT OF INHALED (INSPIRED) AND EXHALED (EXPIRED) AIR

The apparatus in Figure 5.9 can be used to compare the amount of carbon dioxide in inhaled and exhaled air. A person breathes gently in and out through the middle tube. Exhaled air passes out through one tube of indicator solution and inhaled air is drawn in through the other tube. If limewater is used, the limewater in the 'exhaled' tube will turn cloudy before the limewater in the 'inhaled' tube. (If hydrogencarbonate indicator solution is used instead, it will change from red to yellow.)

Safety Note: Breathe in and out slowly – do not blow. Every user must have a clean mouthpiece.

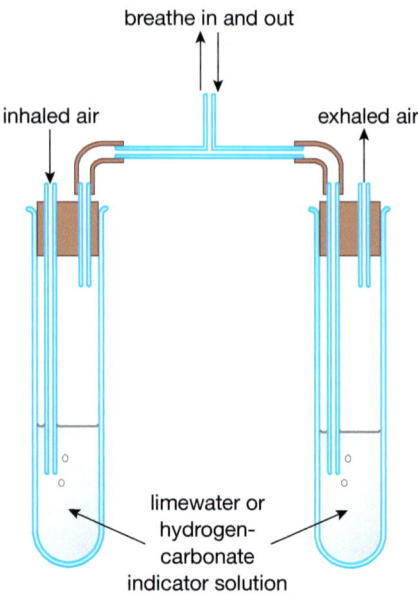

▲ Figure 5.9 Apparatus for Activity 3

ACTIVITY 4

▼ PRACTICAL: INVESTIGATING THE EFFECT OF EXERCISE ON BREATHING RATE

It is easy to show the effect of exercise on a person's breathing rate. They sit quietly for five minutes, making sure they are completely relaxed. They then count the number of breaths they take in one minute, recording their results in a table. They wait a minute and then count their breaths again, recording the result, and repeating if necessary until they get a steady value for the 'resting rate'.

The person then carries out some vigorous exercise, such as running on the spot, for three minutes. When they finish the exercise, they immediately sit down and record their breathing rate as before. They continue to record their breaths per minute, every minute, until they return to their normal resting rate.

The table shows the results of an investigation into the breathing rate of two girls, A and B, before and after exercise.

Time from start of experiment / min	Breathing rate / breaths per min	
	A	B
1	13	13
2	14	12
3	14	12
(3 minutes of vigorous exercise took place here)		
7	28	17
8	24	13
9	17	12
10	14	12

Plot a line graph of these results, using the same axes for both subjects. Join the data points using straight lines, and leave a gap during the period of exercise, when no readings were taken.

- Describe the differences between the breathing rates of the two girls (A and B) after exercise. Which girl is more fit? Explain your reasoning.
- Why does the rate not return to normal as soon as a subject finishes the exercise?
- Why does breathing rate need to rise during exercise? Explain as fully as possible.

MEASURING BREATHING VOLUMES

Breathing volumes can be measured using a piece of apparatus called a **spirometer** (Figure 5.10).

Safety Note: A spirometer should only be used by someone who is trained to operate it.

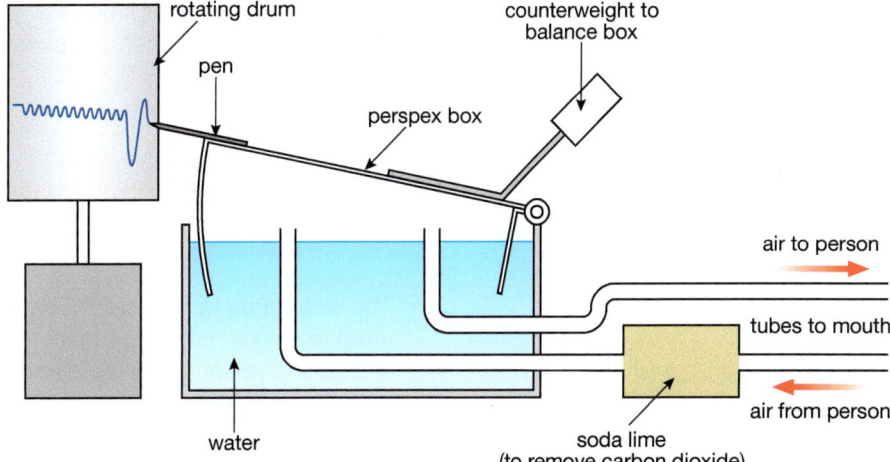

▲ Figure 5.10 A spirometer.

The floating box is filled with air, or with medical-grade oxygen. A person (the 'subject' of the experiment) breathes in and out through a mouthpiece into the floating lid of the spirometer chamber. As they breathe, the lid moves up and down, and a pen records the breathing movements on a rotating drum, which is covered by a piece of graph paper. A canister of soda lime absorbs carbon dioxide from the subject's exhaled air. The speed at which the drum turns is set, so that the number of breaths per minute can be worked out.

DID YOU KNOW?
The breathing volumes depend on various factors, including age, sex of the person, size of the lungs, level of exercise and how healthy the subject is. The vital capacity varies from about 3.0 dm³ in a young woman to 6.0 dm³ in a trained athlete.

Figure 5.11 shows a trace from a subject breathing into a spirometer. The subject began by breathing normally into the spirometer, before taking a deep breath in. He then breathed out as much as possible.

The volume of air breathed in and out with a normal breath is called the **tidal volume**. In a normal healthy adult man, this is about 0.4 dm³. The difference between the maximum breath in and the maximum breath out is about 4.5 dm³. This is called the **vital capacity**.

The spirometer trace does not show the full amount of air in the lungs. There is always about 1.5 dm³ of air left in the lungs, even after a person has breathed out as much as possible. This is called the **residual volume**. The maximum total volume of the lungs is called the **lung capacity**. It is equal to the vital capacity plus the residual volume.

HINT
Notice that the trace in Figure 5.11 is not horizontal, but slowly slopes down to the right. This is because the oxygen in the spirometer is gradually being used up by the person's respiration, and is not replaced by carbon dioxide.

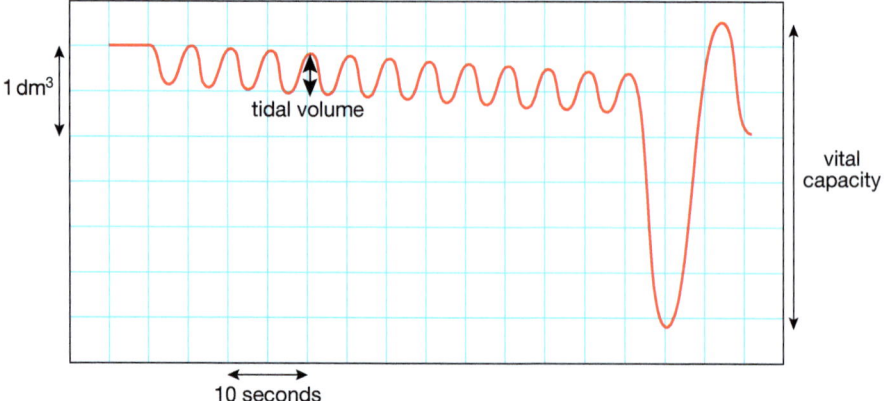

▲ Figure 5.11 A spirometer trace showing breathing volumes.

If you have access to a spirometer, you can use it to measure lung capacity. An alternative, simpler method is shown in Activity 5.

ACTIVITY 5

▼ PRACTICAL: MEASURING LUNG CAPACITY

Safety Note: Use a fresh breathing tube mouthpiece for each person. Avoid over-exertion as this may cause headaches or faintness. Participants with asthma or other respiratory problems should not use this apparatus.

The apparatus in Figure 5.12 can be used to measure lung capacity. It works on the same principle as the spirometer. A large bell jar is marked in dm³ to a volume of at least 6 dm³. It is placed open-end downwards in a large bucket or sink of water. The rubber bung is removed so that the level of water is the same inside and outside the jar, and the position of the bell jar is adjusted until the water level is at the zero mark. The bung is then replaced.

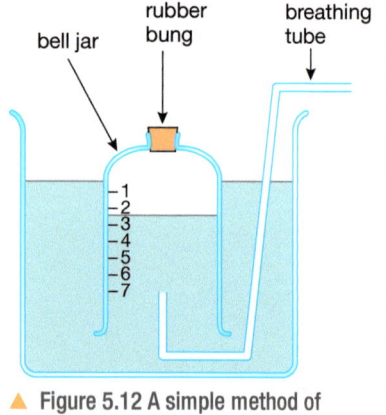

▲ Figure 5.12 A simple method of measuring lung capacity.

A person wearing a nose clip breathes in as deeply as possible and then steadily blows out as much air as they can through a piece of rubber tubing. The air is collected inside the jar and pushes the water down. As the air enters, a second person supports the weight of the bell jar and slowly moves it up so that the water levels inside and outside stay the same.

When the person has finished exhaling, the new volume on the scale is noted. This is the person's vital capacity – see above.

This method needs practice and repeat readings should be taken for reliable results.

Note: The breathing tube needs to have a wide diameter, otherwise it will be difficult to blow through and will give an underestimate of lung capacity. If different people measure their lung capacities, the breathing tube needs to be disinfected and washed before each new person uses it.

CONTROL OF BREATHING RATE

Breathing is automatic and under the control of your brain. When you exercise, your muscles need more oxygen for respiration. They also produce more carbon dioxide, which has to be removed from your body. To bring about these changes, there is an increase in the rate of breathing. As carbon dioxide builds up in the blood, it is detected by receptors in two main places:

- the **medulla** of the brain (see Chapter 9)
- the **aorta** and **carotid arteries** leading from the heart (Chapter 6).

These receptors send messages to the **respiratory centre**, which is also located in the medulla of the brain. The respiratory centre sends messages to the diaphragm and intercostal muscles, causing both the breathing rate and the depth of breathing to increase. This adjusts the levels of carbon dioxide and oxygen in the blood (Figure 5.13).

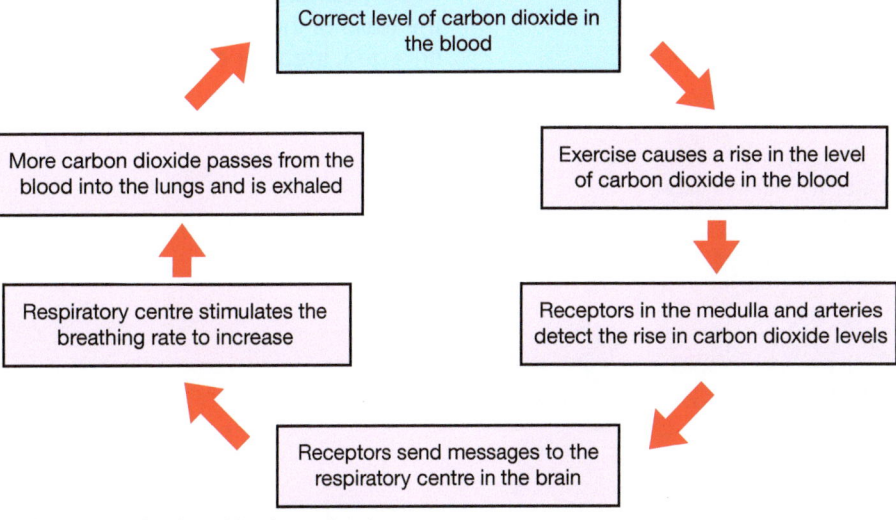

▲ Figure 5.13 How breathing is regulated.

THE EFFECTS OF SMOKING

In order for the lungs to exchange gases properly:

- the air passages must be clear
- the alveoli must be free from dirt particles and bacteria
- the alveoli must have as big a surface area as possible in contact with the blood.

There is one habit that can upset all of these conditions – smoking.

Links between smoking and diseases of the lungs are now a proven fact. Smoking is associated with lung cancer, bronchitis and emphysema. It is also a major contributing factor to other conditions, such as coronary heart disease and ulcers of the stomach and intestine. Pregnant women who smoke are more likely to give birth to underweight babies.

Coronary heart disease will be described in Chapter 6 after you have studied the structure of the heart. Here, we will look at a number of other medical conditions that are caused by smoking.

EFFECTS OF SMOKE ON THE LINING OF THE AIR PASSAGES

Earlier in this chapter, you learnt how the lungs are kept free of dirt and bacteria by the action of mucus and and cilia (see Figure 5.5). In the trachea and bronchi of a smoker, the cilia are destroyed by the chemicals in cigarette smoke.

The reduced numbers of cilia mean that the mucus is not swept away from the lungs. Instead, it remains to block the air passages. This is made worse by the fact that the smoke irritates the lining of the airways, stimulating the cells to secrete more mucus. The sticky mucus blocking the airways is the cause of 'smoker's cough'. Irritation of the bronchial tree, along with infections from bacteria in the mucus can cause the lung disease **bronchitis**. Bronchitis blocks normal air flow, so the sufferer has difficulty breathing properly.

EMPHYSEMA

Emphysema is another lung disease and kills about 20 000 people in Britain every year. Smoking is the cause of one type of emphysema. Smoke damages the walls of the alveoli, which break down and fuse together again, forming enlarged, irregular air spaces (Figure 5.14).

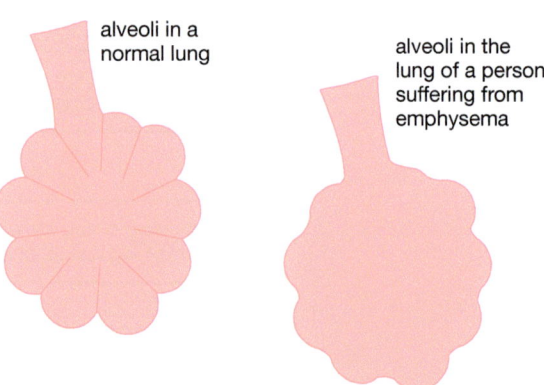

▲ Figure 5.14 The alveoli of a person suffering from emphysema have a greatly reduced surface area. This results in inefficient gas exchange.

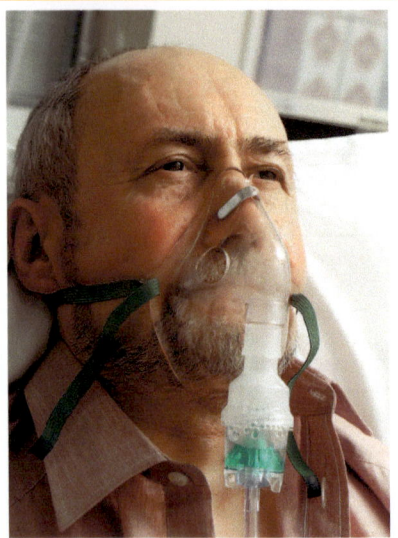

▲ Figure 5.15 Patients with emphysema often need to breathe air enriched with oxygen in order to stay alive.

> **DID YOU KNOW?**
> A person who has chronic (long-term) bronchitis and emphysema is said to be suffering from chronic obstructive pulmonary disease, or COPD. COPD is a progressive disease for which there is no cure.

This greatly reduces the surface area for gas exchange, which becomes very inefficient. The blood of a person with emphysema carries less oxygen. In serious cases, this means the sufferer is unable to carry out even mild exercise, such as walking. Emphysema patients often have to have a supply of oxygen nearby at all times (Figure 5.15). There is no cure for emphysema, and usually the sufferer dies after a long and distressing illness.

LUNG CANCER

Evidence of the link between smoking and lung cancer first appeared in the 1950s. In one study, a number of patients in hospital were asked a series of questions about their lifestyles. They were asked about their work, hobbies, housing and so on, and there was a question about how many cigarettes they smoked. The same questionnaire was given to two groups of patients. Patients in the first group were all suffering from lung cancer. Patients in the second (Control) group were in hospital with various other illnesses, but not lung cancer. To make it a fair comparison, the Control patients were matched with the lung cancer patients for sex, age and so on.

When the results were compared, one difference stood out (Table 5.2). A greater proportion of the lung cancer patients were smokers than in the Control patients. There seemed to be a connection between smoking and lung cancer.

▼ Table 5.2 Comparison of the smoking habits of lung cancer patients and other patients.

	Percentage of patients who were non-smokers	Percentage of patients who smoked more than 15 cigarettes a day
Lung cancer patients	0.5	25
Control patients (with illnesses other than lung cancer)	4.5	13

Although the results did not prove that smoking caused lung cancer, there was a statistically significant link between smoking and the disease. This is called a 'correlation'.

Over 20 similar investigations in nine countries have revealed the same findings. In 1962, a report called 'Smoking and health' was published by the Royal College of Physicians of London. This report warned the public about the dangers of smoking. Not surprisingly, the first people to take the findings seriously were doctors, many of whom stopped smoking. This was reflected in their death rates from lung cancer. In ten years, while deaths from lung cancer among the general male population had risen by 7%, the deaths of male doctors from the disease had *fallen* by 38%.

> **DID YOU KNOW?**
> People often talk about 'yellow nicotine stains'. In fact, it is the *tar* that stains a smoker's fingers and teeth. Nicotine is a colourless, odourless chemical.

Cigarette smoke contains a strongly addictive drug – **nicotine**. It also contains over 7000 chemicals, including carbon monoxide, arsenic, ammonia, formaldehyde, cyanide, benzene, and toluene. More than 60 of these chemicals are known to cause cancer. These chemicals are called **carcinogens** and are contained in the tar that collects in a smoker's lungs. Cancer happens when cells mutate and start to divide uncontrollably, forming a **tumour** (Figure 5.16). If a lung cancer patient is lucky, it may be possible to remove the tumour by an operation before the cancer cells spread to other tissues of the body. Unfortunately, tumours in the lungs usually cause no pain, so in many cases they are not discovered until it is too late. The tumour may be inoperable, or tumours may have developed elsewhere.

If you smoke, you are not *bound* to get lung cancer, but the risk that you will get it is much greater. The more cigarettes you smoke, the more the risk increases (Figure 5.17).

The obvious thing to do is not to start smoking. However, if you are a smoker, giving up the habit will improve your chances of living longer (Figure 5.18). A few years after giving up smoking, the likelihood of dying from a smoking-related disease is almost back to the level of a non-smoker.

▲ Figure 5.16 This lung is from a patient with lung cancer.

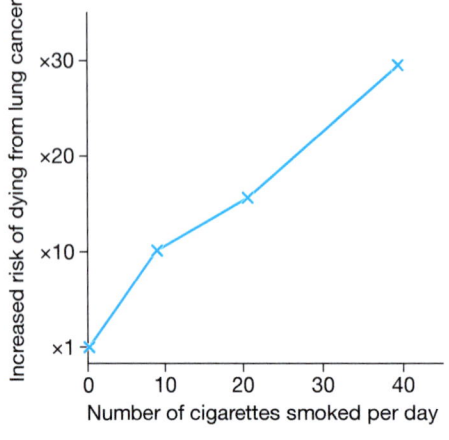

▲ Figure 5.17 The more cigarettes a person smokes, the more likely it is they will die of lung cancer. For example, smoking 20 cigarettes a day increases the risk by about 15 times.

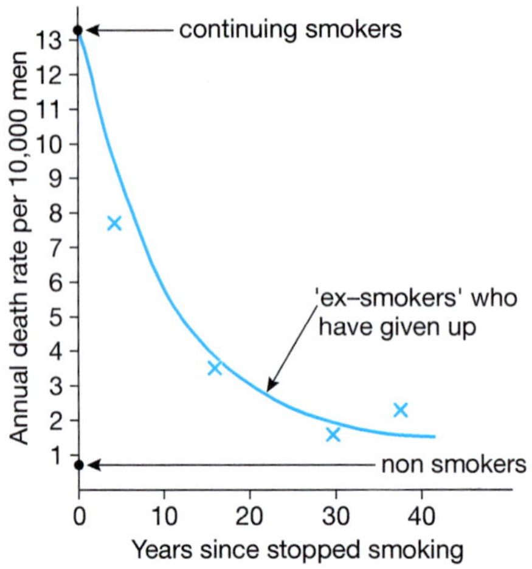

▲ Figure 5.18 Death rates from lung cancer for smokers, non-smokers and ex-smokers.

> **DID YOU KNOW?**
> Studies have shown that the type of cigarette smoked makes very little difference to the smoker's risk of getting lung cancer. Filtered and 'low tar' cigarettes only reduce the risk slightly.

CARBON MONOXIDE IN SMOKE

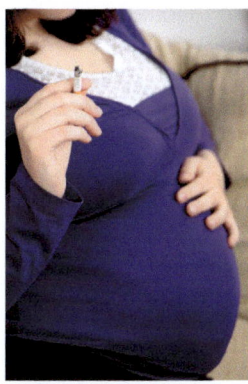

▲ Figure 5.19 Smoking during pregnancy affects the growth and development of the baby.

One of the harmful chemicals in cigarette smoke is the poisonous gas **carbon monoxide**. When this gas is breathed in with the smoke, it enters the bloodstream and interferes with the blood's ability to carry oxygen. Oxygen is carried around in the blood in the red blood cells, attached to a chemical called **haemoglobin** (see Chapter 6). Carbon monoxide can combine with haemoglobin much more tightly than oxygen can, forming a compound called **carboxyhaemoglobin**. The haemoglobin will combine with carbon monoxide in preference to oxygen. When this happens, the blood can carry much less oxygen around the body. Carbon monoxide from smoking is also a major cause of heart disease (Chapter 6).

If a pregnant woman smokes, her unborn **fetus** will be deprived of oxygen (Figure 5.19). This affects the fetus's growth and development. The mass of babies born to mothers who smoke is lower, on average, than the mass of babies born to non-smokers.

SOME SMOKING STATISTICS

- Every year, nearly 6 million people are killed by tobacco-related illnesses. If the current trend continues, this will rise to nearly 8 million deaths per year by 2030 and 80% of these premature deaths will be in developing countries.
- Smoking causes: almost 80% of deaths from lung cancer; 80% of deaths from bronchitis and emphysema; and 14% of deaths from heart disease.
- More than a quarter of all cancer deaths can be linked with smoking. These include cancers of the lungs, mouth, lip, throat, bladder, kidneys, pancreas, stomach, liver and cervix.

[Sources: Action on Smoking and Health (ASH) fact sheets (2015–2016); ASH research reports (2014–2016)]

GIVING UP SMOKING

▲ Figure 5.20 An e-cigarette supplies a smoker with an alternative source of nicotine.

Many smokers would like to find a way to give up the habit. The trouble is that the nicotine in tobacco is a very addictive drug, which causes withdrawal symptoms when people stop smoking. These symptoms include cravings for a cigarette, restlessness and a tendency to put on weight (nicotine reduces the appetite).

Smokers can be helped to give up in various ways. Some methods provide an alternative source of nicotine, without the harmful tar from cigarettes. For example, 'vaping' involves inhaling a vapour containing nicotine from an 'electronic cigarette' or e-cigarette (Figure 5.20). Ex-smokers may also use nicotine patches (Figure 5.21) or nicotine chewing gum. The nicotine is absorbed by the body and reduces the craving for a cigarette. Gradually, the ex-smoker can reduce the nicotine dose until they no longer need it.

There are several other ways that people use to help them give up smoking, including the use of drugs that reduce withdrawal symptoms, acupuncture and even hypnotism.

EXTENSION WORK

You could carry out an internet search to find out about the different methods people use to help them give up smoking. Which methods have the highest success rate? Is there any evidence that e-cigarettes are not safe?

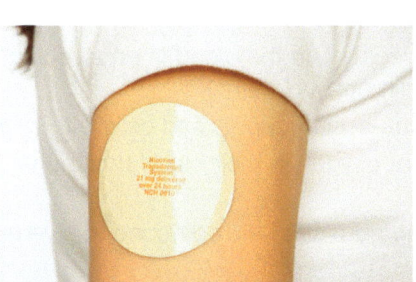

▲ Figure 5.21 A nicotine patch.

CHAPTER QUESTIONS

SKILLS PROBLEM SOLVING

SKILLS CRITICAL THINKING

SKILLS REASONING

SKILLS INTERPRETATION

1. a Write a balanced chemical equation for aerobic respiration.
   b Briefly explain the role of ATP in a cell.
   c List three ways that anaerobic respiration in humans is different from aerobic respiration.
   d Explain why a runner in a 100-metre race develops an oxygen debt. How is the debt repaid?

2. Copy and complete the table, which shows what happens in the thorax during ventilation of the lungs. Two boxes have been completed for you.

	Action during inhalation	Action during exhalation
external intercostal muscles	contract	
internal intercostal muscles		
ribs		move down and in
diaphragm		
volume of thorax		
pressure in thorax		
volume of air in lungs		

SKILLS CRITICAL THINKING

3. A student wrote the following about the lungs.

   When we breathe in, our lungs inflate, sucking air in and pushing the ribs up and out, and forcing the diaphragm down. This is called respiration. In the air sacs of the lungs the air enters the blood. The blood then takes the air around the body, where it is used by the cells. The blood returns to the lungs to be cleaned. When we breathe out, our lungs deflate, pulling the diaphragm up and the ribs down. The stale air is pushed out of the lungs.

   The student does not have a good understanding of how the lungs work. Re-write her description, using correct biological words and ideas.

SKILLS REASONING

4. Sometimes, people injured in an accident such as a car crash suffer from a *pneumothorax*. This is an injury where the chest wall is punctured, allowing air to enter the pleural cavity (see Figure 5.2). A patient was brought to the casualty department of a hospital, suffering from a pneumothorax on the left side of his chest. His left lung had collapsed, but he was able to breathe normally with his right lung.
   a Explain why a pneumothorax caused the left lung to collapse.
   b Explain why the right lung was not affected.

SKILLS CRITICAL THINKING

   c If a patient's lung is injured or infected, a surgeon can sometimes 'rest' it by performing an operation called an *artificial pneumothorax*. What do you think might be involved in this operation?

5. Explain the reasons for the following.
   a The trachea wall contains C-shaped rings of cartilage.
   b The distance between the air in an alveolus and the blood in an alveolar capillary is less than 1/1000th of a millimetre.
   c The lining of the trachea contains mucus-secreting cells and cells with cilia.
   d Smokers have a lower concentration of oxygen in their blood than non-smokers.
   e Nicotine patches and nicotine chewing gum can help someone to give up smoking.
   f The lungs have a surface area of about 60 m² and a good blood supply.

SKILLS CRITICAL THINKING

6 Explain the differences between the lung diseases bronchitis and emphysema.

SKILLS ANALYSIS

7 The graph shows a spirometer trace of the volume of air in a person's lungs during different types of breathing.

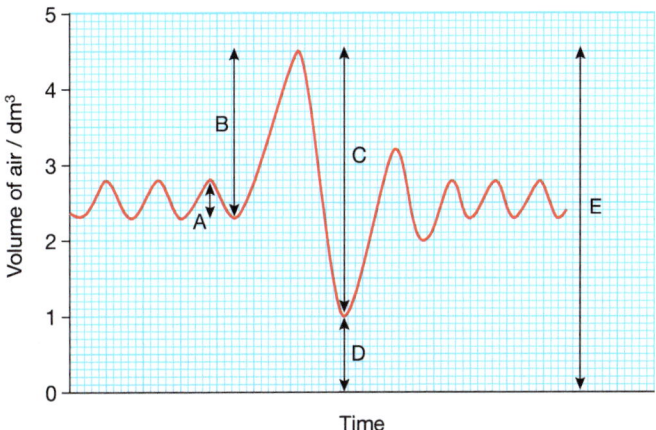

a The tidal volume is shown by the arrow labelled A. Explain what is meant by the term tidal volume and calculate its value from the graph.

b Which letter represents the vital capacity?

SKILLS EXECUTIVE FUNCTION

c 'Boys have a larger vital capacity than girls'. Describe how you could use the spirometer to investigate this hypothesis.

SKILLS CRITICAL THINKING

8 A long-term investigation was carried out into the link between smoking and lung cancer. The smoking habits of male doctors aged 35 or over were determined while they were still alive. Then, the number of deaths and the causes of death among the doctors were monitored over a number of years. (Note that this survey was carried out in the 1950s – very few doctors smoke these days!) The results are shown in the graph.

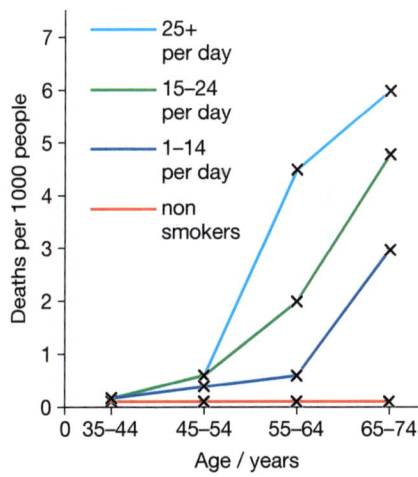

a Write a paragraph to explain what the researchers found out from the investigation.

SKILLS ANALYSIS

b How many deaths from lung cancer would be expected for men aged 55 who smoked 25 cigarettes a day up until their death?

c How many deaths from lung cancer would be expected for men in the same age group smoking 10 cigarettes a day?

d Table 5.2 (page 87) shows the findings of another study linking lung cancer with smoking. Which evidence is more convincing, the results of this investigation or the findings illustrated in Table 5.2?

SKILLS CREATIVITY

9 Design and make a hard-hitting leaflet explaining the link between smoking and lung cancer. It should be aimed at encouraging an adult smoker to give up the habit. Include some smoking statistics, perhaps from an internet search.

6 INTERNAL TRANSPORT

We need a blood system to transport substances to and from the cells of the body. In this chapter, you will find out about the structure and function of the human circulatory system and the composition of blood. You will also read about some medical problems affecting the heart and circulation.

LEARNING OBJECTIVES

- Know the general plan of the circulatory system, including the blood vessels to and from the heart, lungs, liver and kidneys
- Know the structure of the heart and how it functions
- Describe the structure of cardiac muscle as seen through the light microscope
- Explain the terms systolic and diastolic blood pressure
- Explain why the heart rate changes during exercise, and the influence of adrenaline
- Compare the structure of arteries, veins and capillaries and understand their roles, including details of the pulse
- Understand the pulse rate as a measure of heart rate and explain why resting pulse rate can be used as a measure of physical fitness
- Describe how to investigate the effect of exercise on the pulse rate
- Know the role of tissue fluid, explain how it is formed and how it drains into the lymphatic system
- Know the composition of the blood: red blood cells (erythrocytes), white blood cells (phagocytes and lymphocytes), platelets and plasma
- Understand the role of plasma in the transport of carbon dioxide, digested food, urea, hormones and heat energy
- Explain how red blood cells are adapted for the transport of oxygen
- Understand the role of white blood cells, including phagocytosis and antibody production
- Describe the importance of blood clotting, and the role of enzymes in causing the conversion of fibrinogen to fibrin
- Understand the role of ABO blood groups and their importance in blood transfusions
- Understand the causes, prevention and treatment of heart disease, including:
  - the effect of diet and exercise, including cholesterol and atherosclerosis
  - the use of stents, transplants, and artificial hearts
- Understand the problems associated with heart transplants
- Understand the long-term benefits of exercise on the cardiovascular system
- Understand the term aerobic exercise
- Describe the use of statins and plant stanol esters in the treatment and prevention of circulatory disorders
- Understand the role of beta-blockers in the treatment and prevention of circulatory disorders, e.g. heart failure and angina
- Describe the causes, prevention and treatment of hypertension
- Understand the role of ACE inhibitors in the treatment of high blood pressure
- Describe how monoclonal antibodies are produced
- Understand how monoclonal antibodies work to detect and treat diseases such as cancer

UNIT 2 — INTERNAL TRANSPORT

Figure 6.1 shows the human circulatory system. Blood is pumped around a closed circuit made up of the heart and blood vessels. As it travels around the body, it collects materials from some places and unloads them in others. In mammals, blood transports:

- oxygen from the lungs to all other parts of the body
- carbon dioxide from all parts of the body to the lungs
- nutrients from the gut to all parts of the body
- the excretory waste product **urea** from the liver to the kidneys.

Blood also transports hormones, antibodies and many other substances. It distributes heat around the body, and contains cells and cell products that defend the body against microorganisms that cause disease.

HUMANS HAVE A DOUBLE CIRCULATORY SYSTEM

One of the main functions of the circulatory system is to transport oxygen. Blood is pumped to the lungs to load oxygen. It is then pumped to the other parts of the body where it unloads the oxygen. In humans and other mammals, this is carried out by a **double circulation** (Figure 6.2).

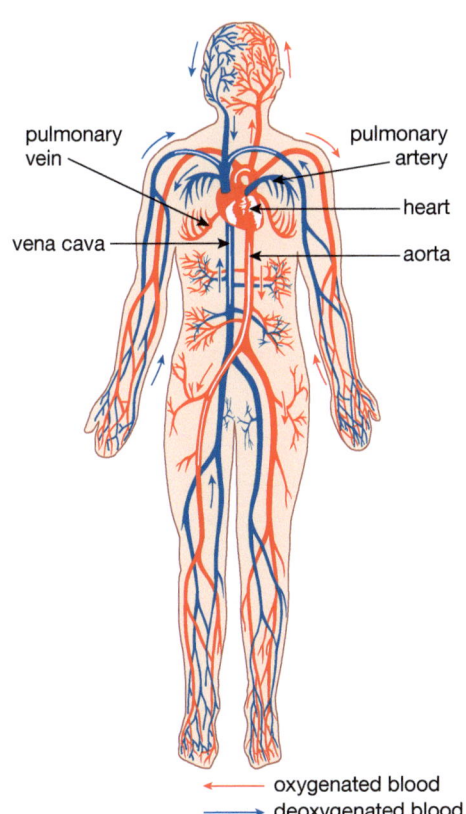

▲ Figure 6.1 The human circulatory system.

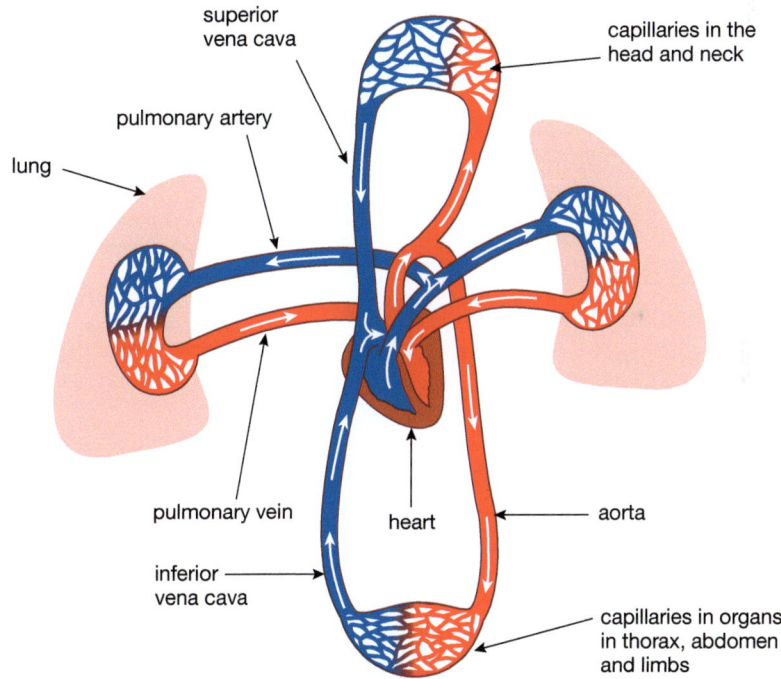

▲ Figure 6.2 The double circulation of blood in the human body. The blood passes through the heart twice during one complete circuit of the body.

KEY TERM
'Pulmonary' means 'to do with the lungs'.

DID YOU KNOW?
There are actually two vena cavae (the plural of vena cava). One (the *superior* vena cava) brings blood back from the head and arms. The other (the *inferior* vena cava) returns blood from the rest of the body.

There are two parts to a double circulatory system:

- The **pulmonary circulation:** Deoxygenated blood leaves the heart through the **pulmonary arteries**. It is circulated through the lungs, where it becomes oxygenated. The oxygenated blood returns to the heart through the **pulmonary veins**.

- The **systemic circulation:** Oxygenated blood leaves the heart through the aorta and is circulated through all other parts of the body, where it unloads its oxygen. The deoxygenated blood returns to the heart through the **vena cava**.

A double circulatory system allows the blood pressure to be different in the pulmonary and systemic circuits. The pressure in the systemic circulation to the body is higher than the pressure in the pulmonary circulation to the lungs.

The human circulatory system is made up of the following parts:

- **The heart:** a muscular pump.
- **Blood vessels:** these carry the blood around the body. **Arteries** carry blood away from the heart and towards other organs. **Veins** carry blood towards the heart and away from other organs. **Capillaries** carry blood through organs, linking the arteries and veins.
- **Blood:** the transport fluid.

Figure 6.3 shows the main blood vessels in the human circulatory system.

> **DID YOU KNOW?**
> The hepatic portal vein is an unusual vein, because it does not return blood directly to the heart. Instead, it carries blood containing the products of digestion from the intestine to the liver. The liver acts as a metabolic 'factory' – it stores some substances, makes others and breaks some down. The *hepatic vein* returns blood from the liver to the heart.

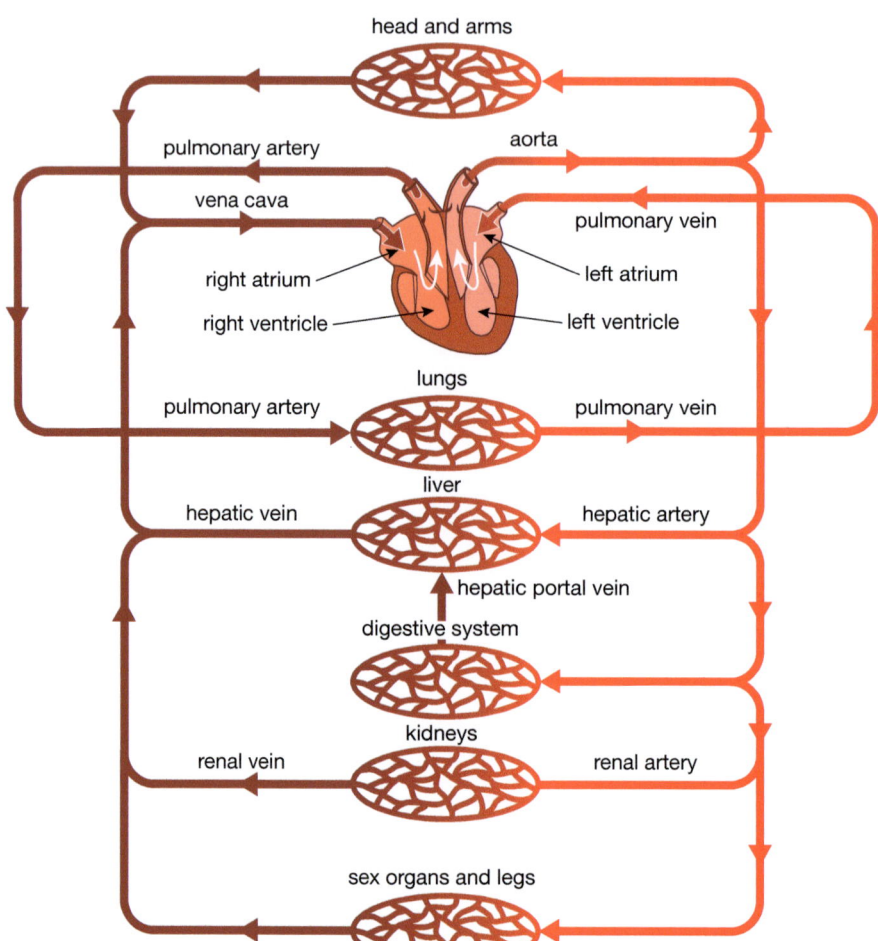

▲ Figure 6.3 The main components of the human circulatory system.

THE STRUCTURE AND FUNCTION OF THE HUMAN HEART

The heart (Figure 6.4) is a muscular pump. It pumps blood around the body at different speeds and at different pressures according to the body's needs. In fact, as you have seen, the heart really consists of *two* pumps – the right side of the heart pumps deoxygenated blood to the lungs and the left side pumps oxygenated blood to the rest of the body.

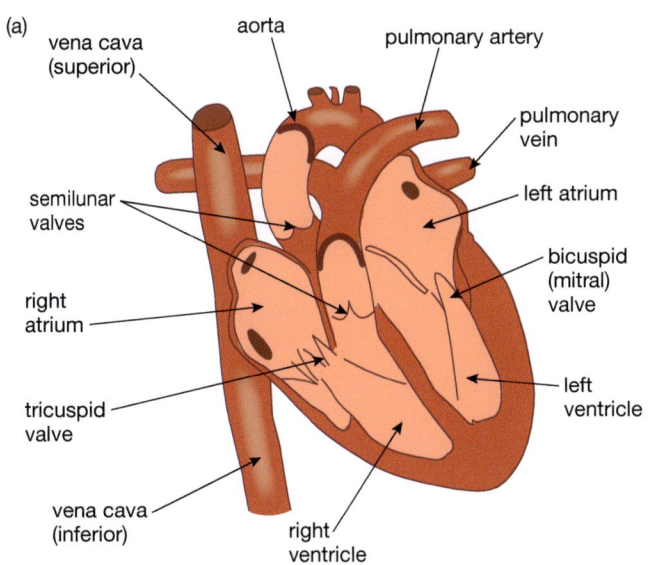
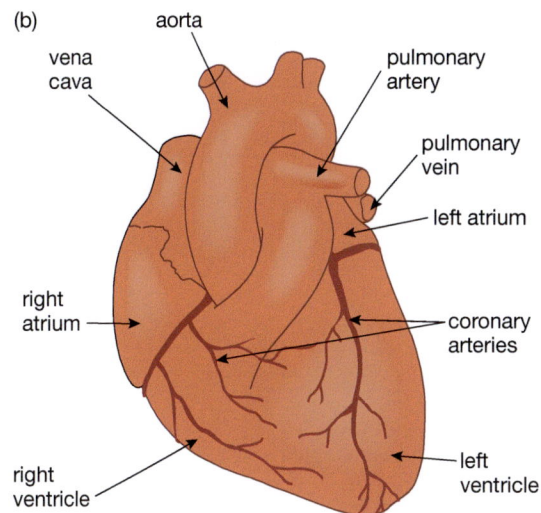

▲ Figure 6.4 The heart: (a) vertical section; (b) external view.

DID YOU KNOW?
The **bicuspid** (mitral) and **tricuspid valves** are both sometimes called atrioventricular valves, as each controls the passage of blood from an atrium to a ventricle.

The wall of the heart is made from a secialised type of muscle called cardiac muscle. You learnt about the different types of muscle in the body in Chapter 1 (see Table 1.2, page 14). Cardiac muscle consists of branching cells called muscle fibres, which form a strong, mesh-like framework. The cells are striated (striped), with several nuclei per cell (Figure 6.5).

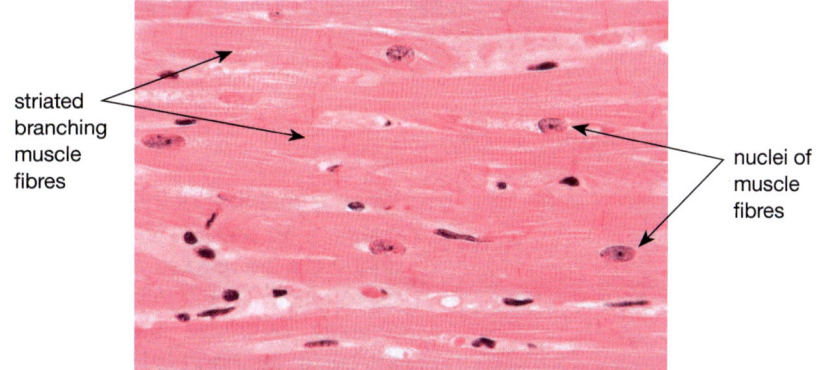

▲ Figure 6.5 Cardiac muscle fibres seen through a light microscope (×300).

Cardiac muscle is unlike any other muscle in the body. It never gets fatigued (tired), unlike skeletal muscle. On average, cardiac muscle fibres contract and relax over and over again, about 70 times a minute. Over a lifetime of 70 years, this special muscle will contract over two billion times – without taking a rest.

Another big difference between heart muscle and skeletal muscle fibres is that heart muscle can contract on its own, without the need for stimulation by nerves. If a sample of heart muscle is removed from a freshly killed mouse and placed on a microscope slide, the muscle fibres will contract rhythmically for as long as they are still alive. In the body, there are nerves to the heart muscle, but they are just there to adjust the rate of contraction. They do not start the contraction process (see the section 'Heart rate').

KEY TERM
'Cardiac' means 'to do with the heart'.

The mesh-like network of muscle fibres in the walls of the heart allows the waves of contraction to spread through the heart in a coordinated way.

Blood is moved through the heart by a series of contractions and relaxations of the muscle in the walls of the four chambers. These events form the **cardiac cycle**. The main stages of this cycle are illustrated in Figure 6.6.

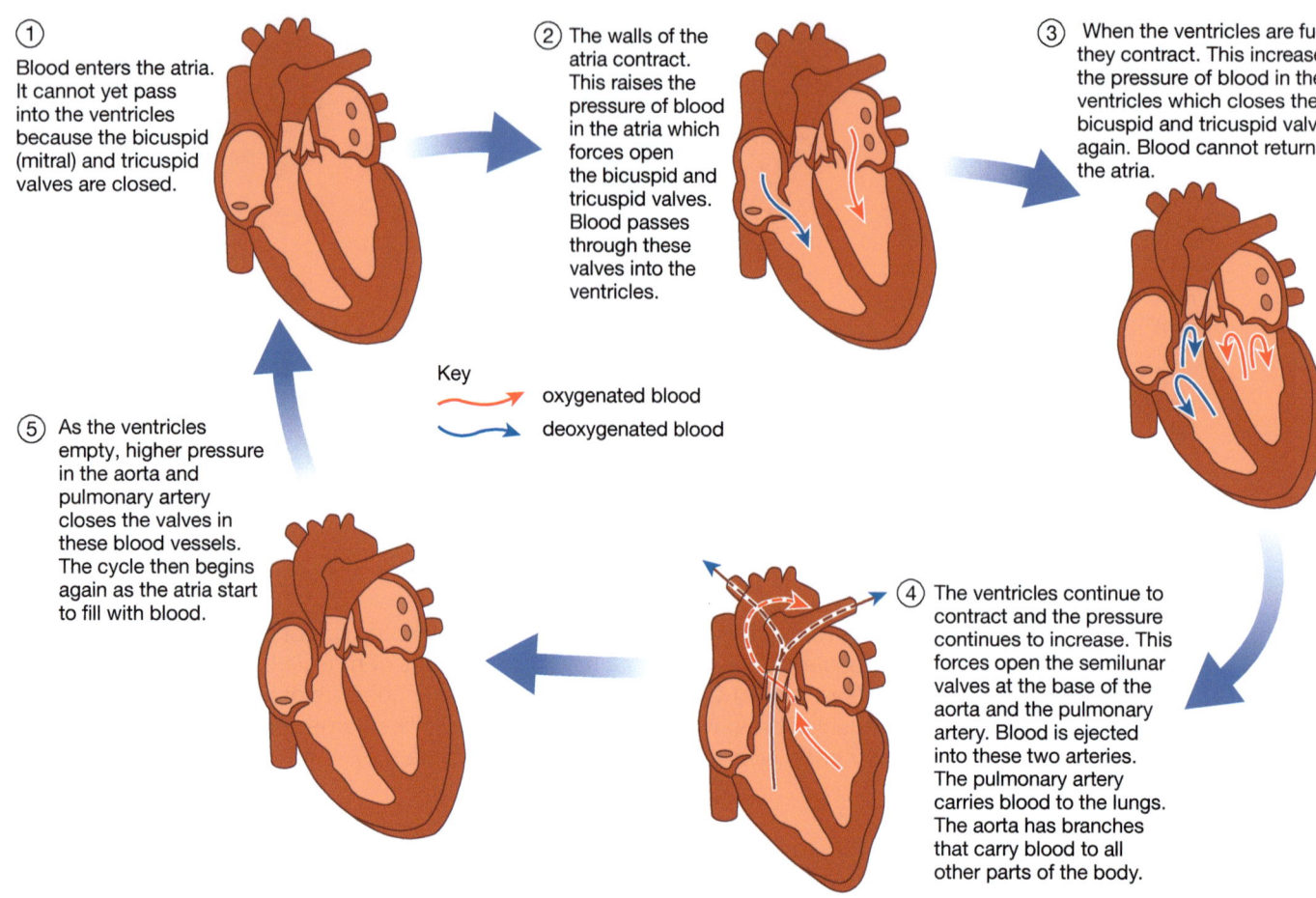

▲ Figure 6.6 The stages in the cardiac cycle.

When a chamber of the heart is contracting, we say it is in **systole**. When it is relaxing, we say it is in **diastole**. In Figure 6.6 stage 1, both upper and lower chambers are relaxed – the whole heart is in diastole. In stage 2, atrial systole takes place. This is followed by ventricular systole, starting at stage 3. Both atria and ventricles relax (diastole) after they have contracted so, by the end of the cycle, the heart is back to a state of diastole.

The structure of the heart is adapted to its function in several ways.

- It is divided into a left side and a right side by a wall of muscle called the septum. The right **ventricle** pumps blood only to the lungs while the left ventricle pumps blood to all other parts of the body. This requires much more pressure, which is why the wall of the left ventricle is much thicker than that of the right ventricle.
- Valves ensure that blood can flow in only one direction through the heart.
- The walls of the **atria** are thin. They can be stretched to receive blood as it returns to the heart but can contract with enough force to push blood through the bicuspid and tricuspid valves into the ventricles.
- The walls of the heart are made of specially adapted cardiac muscle.
- The cardiac muscle has its own blood supply, called the coronary circulation. Blood reaches the muscle via **coronary arteries**. These carry blood to capillaries that supply the heart muscle with oxygen and nutrients. Blood is returned to the right atrium via **coronary veins**.

DID YOU KNOW?
The plural of atrium is atria. Note that during the cardiac cycle both atria contract at the same time (and then relax). After this, both ventricles contract at the same time (and then relax). Students are sometimes confused about this and think that one ventricle contracts, followed by the other.

HEART RATE

The body can change the output of the heart in two ways – one works through the action of nerves and the other is under the control of a hormone.

Normally, the heart beats about 70 times a minute, but this can change according to the needs of the body. When we exercise, muscles must release more energy. They need an increased supply of oxygen for aerobic respiration (see Chapter 5). To deliver this extra oxygen to the muscles, both the number of beats per minute (heart rate) and the volume of blood pumped with each beat (stroke volume) increase.

When we sleep, our heart rate decreases as all our organs are working more slowly. They need to release less energy and so need less oxygen.

These changes in the heart rate are controlled by nerve impulses from the **cardiac centre** in a part of the brain called the medulla (Figure 6.7). When we start to exercise, our muscles produce more carbon dioxide in aerobic respiration. Receptors in the aorta and the carotid artery (the artery leading to the head) detect this increase. They send electrical signals called nerve impulses through the sensory nerve to the medulla. The medulla responds by sending nerve impulses along the accelerator nerve. When carbon dioxide production returns to normal, the medulla receives fewer impulses. It responds by sending nerve impulses along a decelerator nerve. The functions of the medulla are summarised in Chapter 9.

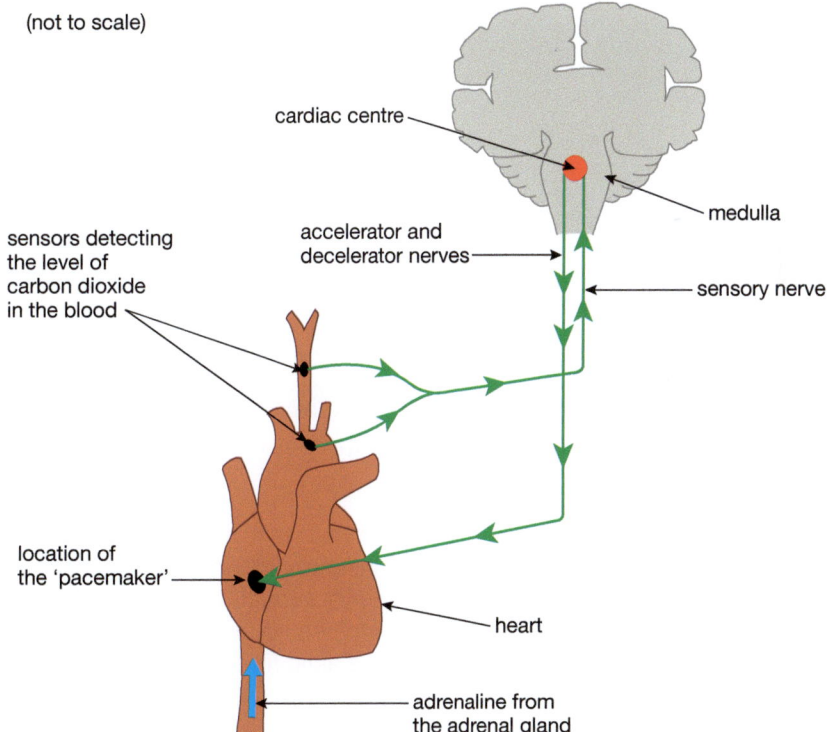

▲ Figure 6.7 How the heart rate is controlled.

The accelerator nerve increases the heart rate. It also causes the heart to beat with more force and so increases blood pressure. The decelerator nerve decreases the heart rate. It also reduces the force of the contractions. Blood pressure then returns to normal.

These controls are both examples of **reflex actions** (see Chapter 9).

> **DID YOU KNOW?**
> Have you noticed a 'hollow' or 'fluttering' feeling in your stomach when you are anxious? This happens because blood that would normally flow to your stomach and intestines has been diverted to the muscles to allow the 'fight or flight' response.

Our heart rate also increases when we are angry or afraid. The increased output supplies extra blood to the muscles, enabling them to release extra energy through aerobic respiration. This allows us to fight or run away and is called the 'fight or flight' response. It is triggered by secretion of the hormone adrenaline from the adrenal glands (see Chapter 9).

ARTERIES, VEINS AND CAPILLARIES

Arteries carry blood from the heart to the organs of the body. This arterial blood is pumped out under high pressure by the ventricles of the heart. When blood leaves the heart through the aorta, valves in the aorta prevent the blood from returning to the heart. The blood leaves in short bursts or 'spurts' of pressure, stretching or distending the wall of the aorta. When the ventricle of the heart relaxes, the stretched section of the aorta recoils, increasing the pressure inside it. A wave of stretching followed by constriction passes along the aorta and through the arteries. This is the **pulse**. The pulse wave passes through the arteries and arterioles, getting weaker as it travels along. It disappears by the time it reaches the capillaries, so there is no pulse in veins. You can find your pulse anywhere an artery can be pressed against a bone, for example, in the wrist. Because the pulse is caused by the surge of blood at each heartbeat, the pulse rate is the same as the heart rate.

The walls of arteries have a thick layer of elastic tissue and smooth muscle fibres so they can stretch and recoil under the pressure wave. The muscle in the walls of smaller arteries can contract to help the blood flow.

The pulse wave travels much faster than the actual blood flow within the arteries. By the time blood reaches the capillaries, it is only travelling at about 1 mm per second.

> **DID YOU KNOW?**
> **Arterioles** are small arteries. They carry blood into organs from arteries. Their structure is similar to the larger arteries, but they have a larger proportion of muscle fibres in their walls. They are also supplied with nerve endings in their walls and they can be made to dilate (become wider) or constrict (become narrower) to allow more or less blood into the organ.
>
> If *all* the arterioles constrict, it is harder for blood to pass through them – there is more resistance. This increases blood pressure. Prolonged stress can cause arterioles to constrict, increasing blood pressure.

> **KEY POINT**
> All arteries carry oxygenated blood, except the pulmonary artery and the umbilical artery of an unborn baby (see Chapter 11). All veins carry deoxygenated blood, except the pulmonary vein and umbilical vein.

Veins carry blood from organs back to the heart. Blood pressure decreases as it flows through the capillaries, so the blood in veins (venous blood) is at a very low pressure – much lower than in the arteries. The walls of veins are thin, with few elastic fibres and smooth muscle. Figure 6.8 shows the structures of a typical artery and a typical vein with the same diameter.

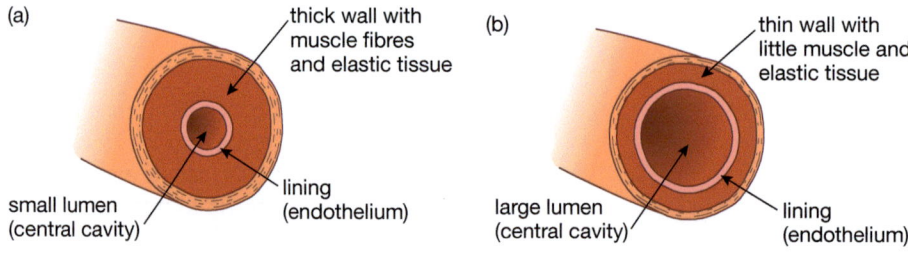

▲ Figure 6.8 The structure of (a) an artery and (b) a vein as seen in cross-section.

Veins have **semilunar** (half-moon shaped) **valves**, which prevent the backward flow of blood. The action of these valves is explained in Figure 6.9.

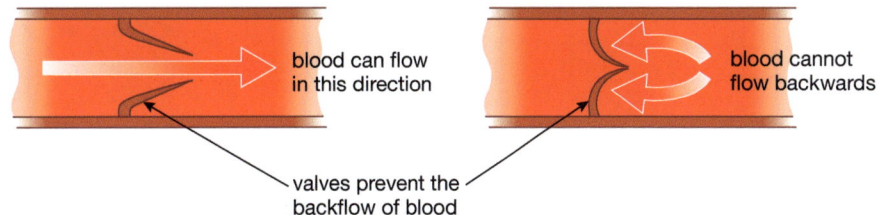

▲ Figure 6.9 Semilunar valves help to maintain a one-way flow of blood in veins.

Capillaries carry blood through organs, bringing the blood close to every cell in the organ. Substances are transferred between the blood in the capillary and the cells. To do this, capillaries must be small enough to 'fit' between cells, and allow materials to pass through their walls easily. Figure 6.10 shows the structure of a capillary and how substances are exchanged between the capillary and nearby cells. The walls of capillaries are one cell thick, which provides a short distance for diffusion of materials into and out of the blood. Capillaries have a tiny **lumen** (central space) which red blood cells can only just fit through. This means the red blood cells are close to the capillary wall, so there is a short distance for oxygen to diffuse.

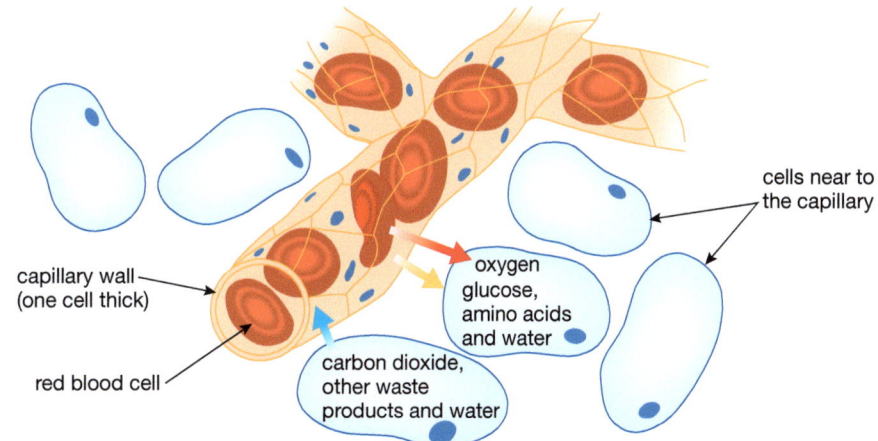

▲ Figure 6.10 Exchange of materials between capillaries and cells.

Figure 6.11 shows a photograph of a cross-section through an artery and a vein.

Between the capillaries and the cells is a watery liquid called **tissue fluid**. The squamous (flattened) epithelium cells of the capillary wall are leaky, so the blood pressure causes fluid to leak out of the capillaries. Tissue fluid is similar in composition to blood plasma, except that it lacks proteins. These are too large to pass through the capillary cells. Tissue fluid forms a pathway for diffusion of substances between the capillaries and the cells. Most of the water from tissue fluid re-enters capillaries by osmosis, and some tissue fluid passes into another system of vessels called the lymphatic system (Figure 6.12). The lymphatic system consists of vessels similar to blood capillaries, which are sometimes called lymphatic capillaries. They transport the fluid, called lymph, back to the blood (Figure 6.13). The lacteals of the small intestine are part of the lymphatic system (see Chapter 4).

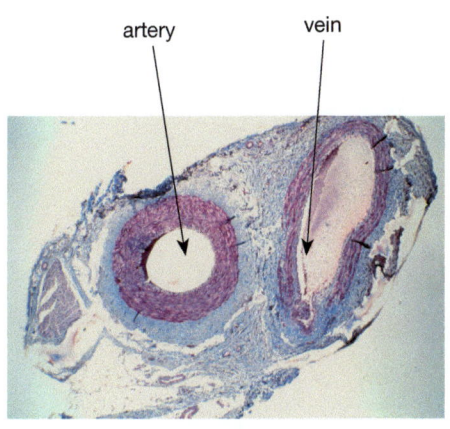

▲ Figure 6.11 A section through an artery and a vein. Notice that the artery has a thick muscular wall and small lumen compared with the vein.

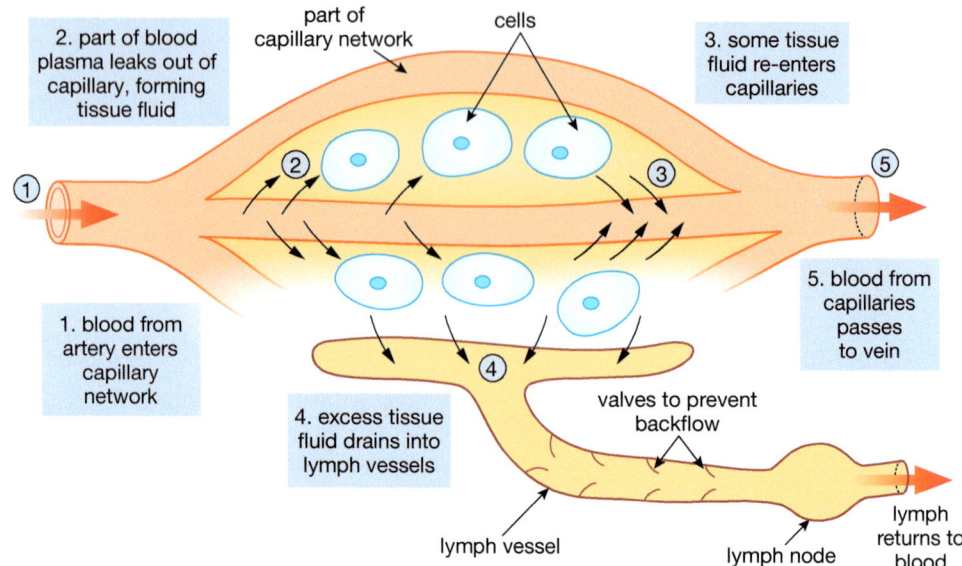

▲ Figure 6.12 The relationship between blood capillaries, tissue fluid and lymph.

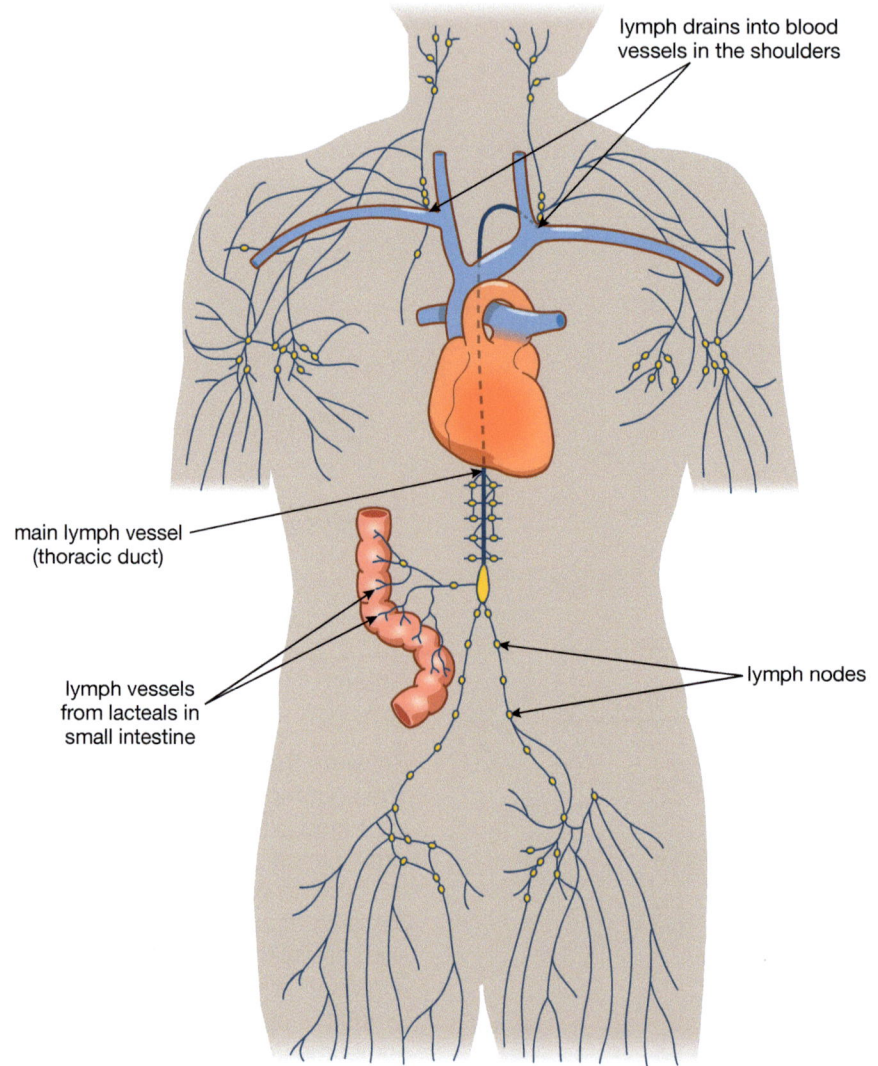

▲ Figure 6.13 The lymphatic system. Notice the large numbers of lymph nodes in the neck, armpits and groin. These may swell up if a person has an infection.

Before lymph passes back into the blood, it is filtered to remove dead cells and bacteria. This takes place in swellings called **lymph nodes**. Lymph nodes contain white blood cells that are important in destroying harmful bacteria (see later in this chapter). When a person gets an infection, one of the first signs may be a swelling of the lymph nodes, often referred to as 'swollen glands'.

THE PULSE RATE AND PHYSICAL FITNESS

Normal resting heart rate ranges from around 40 beats per minute (bpm) to 100 bpm. It depends upon sex, age and physical fitness. The average resting heart rate for a man is 70 bpm and for a woman it is 75 bpm.

If you are in training and trying to improve your physical fitness, you should notice a drop in your resting heart rate. Generally, a lower rate is a sign of improved fitness. This is because the heart becomes more efficient at pumping blood around the body, so more blood is pumped with each beat (the stroke volume increases).

It is not just the resting heart rate that matters. What is also important is how quickly your heart recovers after exercise. You can carry out an investigation into this (Activity 6).

> **DID YOU KNOW?**
> The world record for the lowest resting heart rate is 27 bpm. This was recorded on a 36-year-old man in Guernsey, UK in 2005.

ACTIVITY 6

▼ PRACTICAL: INVESTIGATING THE EFFECT OF EXERCISE ON THE PULSE RATE

The pulse is a short, sudden increase in blood pressure in an artery, caused by the heartbeat. If the artery is near the surface of the skin, e.g. in the wrist or neck, you can feel the pulse and measure the pulse rate.

A person (the 'subject' of the investigation) finds their pulse in their wrist or neck and records the number of pulse 'beats' in 15 seconds. This number is multiplied by four to give beats per minute (bpm). This is repeated three times and an average resting pulse rate is calculated.

The subject then carries out a standard form of exercise, such as 'running on the spot' for five minutes, or a hundred 'step-ups'. (The type of exercise is standardised so that you can compare different people's results.)

Immediately after the exercise, the subject finds their pulse again and measures the number of beats in 15 seconds, converting this value to bpm. They continue to measure their pulse rate like this, taking a reading every 30 seconds until the rate returns to the resting value.

If you are able to do this investigation, use your results to plot a line graph of pulse rate against the time after finishing the exercise. Draw a horizontal line on the graph to show your resting pulse rate.

Compare your results with those of other students. How much variation is there in the recovery time after a standard period of exercise?

THE COMPOSITION OF BLOOD

Blood is a complex tissue composed of several types of cell suspended in a liquid called **plasma**. Plasma is a watery fluid containing many dissolved substances. Carbon dioxide is carried in the plasma, mainly as hydrogen carbonate ions (HCO_3^-). Plasma also contains nutrients from the digestion of food, such as glucose and amino acids. It transports hormones (see Chapter 9) and waste products such as urea (see Chapter 10).

Figure 6.14 shows the main types of component found in blood.

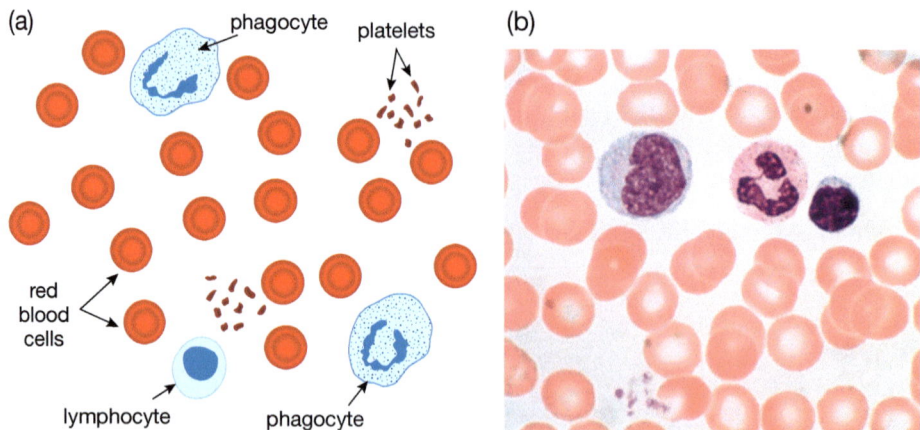

▲ Figure 6.14 The different types of blood cell. (a) Diagram of the different cells. (b) A blood smear seen through a light microscope. The smear shows many red blood cells and three different kinds of white blood cell.

Table 6.1 summarises the functions of the different parts of the blood.

▼ Table 6.1 Functions of the different components of blood.

Component of blood	Description of component	Function of component
plasma	liquid part of blood: mainly water	carries the blood cells around the body; carries dissolved nutrients, hormones, carbon dioxide and urea; distributes heat around the body
red blood cells (erythrocytes)	biconcave, disc-like cells with no nucleus; millions in each mm^3 of blood	transport oxygen – contain mainly haemoglobin, which loads oxygen in the lungs and unloads it in other regions of the body
white blood cells: lymphocytes	about the same size as red blood cells with a large round nucleus	produce antibodies to destroy microorganisms – some lymphocytes remain in our blood after infection and give us immunity to specific diseases
white blood cells: phagocytes	much larger than red blood cells, with a large spherical or lobed nucleus	digest and destroy bacteria and other microorganisms that have infected our bodies
platelets	the smallest cells – are really fragments of other cells	release chemicals to make blood clot when we cut ourselves

RED BLOOD CELLS

The red blood cells or **erythrocytes** are highly specialised cells made in the bone marrow. They have a limited life span of about 100 days; after this, they are destroyed in the spleen. They have only one function – to transport oxygen. Several features enable them to carry out this function very efficiently.

Red blood cells contain an iron-containing protein called haemoglobin. When there is a high concentration of oxygen in the surroundings, haemoglobin associates (combines) with oxygen to form **oxyhaemoglobin**. We say that the red blood cell is loading oxygen. When the concentration of oxygen is low, oxyhaemoglobin turns back into haemoglobin and the red blood cell unloads its oxygen.

$$\text{haemoglobin} + \text{oxygen} \underset{\text{low } O_2 \text{ (tissues)}}{\overset{\text{high } O_2 \text{ (lungs)}}{\rightleftharpoons}} \text{oxyhaemoglobin}$$

As red blood cells pass through the lungs, they load oxygen. As they pass through active tissues, they unload oxygen.

Red blood cells do not contain a nucleus. It is lost during their development in the bone marrow. This means that more haemoglobin can be packed into each red blood cell so more oxygen can be transported. Red blood cells are concave on both sides (biconcave). This shape allows efficient exchange of oxygen into and out of the cell. Each red blood cell has a high surface area to volume ratio, giving a large area for diffusion. The thin shape of the cell results in a short diffusion distance to the centre of the cell. Red blood cells are also flexible, so they can squeeze through the narrow capillaries.

WHITE BLOOD CELLS

There are several types of white blood cell. Their main role is to protect the body against invasion by disease-causing microorganisms (called **pathogens**) such as bacteria and viruses. They do this in two main ways: **phagocytosis** and the production of **antibodies**.

About 70% of white blood cells can take up and destroy microorganisms such as bacteria. This process is called phagocytosis, and the cells are **phagocytes**. They do this by changing their shape to produce extensions of their cytoplasm, called pseudopodia. The pseudopodia surround and enclose the microorganism in a vacuole. Once the microorganism is inside, the phagocyte secretes enzymes into the vacuole to break the microorganism down (Figure 6.15). Phagocytosis means 'cell eating' – you can see why it is called this.

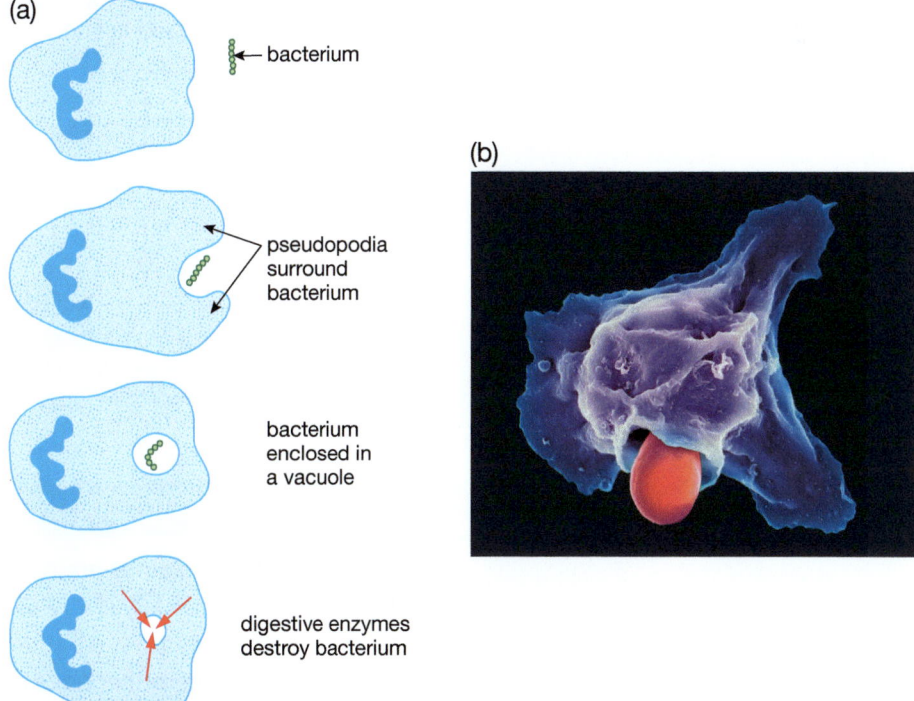

▲ Figure 6.15 (a) Phagocytosis by a white blood cell. (b) A phagocyte ingesting a yeast cell.

About 25% of white blood cells are **lymphocytes**. Their function is to make chemicals called antibodies. Antibodies are soluble proteins that pass into the plasma. Their role is to destroy disease-causing organisms such as bacteria and viruses. Antibodies recognise other chemicals, called antigens, on the surface of these pathogens. Production of antibodies against particular antigens is known as the **immune response**. This is described in more detail in Chapter 13.

> ### DID YOU KNOW?
> When we are exposed to a particular disease-causing microorganism for the first time, certain lymphocytes retain a 'memory' of their antigens. This memory may last for many years or even a lifetime. If the same pathogen re-infects us, these *memory cells* can reproduce and make more antibodies. This *secondary* immune response is faster and stronger than the first immune response, and gives us *immunity* against the invading organism, so that it is killed before it can multiply to a level where it would cause the disease.

PLATELETS

Platelets are not whole cells, but fragments of large cells made in the bone marrow. If the skin is cut, exposure to the air stimulates the platelets to start a chain of reactions leading to the formation of a blood clot. The final part of the chain is shown in Figure 6.16. A soluble protein in the plasma called **prothrombin** is converted into an enzyme called **thrombin**. This enzyme causes the soluble plasma protein **fibrinogen** to change into insoluble fibres of another protein, **fibrin**. The fibrin forms a network across the wound, in which red blood cells become trapped. This is the blood clot, which prevents further loss of blood and stops pathogens entering the wound. The clot develops into a scab, a rough protective crust that forms over the wound. The scab protects the damaged tissue while new skin grows.

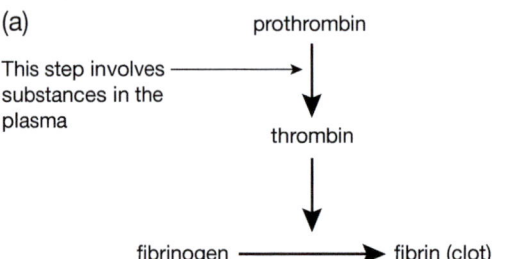
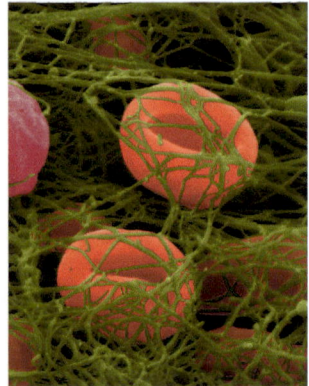

▲ Figure 6.16 (a) The final steps in the formation of fibrin (a blood clot). (b) A blood clot seen through an electron microscope, showing red blood cells trapped in a meshwork of fibrin.

BLOOD GROUPS AND TRANSFUSIONS

Successful blood transfusions are only possible because of our knowledge of blood grouping (Figure 6.17). Blood of the wrong group will be 'rejected' by the body.

It is not only bacterial cells that have antigens on the surface. All animal cells have them, including blood cells. Your blood group depends on the antigens on the surface of your red blood cells. Two types of antigen are important in

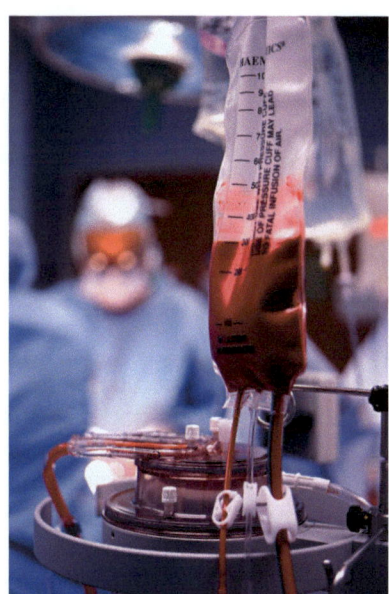

▲ Figure 6.17 A patient receiving a transfusion of blood during an operation.

KEY POINT

Agglutination is not the same process as blood clotting (see above). That is why we use the word 'clumping' to describe what happens during agglutination, rather than 'clotting'.

DID YOU KNOW?

These antibodies differ in one important way from the antibodies we make against organisms that cause disease. They are present in our blood all the time, so we do not need to be exposed to the antigen to make the antibody.

blood grouping. They are called the A and B antigens. The four possible blood groups, based on the presence of these antigens, are called A, B, AB and O.

Each person also has antibodies in their plasma. These antibodies will destroy red blood cells with a particular antigen by making them **agglutinate**. **Agglutination** means that they stick together in clumps. Anti-A antibody agglutinates any red blood cells with the A antigen. Anti-B antibody agglutinates any red blood cells with the B antigen (Figure 6.18).

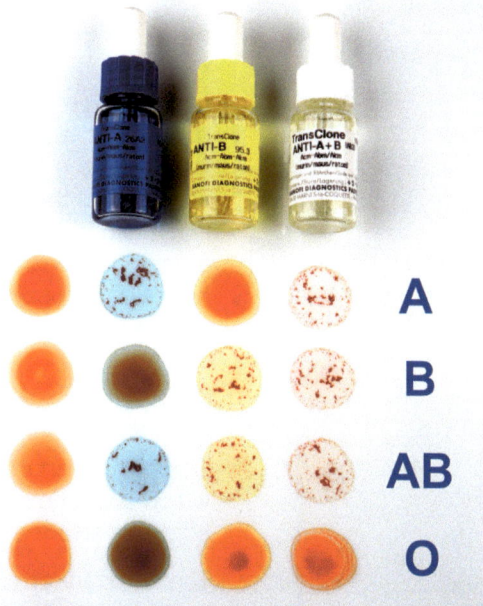

▲ Figure 6.18 A test for ABO blood groups. The bottles hold serum containing anti-A, anti-B and (anti-A + anti-B) antibodies. The spots below each bottle show the effects of the serum on different samples of blood. For example, blood group A has agglutinated with anti-A and with (anti-A + anti-B). No agglutination happens with blood group O. The four spots of blood on the left are Controls – no serum has been added to them.

Table 6.2 summarises the presence or absence of antigens and antibodies in the different blood groups. As a general principle, remember that a person never carries the antibodies that would react with the antigens on their own red blood cells.

▼ Table 6.2 Blood groups and their antigens and antibodies.

Blood group	Antigens on red cells	Antibodies in plasma
A	A	anti-B
B	B	anti-A
AB	AB	neither
O	neither	anti-A and anti-B

If a doctor gives a blood transfusion of type A blood to a person with type B blood, the patient may die. The person receiving the blood has anti-A antibodies in their plasma that will make the red cells of the transfused blood agglutinate. Blood cells will clump inside the blood vessels and block them. The safety of a transfusion depends on the antigens on the red cells of the donated blood (blood from another person) and the antibodies in the plasma of the person receiving the blood. If the antigens and antibodies can react (e.g. antigen A and anti-A antibody), the transfusion will be unsafe. Table 6.3 shows safe and unsafe transfusions.

> **DID YOU KNOW?**
> It is the *donated* antigens (on the red cells) that are important in matching donor and recipient. You might think that the antibodies in the donated plasma have to be taken into consideration, but they do not – the volume of donated blood is small compared with the total volume of blood in the body, so any antibodies in the donated blood are too diluted to matter.

Table 6.3 Blood transfusions and blood groups (✓ = safe transfusion, ✗ = unsafe transfusion).

Blood group of donor		Blood group of recipient (antibodies present shown in brackets)			
Group	Antigen	A (anti-B)	B (anti-A)	AB (neither)	O (anti-A + anti-B)
A	A	✓	✗	✓	✗
B	B	✗	✓	✓	✗
AB	A+B	✗	✗	✓	✗
O	neither	✓	✓	✓	✓

Blood group O is sometimes called the 'universal donor' because, in theory, it can be given to any other blood group. There are no antigens on the surface of the red blood cells, so there can be no reaction with any antibodies in the plasma of the person receiving the blood (the recipient). Similarly, blood group AB is called the 'universal recipient'. Because there are no antibodies in the plasma, antigens in the donated blood cannot cause a reaction, so in theory a person who is blood group AB can receive any blood type. In practice, however, a hospital would only use blood of a patient's own group when giving them a transfusion.

CORONARY HEART DISEASE

Coronary heart disease (CHD) is caused by blockage of the coronary arteries, which supply blood to the heart muscle (Figure 6.19). It is also known as coronary artery disease.

The coronary arteries are among the narrowest arteries in the body. They are easily blocked by a build-up of fatty substances such as cholesterol in their walls. This is called 'hardening of the arteries' or **atherosclerosis**. If this happens, the oxygen supply to the heart muscle is reduced (Figure 6.20). The first symptom of a blockage may be chest pain when the person exercises. This is called **angina**. The fatty deposits are called a *plaque*, and can build up to block the coronary artery completely. They may also lead to the formation of blood clots, which can cause a blockage. If the artery is completely blocked, the oxygen supply to the heart muscle may be cut off altogether and the muscle will stop contracting – this is a **heart attack**.

If a large part of the heart muscle is affected, a heart attack is often fatal. However, many heart attacks are less severe and, with treatment, the person can recover.

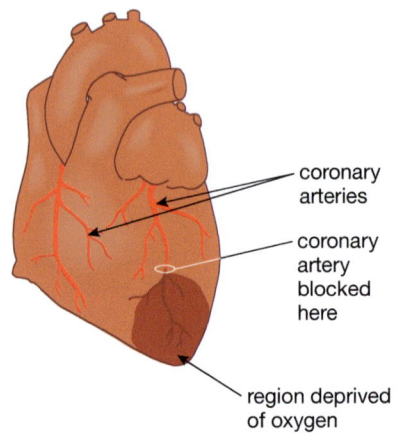

▲ Figure 6.19 A blockage of a coronary artery cuts off the blood supply to part of the heart muscle.

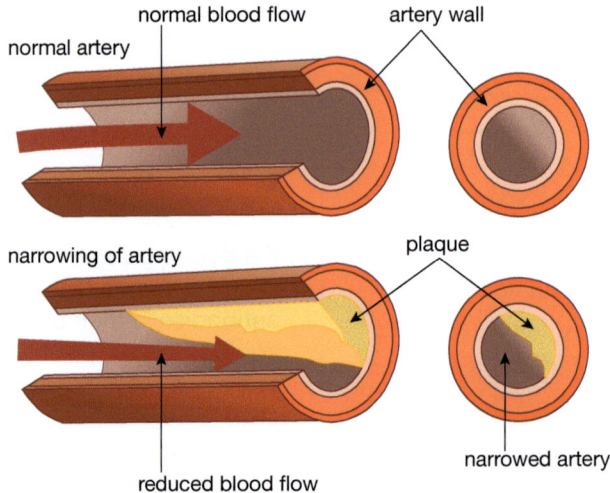

▲ Figure 6.20 Atherosclerosis and formation of a plaque in a coronary artery.

A number of factors make coronary heart disease more likely.

- **Heredity:** Some people inherit genes that make them more likely to suffer from heart disease.
- **High blood pressure:** Permanently high blood pressure is called **hypertension**. The heart has to work harder to pump the blood, which puts more strain on the heart muscle.
- **Diet:** A diet that is high in fat, especially saturated fat, causes raised blood cholesterol. High cholesterol in the blood combines with other lipids to form a plaque.
- **Smoking:** Nicotine in tobacco smoke has several harmful effects. It constricts blood vessels, raising blood pressure, speeds up the heart rate, and increases blood cholesterol. Smoke also contains carbon monoxide, which reduces the ability of the blood to carry oxygen (see Chapter 5).
- **Stress:** Hormones released during stress constrict blood vessels, raising blood pressure.
- **Lack of exercise:** Regular exercise helps to reduce blood pressure and strengthens the heart.
- **Diabetes:** This disease results in high blood glucose levels (see Chapter 9). This has been shown to lead to an increase in the fatty substances deposited in plaques.

Clearly, you cannot do anything to avoid some risk factors, such as heredity. Most of them, however, are lifestyle choices. There are four main ways to avoid coronary heart disease:

- do not smoke
- exercise regularly
- eat a healthy balanced diet
- stay at a healthy weight.

A person who has any other health problems that raise the risk of a heart attack – including diabetes, high blood pressure and high cholesterol – should also manage these problems with the help of health professionals and medical care.

If a person has an advanced case of coronary heart disease, or has suffered from a heart attack, this can be treated by a medical procedure called an **angioplasty**. This widens the artery, so that normal blood flow is returned to the heart muscle.

An angioplasty is carried out using a long, thin tube called a catheter. First, the doctor inserts the catheter into a blood vessel in the arm or groin. The catheter has a very thin guide wire, which the doctor uses to guide the catheter through blood vessels to the blocked region of the coronary artery. The doctor watches the movement of the catheter on an X-ray screen.

At the end of the catheter is a small inflatable balloon, with an expandable mesh tube called a **stent** around the balloon (Figure 6.21). Once the catheter has reached the blocked region of the artery, the balloon is inflated, pushing the plaque against the wall of the artery. The inflated balloon also expands the stent. Next, the balloon is deflated, and the balloon, guide wire and catheter are removed. The stent remains in the blood vessel, allowing the blood to flow normally again. Sometimes inflation of the balloon is enough to widen an artery, without the need for a stent. Stents can also be used to treat arteriosclerosis (thickening and hardening of the artery walls, often due to old age) in other arteries, such as the carotid arteries leading to the head.

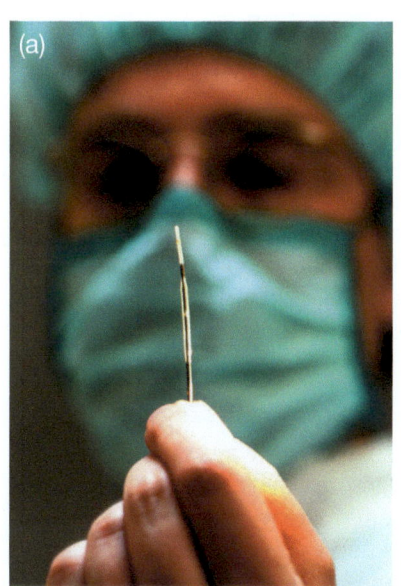
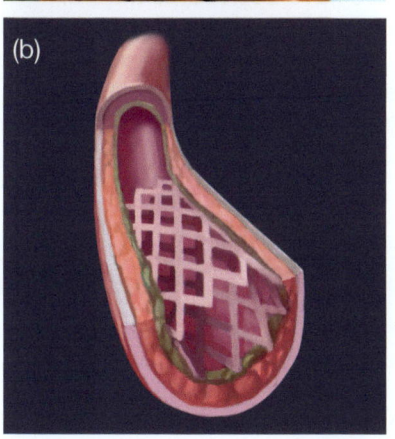

▲ Figure 6.21 (a) A surgeon holding a stent. (b) A stent in place in a clogged artery.

HEART TRANSPLANTS

A **heart transplant** is a surgical procedure carried out on a patient with heart failure or severe disease of the coronary arteries. A functioning heart is removed from the body of a person who has recently died (the organ donor) and implanted into the patient. Usually, the patient's heart is removed and replaced with the new heart. In some operations, however, the old heart is left in place to support the new one. The technique was first performed in 1967, and today about 3500 heart transplants are performed around the world every year. Over the years, the success rate has improved, so that the average length of time a patient survives after the operation is now 15 years.

Heart transplants cannot be used on all patients with heart failure. Patients with certain diseases (such as advanced lung, liver or kidney disease; cancer; or severe diabetes) are less likely to be suitable for organ transplant.

One of the problems that can happen with any transplant is **organ rejection**. All cells have surface antigens, and each of us has our own unique set of antigens on the cells of our bodies. Lymphocytes will not produce antibodies against these because they are 'self' antigens. The antigens on bacteria that infect us are 'non-self', so our lymphocytes recognise them and produce antibodies to destroy them. The same is true of organs transplanted from other people. If a heart or other organ is used for a transplant, the antigens on the cells of that organ will be recognised in the recipient's body as 'non-self'. Lymphocytes will produce antibodies to destroy these 'foreign' cells. This is what is meant by organ rejection.

When looking for a donor organ, it is important to find one with antigens on the cells that match the patient's antigens as closely as possible. Our antigens are determined by our genes, so organs from family members are often a good match. Those of identical twins are particularly closely matched. Finding an organ with antigens similar to those of the person needing the transplant is called tissue typing. The risk of rejection can be reduced considerably by matching the antigens closely and using immunosuppressive drugs.

> **DID YOU KNOW?**
> Immunosuppressive drugs lower our immune response. They reduce the risk of rejection of a transplanted organ, but they also reduce our ability to fight disease.

ARTIFICIAL HEARTS

The demand for donor organs for heart transplants is always greater than the supply. This was one of the reasons that led to the development of **artificial hearts**. An artificial heart is a synthetic device, made mainly of plastics and some metal alloys. Their main purpose is to keep people alive in the time leading up to a heart transplant, although there have been some trials of artificial hearts as permanent replacements for a diseased heart. As yet, 'total' artificial hearts have not been successful as long-term replacements. However, research in this area has produced a number of *ventricular assist devices*, which can be used in patients with some remaining heart function. They do not replace the heart, but help it to function.

EXERCISE AND THE CARDIOVASCULAR SYSTEM

The heart and blood vessels are sometimes called the **cardiovascular system**. As soon as we start to exercise, our cardiovascular system has to work harder (Figure 6.22). Our muscle cells respire more quickly to release more energy and produce more carbon dioxide. This produces a reflex action to increase heart rate and blood pressure. More intense exercise results in a greater increase in heart rate and blood pressure.

UNIT 2 — INTERNAL TRANSPORT

During a period of exercise, the heart rate and blood pressure increase to a maximum to deliver the extra oxygen needed by the muscles. They remain at this level during the period of exercise, then begin to decrease to the pre-exercise levels as soon as the exercise ends. Figure 6.23 shows these changes.

The period when heart rate and blood pressure are returning to normal following exercise is called the *recovery period*. Heart rate and blood pressure do not drop back to pre-exercise levels straight away; instead, they decrease gradually during this period. This is because, during exercise, lactate is formed in the muscles by anaerobic respiration (see Chapter 5). Lactate must be oxidised before it can be removed from the muscles, so as long as there is any lactate left in the muscles, extra oxygen will be needed to oxidise it. The heart must beat faster and with more force to deliver this extra oxygen. As the amount of lactate drops, so do the heart rate and blood pressure. When all the lactate has been oxidised, both heart rate and blood pressure return to pre-exercise levels. Figure 6.24 shows the changes in the heart rate and level of lactate in the blood following exercise. The length of recovery time is a rough measure of how healthy your heart is. If the heart can pump enough oxygen to the muscles during exercise, only a little lactate will be formed and the recovery period will be short.

▲ Figure 6.22 As we exercise, our cardiovascular system works harder.

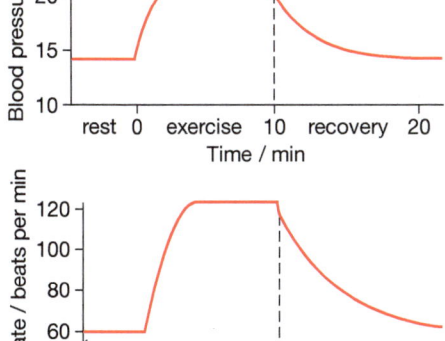

▲ Figure 6.23 How the heart rate and blood pressure change during exercise.

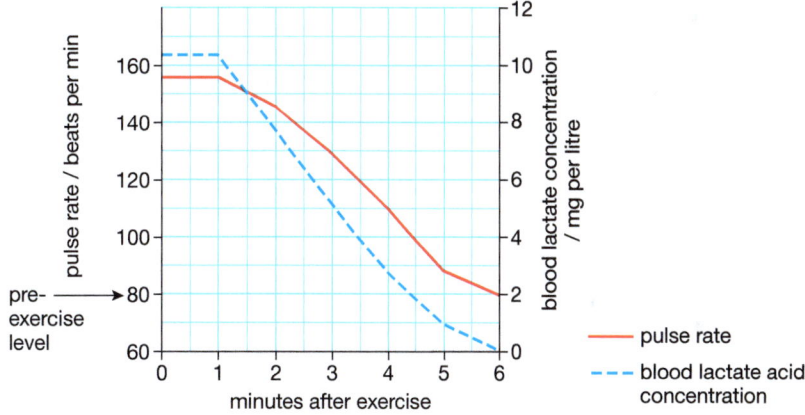

▲ Figure 6.24 The changes in pulse rate and lactate levels in the period following exercise.

THE BENEFITS OF AEROBIC EXERCISE

KEY POINT

Compare aerobic exercise with anaerobic exercise. Anaerobic exercises such as weight lifting or sprinting are short duration, high intensity exercises, which result in a build-up of lactate in the muscles and blood.

Aerobic exercise is physical exercise that depends mainly on ATP supplied by aerobic respiration (see Chapter 5). It consists of light to moderate activities that can be performed for extended periods of time, such as walking, jogging, swimming and cycling.

Regular aerobic exercise strengthens the heart muscle, and reduces the resting heart rate and blood pressure. It has also been shown to increase the number of red blood cells in the body, helping with the transport of oxygen. Aerobic exercise does not only benefit the cardiovascular system; it also improves a person's overall fitness in many ways. It:

- helps maintain a healthy body weight, to prevent obesity
- reduces the levels of lipids, including cholesterol, in the blood
- builds skeletal muscle, increasing the mass and 'tone' of muscles (see Chapter 7)
- improves the strength of tendons and ligaments

- strengthens the diaphragm and intercostal muscles and increases the vital capacity of the lungs, so that breathing is more efficient
- stimulates the immune system
- helps maintain the level of glucose in the blood, which reduces the risk of type 2 diabetes (see Chapter 9)
- reduces the risk of contracting certain cancers, such as colon cancer
- makes people feel happier and more satisfied with life – some studies have even shown that regular exercise is as good as many medicines in treating depression.

SOME OTHER DISORDERS OF THE CARDIOVASCULAR SYSTEM AND THEIR TREATMENT

Atherosclerosis is not just a problem in the coronary arteries. It can also happen in any of the other arteries of the body, blocking blood flow to the brain, pelvis, legs, arms or kidneys. A piece of a plaque may break off, or a blood clot may form on the surface of the plaque. If an artery is blocked (for example, by a piece of a plaque or a blood clot), this may cause a heart attack or stroke.

> **DID YOU KNOW?**
> A stroke cuts off blood supply to part of the brain. If it is not caught early, it can cause permanent brain damage or death.

STATINS

Statins are chemicals that lower blood cholesterol. They are very effective in preventing cardiovascular disease caused by atherosclerosis. Statins have been found to be especially effective for treating people who have high levels of blood cholesterol but no previous history of heart disease. In other words, they help to reduce one of the main causes of cardiovascular disease.

The main way in which statins work is by inhibiting an enzyme that is involved in the synthesis of cholesterol. This is important, because most cholesterol is made in our liver, rather than gained from our diet. Statins also affect various other aspects of lipid metabolism in the cells of the body.

Statins are the most commercially successful drugs ever made, earning drug companies tens of billions of dollars every year.

PLANT STANOL ESTERS

Another group of substances known to lower cholesterol is plant **stanol esters**. These are naturally occurring molecules found in most plants; they are particularly concentrated in vegetable oils. Stanol esters have a structure that is very similar to cholesterol.

Certain foods – mainly dairy products, such as yoghurts and spreads – can be enriched with stanol esters. These foods have been shown to lower blood cholesterol in clinical trials, but have not been found to affect the development of cardiovascular disease itself.

BETA BLOCKERS

Beta blockers (β-blockers) are a class of drugs that are mainly used to manage unusual heart rhythms and to protect against a second heart attack after a person has already had one attack.

Beta blockers work on the nervous system. Nerves communicate with each other by secreting chemicals called **neurotransmitters**, one of which is the substance **adrenaline**. (Adrenaline may be used where it is made or released into the blood as a hormone – see Chapter 9).

Beta blockers block the action of adrenaline at its receptor sites in various parts of the body, including the heart. In doing this, they reduce heart rate and stroke volume, which means the heart has to work less hard and needs less oxygen. This is why beta blockers are a powerful treatment for angina, the pain associated with coronary heart disease.

HYPERTENSION

Hypertension or high blood pressure is a long-term medical condition in which the blood pressure in the arteries is persistently higher than normal. It does not cause many symptoms of its own but it is a major risk factor in coronary heart disease, strokes, heart failure, kidney failure and several other diseases.

Blood pressure is expressed as two measurements, the systolic pressure and the diastolic pressure (see the heart cycle earlier in this chapter). The systolic pressure results from contraction of the left ventricle. The diastolic pressure is the pressure in the arteries when the ventricles relax. Systolic pressure is higher than diastolic pressure.

Blood pressure is usually measured in an old-fashioned unit called 'millimetres of mercury', which is taken from the measurement of pressure on a mercury manometer. The symbol for the element mercury is Hg, so this unit is written as mmHg. Blood pressure depends on sex, age, health and whether a person is at rest or exercising. However, the normal range is 100–140 mmHg (systolic) and 60–90 mmHg (diastolic). A doctor who records a man's systolic pressure as 120 mmHg and diastolic pressure as 70 mmHg would write this down as 120/70. Generally, anyone with a reading that is persistently higher than 140/90 has mildly high blood pressure. Someone with a reading over 160/100 has high blood pressure.

> **DID YOU KNOW?**
> Mercury manometers are no longer used to measure blood pressure. Instead, doctors use digital (electric) blood pressure meters. These still use the old mmHg units.

Many factors contribute to hypertension, including:

- genetic factors (35 genes affecting blood pressure have been identified)
- high salt content in the diet
- lack of exercise
- being overweight
- smoking
- drinking alcohol
- pregnancy
- many diseases, such as kidney disease.

You can help yourself to avoid developing hypertension by avoiding the preventable risk factors. Eat healthily, exercise regularly, maintain a healthy weight, and do not smoke or drink.

A number of drugs are used to treat hypertension, including **ACE inhibitors**. ACE stands for angiotensin converting enzyme and is part of the **renin–angiotensin system**. This system controls blood pressure (Figure 6.25).

When the blood pressure falls, the kidneys produce an enzyme called **renin**. Renin is released into the blood, where it cuts a plasma protein called angiotensinogen into a peptide made up of 10 amino acids, called angiotensin I. ACE then converts angiotensin I into another peptide of 8 amino acids, called angiotensin II. Angiotensin II is a hormone that acts on blood vessels, causing them to constrict (narrow). This increases blood pressure.

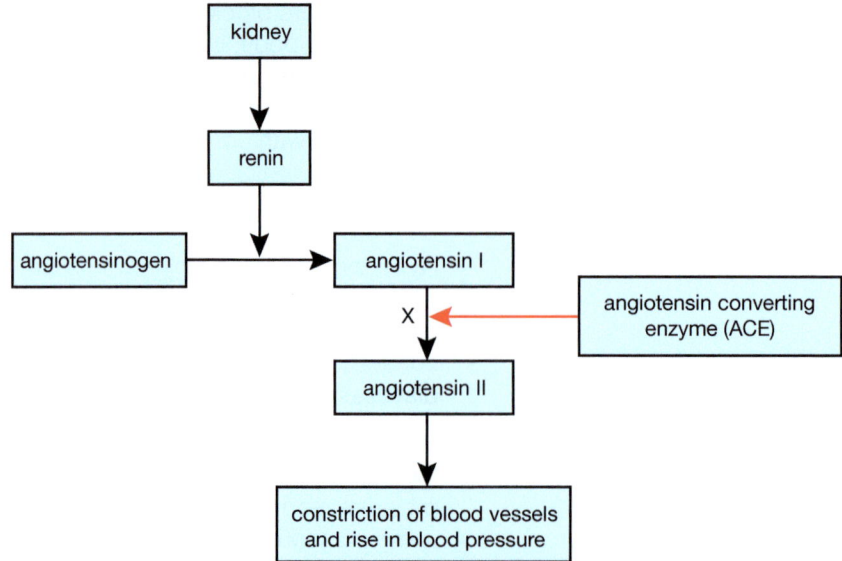

▲ Figure 6.25 The renin–angiotensin system and control of blood pressure.

ACE inhibitors act at point X in Figure 6.25. They inhibit the enzyme so that angiotensin II is not produced and there is no increase in blood pressure. People with high blood pressure take ACE inhibitors to control their hypertension.

MONOCLONAL ANTIBODIES

Monoclonal antibodies are antibodies that are made in the laboratory. They can be made on a large scale and used for the diagnosis and treatment of disease, and for medical research. They are made by a single clone of cells, so each antibody is specific to a particular antigen.

Monoclonal antibodies are produced by special cells called **hybridoma cells**. Hybridoma cells are made by fusing (joining) antibody-producing lymphocytes with certain cancer cells. The lymphocytes make antibodies but cannot divide, while the cancer cells divide indefinitely. The hybridoma cell can divide *and* make antibodies.

The method of production of monoclonal antibodies is shown in Figure 6.26.

Firstly, an antigen (e.g. from a disease-causing bacterium) is injected into a laboratory mouse. This stimulates the mouse to make antibody-producing lymphocytes. Several types of lymphocyte are made against the antigen, because receptor proteins on the surface of the lymphocytes bind to different parts of the antigen molecule.

Next, lymphocytes are taken from the spleen of the mouse and placed together with cancer cells in a special medium that encourages the two types of cell to fuse together. This produces hybridoma cells. Some unfused cells remain in the medium, but the conditions of the culture only allow the hybridoma cells to grow. Individual hybridoma cells are then placed in wells in a tray, where they grow into clones.

The clones produce different antibodies. The clone that makes the required antibody is identified by testing its ability to bind with the antigen. Once isolated, the clone will divide indefinitely, producing unlimited supplies of the antibody.

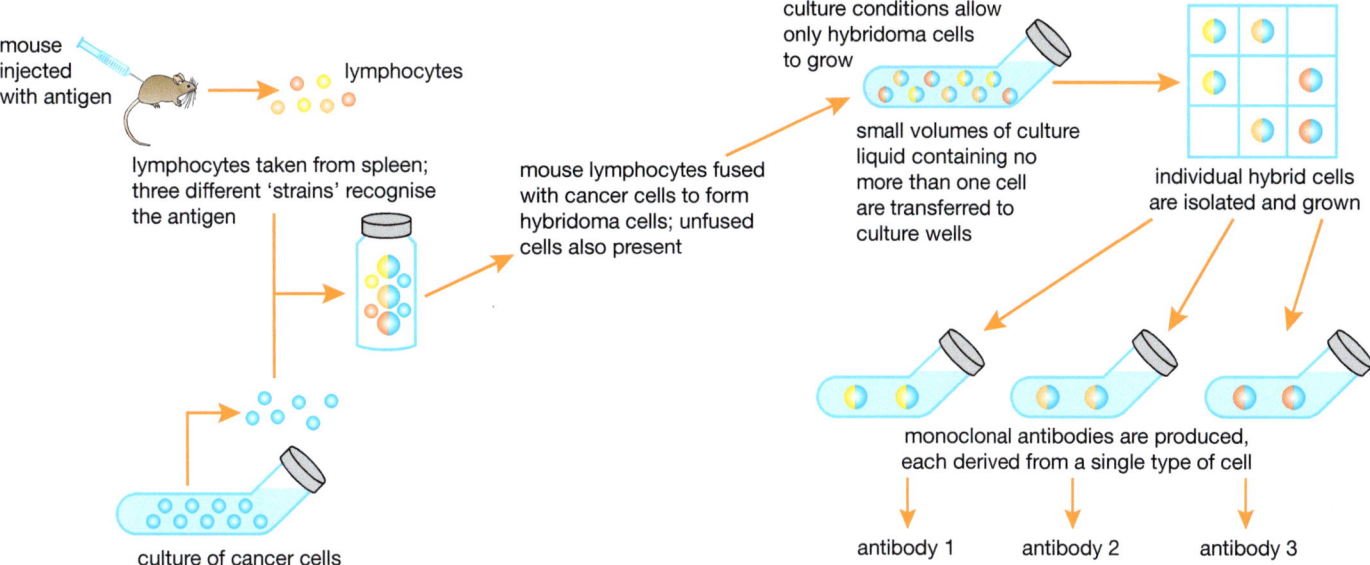

▲ Figure 6.26 How monoclonal antibodies are produced.

Monoclonal antibodies have several important uses, including detecting and treating cancer. To detect cancer, monoclonal antibodies are produced using antigens from cancer cells. These antibodies can then be used to detect the cancer antigen in tissue samples from other patients. The antibody must be 'labelled' so that it shows up in the test. This can be done in a number of ways, for example, by binding the antibody to a coloured dye, to a radioactive chemical or to a fluorescent substance (Figure 6.27).

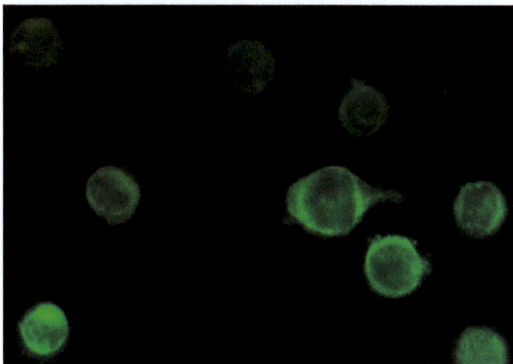

▲ Figure 6.27 Fluorescent monoclonal antibodies binding to the surface of brain cancer cells taken from a patient with a brain tumour.

Monoclonal antibodies can also be used to treat various diseases, including cancers. The antibodies bind to particular antigens on the infected or cancerous cells and act in various ways. Sometimes they kill the cells. In other cases, they interfere with signalling pathways in the cells.

For example, a monoclonal antibody called Herceptin™ is used to treat some types of breast cancer. In the cell membrane of many human cells is a receptor protein called HER2 (human epidermal growth factor receptor 2). HER2 responds to chemicals in the blood that switch on mitosis. Some cells (e.g. in breast tissue) have too many HER2 receptors. This results in uncontrolled cell division that leads to cancer. Herceptin™ binds to these receptors and prevents their action, as well as marking them for destruction by the woman's immune system.

DID YOU KNOW?

Partial pressure is a measure of the pressure of a gas in a mixture of different gases (such as the air). It is measured in kilopascals (kPa). Gases diffuse according to their partial pressures.

LOOKING AHEAD – HAEMOGLOBIN

In this chapter, you learnt that haemoglobin loads oxygen in the lungs and unloads it at the respiring tissues. This can be shown as a graph called an oxygen dissociation curve (Figure 6.28). In the lungs, there is a high partial pressure of oxygen. Haemoglobin takes up oxygen and becomes more and more saturated with it, until it is fully saturated. At the respiring tissues, the opposite happens. The partial pressure of oxygen is low, and the haemoglobin becomes less saturated with oxygen – it gives up its oxygen to the tissues for respiration.

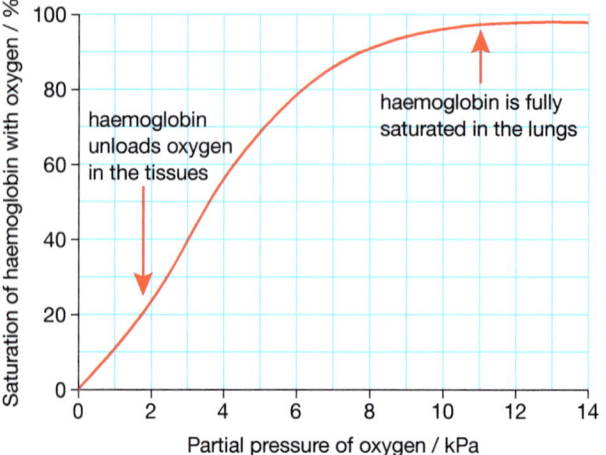

▲ Figure 6.28 The oxygen dissociation curve for haemoglobin.

Carbon dioxide also affects the oxygen dissociation curve. If the partial pressure of carbon dioxide is high (as it is at the respiring tissues), the dissociation curve shifts to the right (Figure 6.29). This is called the *Bohr effect*. This means that the affinity for oxygen is reduced, so more oxygen is released to the tissues.

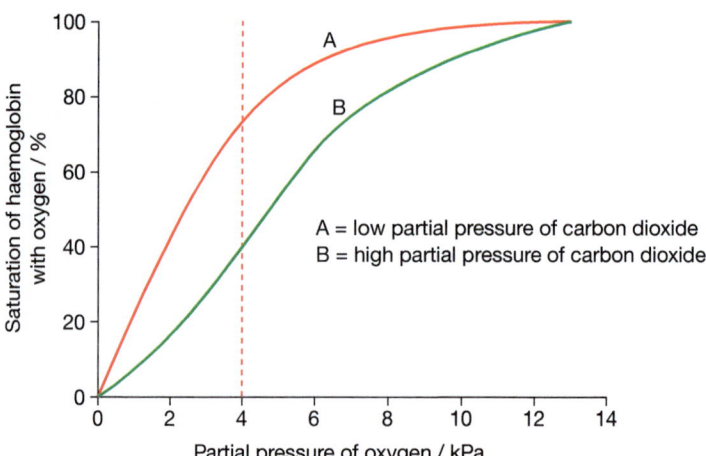

▲ Figure 6.29 The Bohr effect. An increased partial pressure of carbon dioxide lowers the affinity of the blood for oxygen. For example, if the partial pressure of oxygen is 4 kPa (shown by the dotted line on the graph), the blood is about 73% saturated with oxygen at a low partial pressure of carbon dioxide, but only 40% saturated at a higher partial pressure of carbon dioxide.

UNIT 2 — INTERNAL TRANSPORT

CHAPTER QUESTIONS

SKILLS ANALYSIS

1 The circulation system carries nutrients, oxygen and carbon dioxide around the body.

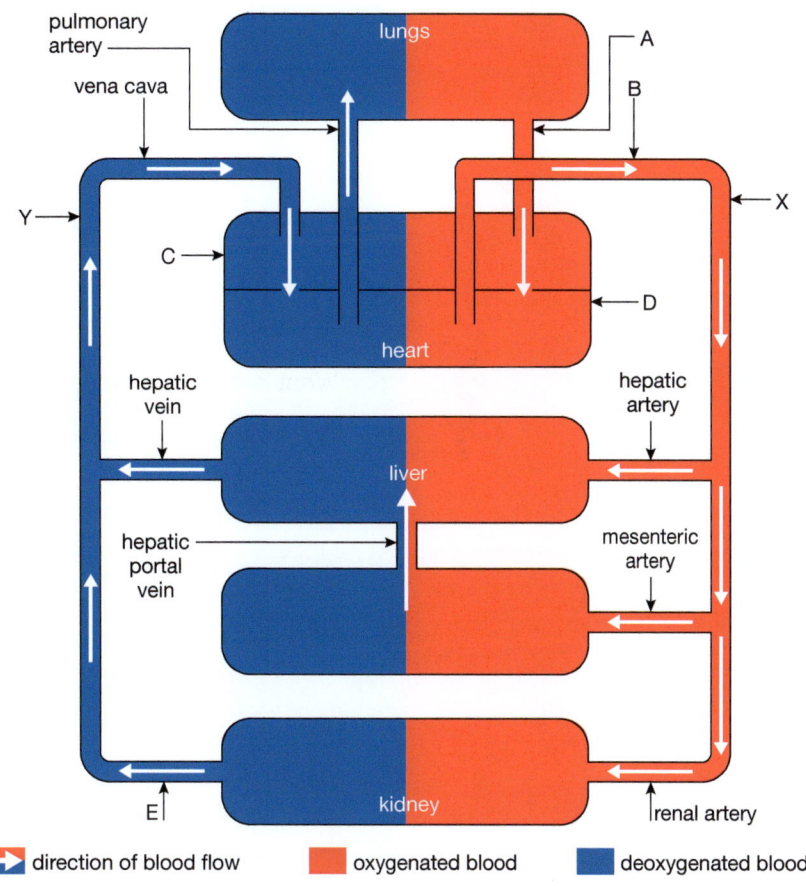

a Write down the names of the structures labelled A to E.
b Give two differences between the blood vessels at point X and point Y.
c Which blood vessel contains the highest concentration of urea?

SKILLS CRITICAL THINKING

2 Blood transports oxygen and carbon dioxide around the body. Oxygen is transported by the red blood cells.
  a State three ways in which a red blood cell is adapted to its function of transporting oxygen.
  b Describe how oxygen:
    i enters a red blood cell from the alveoli in the lungs
    ii passes from a red blood cell to an actively respiring muscle cell.
  c How is carbon dioxide transported around the body?

3 Capillaries are the smallest blood vessels in the body.
  a Describe two ways in which the structure of a capillary is adapted for its function.
  b Briefly describe how tissue fluid is formed.
  c What is the function of tissue fluid?

SKILLS ANALYSIS

4 The diagram shows a section through a human heart.

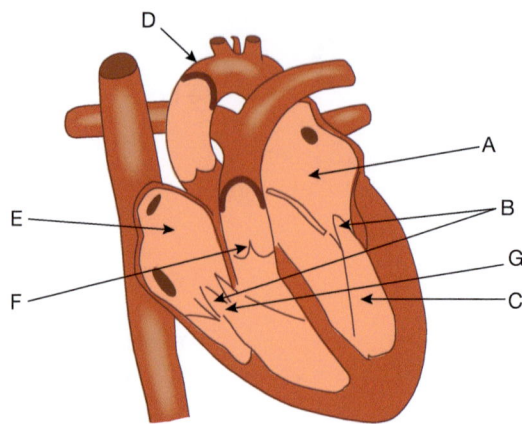

a Name the structures labelled A to E.
b What is the importance of the structures labelled B and F?
c Which letters represent the chamber of the heart to which blood returns:
   i from the lungs
   ii from all the other organs of the body?

5 The diagram shows three types of cell found in human blood.

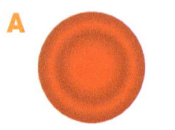  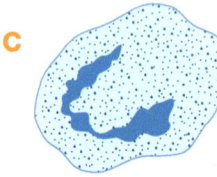

a Giving a reason for each answer, identify the blood cell which:
   i engulfs and digests bacteria.
   ii transports oxygen around the body
   iii produces antibodies to destroy bacteria

SKILLS CRITICAL THINKING

b Name one other component of blood found in the plasma and state its function.

SKILLS ANALYSIS

6 The graph shows changes in a person's heart rate over a period of time.

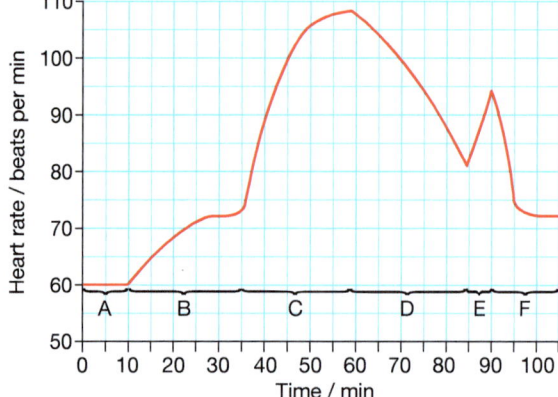

Give the letter of the time period when the person was:
a running
b frightened by a sudden loud noise
c sleeping
d waking.

In each case give a reason for your answer.

SKILLS ANALYSIS

7 The graph shows the changes that take place in heart rate before, during and after a period of exercise.

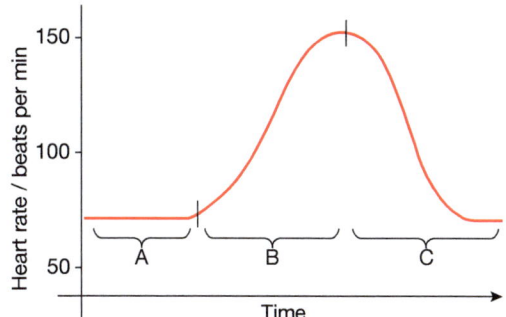

a Describe and explain the heart rates found:
   i at rest, before exercise (period A)
   ii as the person begins exercising (period B)
   iii as the person recovers from the exercise (period C).

SKILLS CRITICAL THINKING

b How can the recovery period (period C) be used to assess a person's fitness?

c Write down four of the benefits of taking regular exercise.

8 A study was carried out to find the effect of exercise and diet on the blood pressure of patients with hypertension. One group of patients carried out 45 minutes of aerobic exercise five times a week. A second group of patients was put on a low energy and low fat diet. A third group was given a 'combination' treatment of aerobic exercise and the special diet. The blood pressures of each group were measured before and after the treatments and the reduction in pressure was compared with a Control group of patients.

The results are shown in the graph.

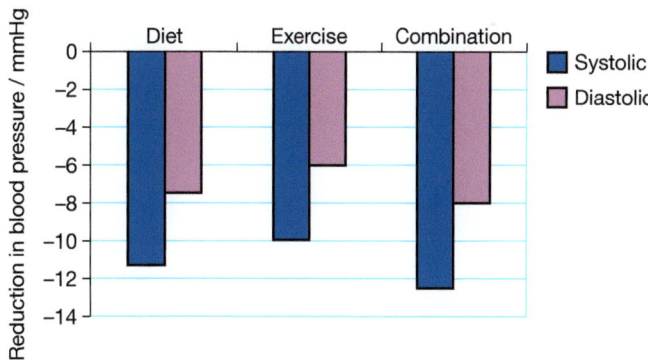

a What is hypertension?
b What is meant by aerobic exercise?
c What treatment would the Control group of patients have received?
d The graph shows the systolic and diastolic blood pressures. Explain the difference between these.

SKILLS ANALYSIS

e Describe the effect of (i) exercise alone and (ii) diet alone on the patients' blood pressures.

SKILLS DECISION MAKING

f The patients receiving the combination treatment showed a drop in blood pressure compared with the Control patients. However, there was no significant difference between this reduction and the drop shown by either treatment alone. One explanation for this is that the results may not be reliable. How could the scientists improve the reliability of their results? Explain your answer.

SKILLS CRITICAL THINKING

9 Coronary heart disease is caused by atherosclerosis in the coronary arteries that supply blood to the heart muscle. Without treatment, this can result in a heart attack, which can be fatal. One treatment is to place a stent in a coronary artery.

   a What is atherosclerosis? Explain how it can lead to a heart attack.

   b What is a stent? Explain how it is inserted into a coronary artery.

SKILLS INTERPRETATION

10 a Construct a flow chart to show the steps in the production of monoclonal antibodies. The flow chart has been started for you.

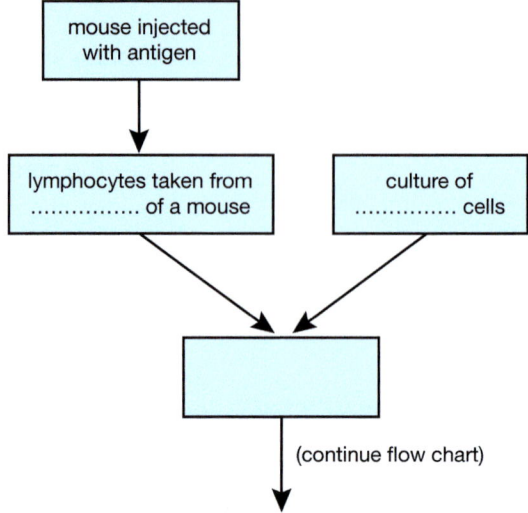

(continue flow chart)

SKILLS CRITICAL THINKING

   b Explain how monoclonal antibodies can be used to detect cancer.

UNIT QUESTIONS

SKILLS CRITICAL THINKING

1. Which of the following statements about digestion is *not* true?
   A Digestion produces fatty acids and glycerol.
   B Digestion converts insoluble molecules into soluble molecules.
   C Digestion changes proteins into amino acids.
   D Digestion releases energy from food. **(Total 1 mark)**

2. Which of the following shows the correct order in which food passes through the gut?
   A duodenum → ileum → colon → oesophagus
   B oesophagus → duodenum → ileum → colon
   C duodenum → colon → oesophagus → ileum
   D colon → ileum → duodenum → oesophagus **(Total 1 mark)**

3. Which of these organs does *not* produce digestive enzymes?
   A salivary gland
   B gall bladder
   C stomach
   D pancreas **(Total 1 mark)**

SKILLS PROBLEM SOLVING

4. A man with a height of 1.72 m weighs 65 kg. What is his BMI?
   A 17
   B 22
   C 26
   D 38 **(Total 1 mark)**

SKILLS CRITICAL THINKING

5. The structures below are found in the human bronchial tree:
   1. alveoli
   2. trachea
   3. bronchioles
   4. bronchi

   Which of the following shows the route taken by air after it is breathed in through the mouth?
   A 2 → 3 → 4 → 1
   B 1 → 4 → 3 → 2
   C 2 → 4 → 3 → 1
   D 4 → 1 → 2 → 3 **(Total 1 mark)**

SKILLS CRITICAL THINKING

6   Which of the following is *not* a feature of an efficient gas exchange surface?

   A thick walls

   B moist lining

   C close to blood capillaries

   D large surface area

   (Total 1 mark)

7   Which row in the table shows the correct percentage of oxygen in atmospheric and exhaled air?

	Atmospheric air / %	Exhaled air / %
A	78	21
B	21	16
C	16	4
D	4	0.04

   (Total 1 mark)

8   Chemicals in cigarette smoke lead to the breakdown of the walls of the alveoli. What is the name given to this disease?

   A bronchitis

   B emphysema

   C coronary heart disease

   D lung cancer

   (Total 1 mark)

9   After a period of exercise, which of these blood vessels will contain the highest concentration of carbon dioxide?

   A aorta

   B pulmonary vein

   C hepatic artery

   D vena cava

   (Total 1 mark)

10 When the right ventricle contracts, where does the blood flow next?

   A aorta

   B left atrium

   C pulmonary artery

   D left ventricle

   (Total 1 mark)

UNIT 2   UNIT QUESTIONS   121

SKILLS ANALYSIS

11 The diagram below shows sections through three blood vessels (not drawn to scale).

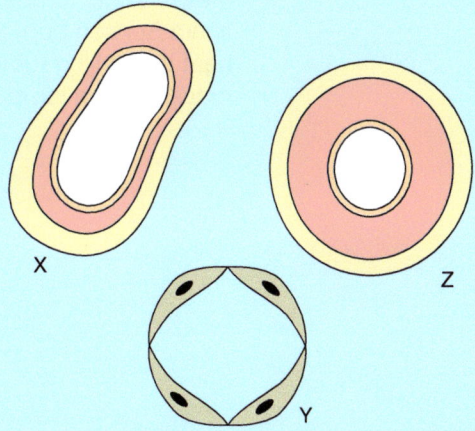

Which row in the table shows the correct names of vessels X, Y and Z?

	X	Y	Z
A	vein	capillary	artery
B	artery	capillary	vein
C	vein	artery	capillary
D	capillary	vein	artery

(Total 1 mark)

12 The diagram shows the results of an ABO blood group test. What is the blood group of the person being tested?

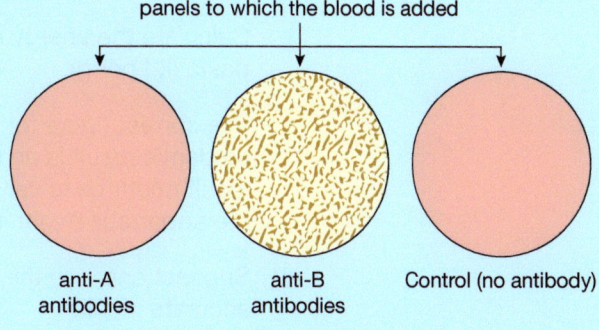

A A
B B
C AB
D O

(Total 1 mark)

SKILLS PROBLEM SOLVING   **13** The diagram shows a method that can be used to measure the energy content of some types of food. A student placed 20 cm³ of water in a boiling tube and measured its temperature. He weighed a small piece of pasta, and then held it in a Bunsen burner flame until it caught alight. He used the burning pasta to heat the boiling tube of water, until the pasta had finished burning. Finally, he measured the temperature of the water at the end of the experiment.

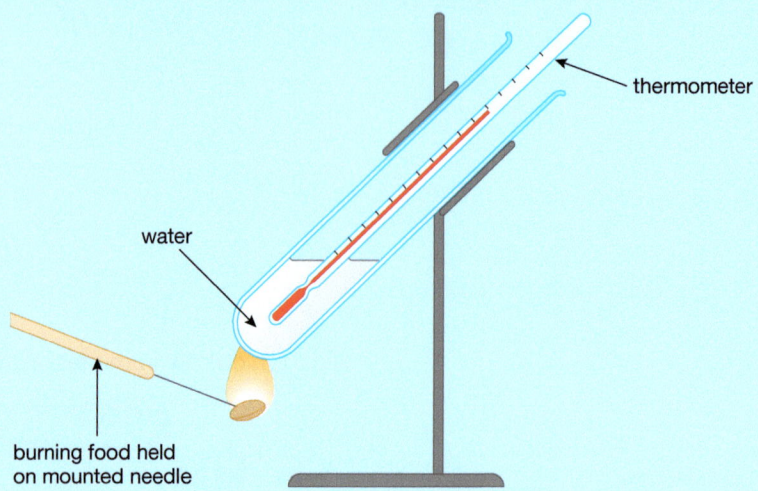

To answer the questions that follow, use this information:

- The density of water is 1 g per cm³.
- The pasta weighed 0.22 g.
- The water temperature at the start was 21 °C and at the end was 39 °C.

The heat energy supplied to the water can be calculated using the formula:
energy (in joules) = mass of water × temperature change × 4.2

a Calculate the energy supplied to the water in the boiling tube in joules (J). Convert this to kilojoules (kJ) by dividing by 1000. (2)

b Calculate the energy released from the pasta as kilojoules per gram of pasta (kJ per g). (1)

SKILLS DECISION MAKING

c The correct figure for the energy content of pasta is 14.5 kJ per g. The student's result is an underestimate. Write down three reasons why this result might be lower than expected. (Hint: think about how the design of the apparatus might introduce errors.) (3)

d Suggest one way the apparatus could be modified to achieve a more accurate result. (1)

SKILLS REASONING

e The energy in a peanut was measured using the method described above. The peanut was found to contain about twice as much energy per gram as the pasta. Explain why this is the case. (2)

(Total 9 marks)

SKILLS EXECUTIVE FUNCTION **14** A piece of meat is made of muscle fibres. Muscle fibres use ATP when they contract. Describe an investigation to find out if a solution of ATP will cause the contraction of muscle fibres.

(Total 6 marks)

UNIT 2 — UNIT QUESTIONS

SKILLS INTERPRETATION

15 The immune system responds to infections using white blood cells. A phagocyte is one type of white blood cell.

SKILLS CRITICAL THINKING

a Draw and label a phagocyte. (3)

b State one way that the structure of a phagocyte differs from that of a red blood cell. (1)

c Phagocytes carry out phagocytosis. Describe this process. (3)

d Describe how other white blood cells are involved in the immune response. (3)

(Total 10 marks)

SKILLS INTERPRETATION

16 Read the passage below. Use the information in the passage and your own knowledge to answer the questions that follow.

> Most humans live near sea level, but our bodies are able to adapt so we can survive at altitudes of several thousand metres above sea level. At high altitude, the air pressure is low compared to the pressure at a low altitude, so less oxygen is taken in with each breath. This leads to a drop in the percentage oxygen saturation of haemoglobin in the blood.   5
>
> In the short term (minutes to hours), the lack of oxygen at high altitude is sensed by oxygen receptors in the arteries. This leads to an increase in breathing rate, called hyperventilation. However, hyperventilation increases the rate of removal of carbon dioxide from the body, which leads to a decrease in breathing rate. The low oxygen and low carbon dioxide 'fight' against each   10 other to control the breathing rate. The oxygen receptors also stimulate the heart rate to increase while stroke volume decreases. These and other changes result in an unpleasant medical condition called 'altitude sickness'.
>
> Eventually, over days or weeks, the body adapts to high altitude in a process called acclimatisation. Changes in the composition of the urine counteract the   15 loss of carbon dioxide due to hyperventilation. The liver is stimulated to make more red blood cells and, at the same time, the blood volume of the body decreases, producing an increase in haemoglobin concentration.
>
> During acclimatisation, the body undergoes several other changes, including: growth of capillaries in the muscles; an increase in the number of   20 mitochondria in muscle cells; and an increase in the size of the right ventricle of the heart. The time taken (in days) for acclimatisation to take place can be found by multiplying the altitude in kilometres by 11.4. For example, to adapt to an altitude of 4000 metres (4 km) takes 45.6 days.

a Explain how removal of carbon dioxide from the body brings about a decrease in breathing rate (line 8). (3)

b Explain how the following will help a person to acclimatise to high altitude;

  i an increase in the haemoglobin concentration of the blood (line 18) (2)

  ii growth of capillaries in the muscles (line 20) (1)

  iii an increase in the number of mitochondria in muscles cells (line 20) (2)

  iv an increase in the size of the right ventricle of the heart (line 21). (2)

SKILLS PROBLEM SOLVING

c Calculate the time taken for a body to acclimatise to an altitude of 1750 m above sea level (line 22). (2)

(Total 12 marks)

BONES, MUSCLES AND JOINTS 125 | SENSORY RECEPTORS – THE EYE AND THE EAR 136 | COORDINATION 149

HOMEOSTASIS AND EXCRETION 171

UNIT 3

The human body can move from place to place. This process is known as locomotion. In Chapter 7, you will learn about the systems of bones, muscles and joints that make movement and locomotion possible. The body is also able to detect changes in its surroundings and respond to them. It can do this because it has a number of sense organs. In Chapter 8, you will study these organs, focusing in particular on the structure and function of the eye and the ear. Stimulus and response are linked by coordination systems. The body has two of these systems: the nervous system and the hormonal or endocrine system. They are described in Chapter 9. The final chapter in Unit 3 looks at the principles of homeostasis and excretion. Homeostasis means keeping conditions in the body constant, while excretion means getting rid of waste products. In Chapter 10, you will learn about these body functions, especially the working of the kidneys.

7 BONES, MUSCLES AND JOINTS

Large animals such as humans need a system of support for their body – a skeleton. The skeleton also forms a framework for muscles so they can pull on our bones, allowing movement and locomotion. Together, our bones, muscles and joints, along with tendons, ligaments and cartilage, make up the musculoskeletal system.

LEARNING OBJECTIVES

- Describe the structure and function of the main parts of the skeleton:
  - axial skeleton (vertebral column, ribcage and skull)
  - appendicular skeleton (scapula, clavicle, pelvis and limb bones)
  - the structure of a long bone, including the distribution of spongy bone, compact bone and epiphysis
- Describe the structure of bone as seen through a light microscope
- Understand the causes and symptoms of osteoporosis
- Describe the structure of a synovial joint
- Explain the functions of joints, using the elbow, shoulder and cartilaginous intervertebral joint as examples
- Explain the relationship between voluntary muscles and bones to bring about movement, illustrated by the biceps and triceps muscles and associated bones in the arm and shoulder
- Understand the dietary factors controlling the healthy development of muscle and bone
- Describe the structure of voluntary muscle as seen through the light microscope

Humans are **vertebrates** – the human body is supported by an internal skeleton made of bone, which has a central 'axis' comprising the **vertebral column**, or backbone. Attached to the bones of the skeleton are voluntary muscles. These muscles are stimulated to contract by the nervous system, under conscious control by the brain. Contraction of the muscles pulls on bones, which produces movement.

THE SKELETON

The human skeleton (Figure 7.1) has several functions. It protects many important organs. For example, the **cranium** (skull) protects the brain, eyes and ears, the **vertebrae** protect the spinal cord, and the ribcage protects the heart and lungs. Bone also makes some components of the blood. Red blood cells and platelets are made in the **bone marrow** of larger bones, such as the **sternum**, **femur** and **pelvis** (Figure 7.2). The skeleton's most obvious roles, however, are in support and movement.

UNIT 3 — BONES, MUSCLES AND JOINTS

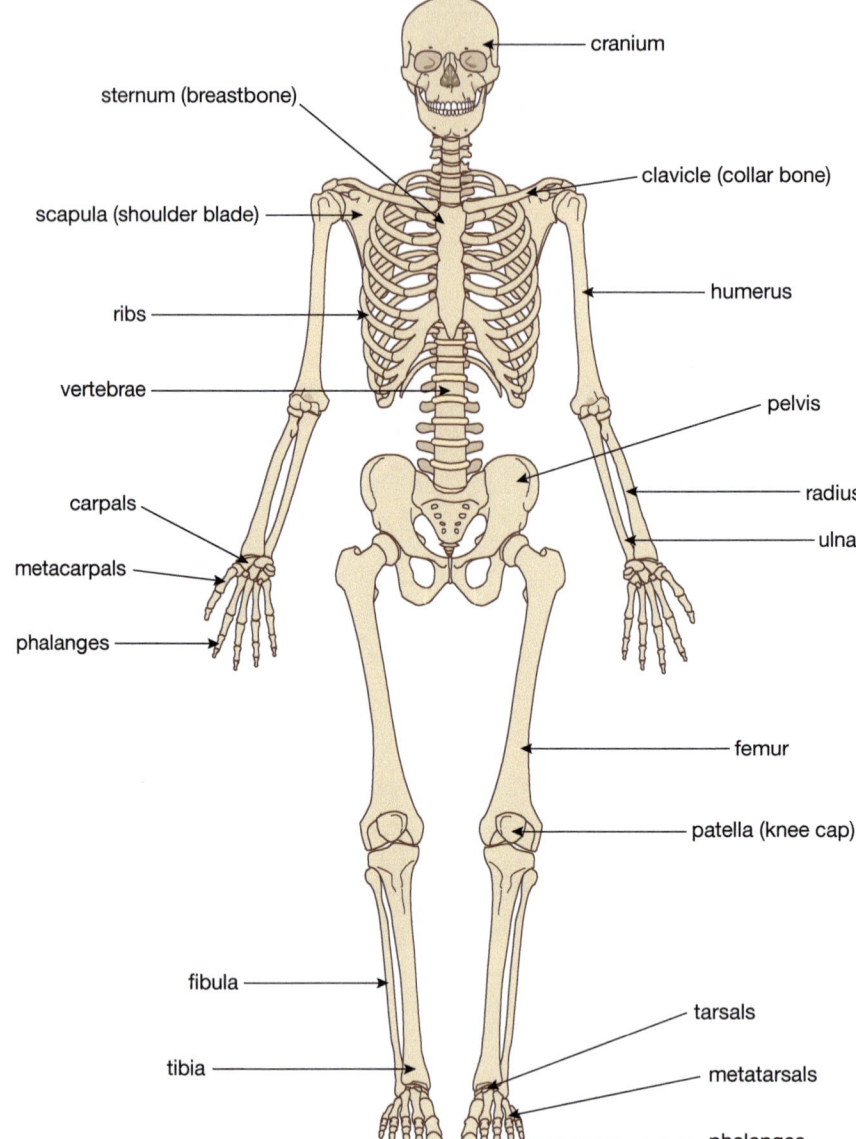

▲ Figure 7.1 The human skeleton.

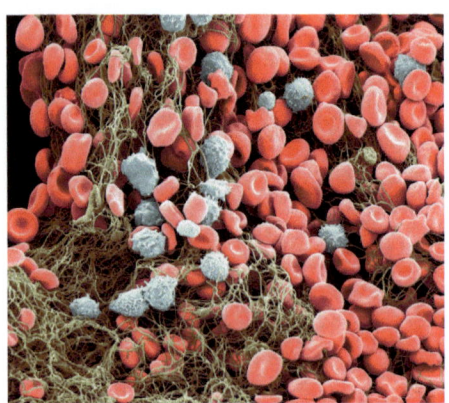

▲ Figure 7.2 An electron microscope photograph of bone marrow (magnification ×2500). Bone marrow contains stem cells that generate red blood cells (coloured red), white blood cells (coloured grey) and cells that will form blood platelets.

The vertebral column, cranium, ribcage and sternum (breastbone) comprise the **axial skeleton**. The other bones are attached to this axis – the scapulas, clavicles, pelvis and limb bones. These form the **appendicular skeleton**.

BONE

Bone is a hard substance because it contains calcium salts, mainly calcium phosphate. This results in a rigid material that can resist bending and compression (squashing) forces. Outside the body, a bone looks dead. In the body, however, it is a living organ, made of cells called **osteocytes**. These cells, along with protein fibres, stop the bone being too brittle and easily broken. The presence of living cells and blood vessels in the bone also means that bones can repair themselves if they are broken. There are many different shapes and sizes of bone in the human body, from the tiny bones (ossicles) in the inner ear to long bones such as the femur. Figure 7.3 shows a section through a long bone and Figure 7.4 is a photomicrograph of a section through the outer part of a bone.

DID YOU KNOW?

If a long bone is left in a beaker of acid, the calcium salts will dissolve out of the bone. The fibres that are left are tough but flexible so the bone will bend and can even be tied in a knot!

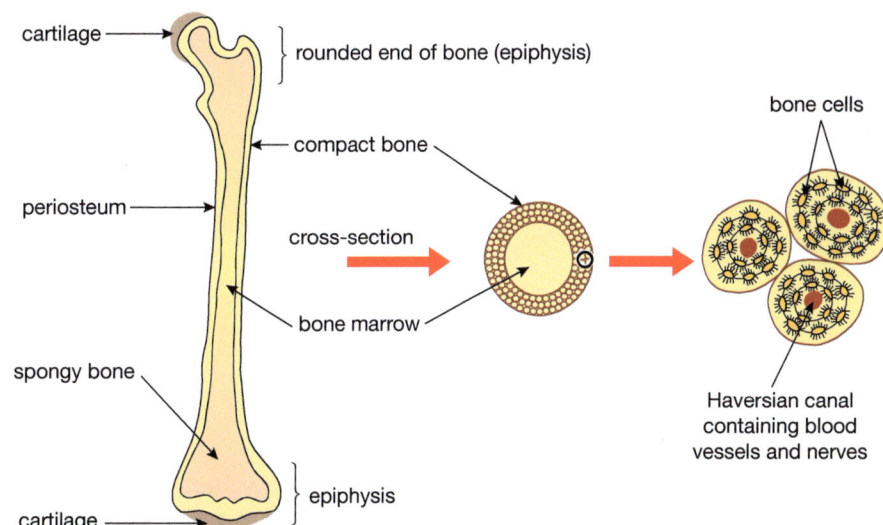

▲ Figure 7.3 Section through a long bone and the microscopic structure of compact bone.

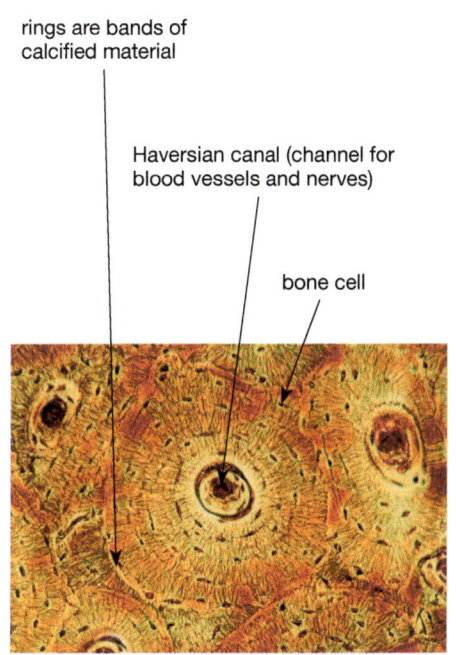

▲ Figure 7.4 This is a cross-section through a normal bone. The rings are calcified material, made mainly of calcium phosphate.

DID YOU KNOW?

The arrangement of the bone matrix in concentric rings makes a long bone very good at resisting compression forces due to the weight of the body. It still allows the bone a certain amount of side-to-side flexibility.

The middle of a bone is composed of **spongy bone**, with fewer calcium salts. This contains spaces like a sponge, making it less hard. The spaces are filled with bone marrow. Some larger bones have a hollow central cavity containing bone marrow, which makes them lighter. Marrow stores fats and produces blood cells.

The outside of the bone is made of harder material, called **compact bone**. In the embryo, bones start off made of cartilage; as the embryo grows, the cartilage is gradually replaced by bone. This is a process called **ossification**. This process is nearly complete by the time a baby is born. Osteocytes arrange themselves in rings, called **Haversian systems**, around canals containing blood vessels and nerves. The osteocytes secrete calcium phosphate salts which, along with protein fibres, make up the bone 'filling' or matrix.

Cartilage remains present at the ends of long bones, where it acts as a cushion between two bones at a joint (see below). Cartilage is a tough but flexible tissue containing cells called **chondrocytes**. They secrete a matrix containing various types of protein fibre.

Covering the outer surface of the bone is a tough membrane called the **periosteum**.

A person's long bones stop growing in their late teens. During childhood, a healthy balanced diet is essential for bone growth. In particular, calcium is needed for the calcium salts in the bone matrix, and vitamin D is needed to allow calcium to be absorbed by the body. Lack of vitamin D causes the bone disease called rickets (see Chapter 4). Although bones stop growing in length by adulthood, their internal structure continues to change throughout a person's life, under the influence of hormones, diet, exercise and general health.

Bones do not grow in length at their ends. In a child, a long bone has regions of growth close to the ends of the bone. These regions are made of cartilage. The bone gets longer because the regions of cartilage grow and then become ossified (Figure 7.5).

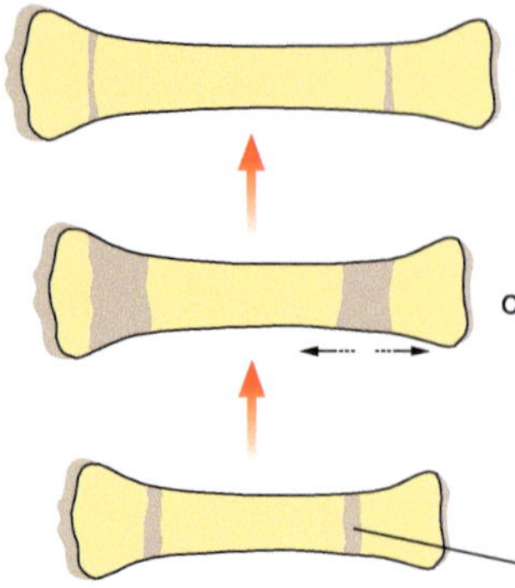

▲ Figure 7.5 How a long bone grows.

OSTEOPOROSIS

Osteoporosis is a medical condition that affects elderly people. Their bones lose calcium salts and become porous and less dense, so that they break easily (Figure 7.6). This type of bone loss is normal in all people from middle age; by the time a person is about 70 years old, they will have lost about a third of their bone mass. However, some individuals will suffer greater bone loss. Women are particularly vulnerable after the menopause, due to changes in hormone levels. There is no cure for osteoporosis but it can be helped by good diet, calcium and vitamin D supplements, and treatment with hormones.

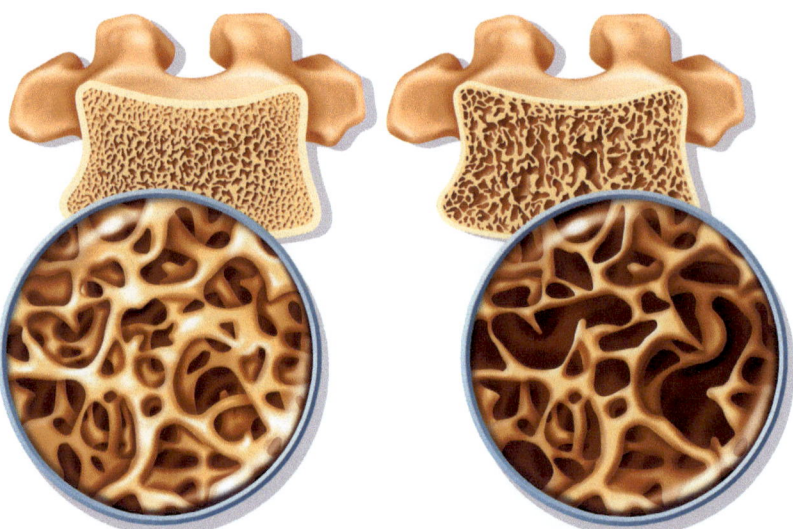

▲ Figure 7.6 Osteoporosis in a vertebra. The section of bone on the left contains normal spongy bone. The one on the right is affected by osteoporosis and is more porous, with more spaces in the spongy tissue. This makes the bone weaker and more likely to fracture.

JOINTS

When humans move, their bones move in relation to each other. The point where two bones meet is called a **joint**. We say that the bones *articulate* at joints. A movable joint, such as the hip or elbow, needs to have certain features. These include:

- a way to keep the ends of the bones together, so that they do not separate (dislocate)
- a means of reducing friction between the ends of the moving bones
- a shock-absorbing surface between the two bones.

You can see the structures that do these things in Figure 7.7.

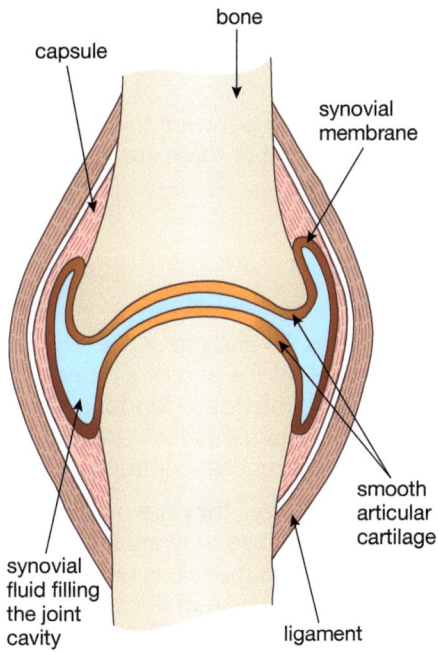

▲ Figure 7.7 The structure of a synovial joint. The size of the space filled with synovial fluid has been exaggerated in the diagram.

Movable joints are called synovial joints. They contain a liquid called **synovial fluid**. This is secreted by the **synovial membrane** which lines the space in the middle of the joint. Synovial fluid is oily and acts as a lubricant, reducing the friction between the ends of the bones. The end of each bone has an articulating surface covered with a smooth layer of cartilage. Cartilage is a strong material but it is not brittle. It acts as a shock absorber between the ends of the bones, rather like a rubber gym mat compresses to absorb the shock when you fall over.

The joint is surrounded by a tough fibrous **capsule**, and held together by **ligaments** which run from one bone to the other across the joint. Ligaments are composed of fibres that make them very tough. They have great strength to resist stretching, called tensile strength, but they also have some elasticity. Ligaments can allow movement but prevent the bones becoming dislocated.

> **DID YOU KNOW?**
> The scientific definition of an 'elastic' material is one which can be bent or stretched but will return to its original shape afterwards. In this chapter, we use the word 'elastic' to mean 'easily stretched'.

There are various kinds of joint in the body. They are classified into three groups, according to the amount of movement they allow:

- freely movable joints
- partially movable joints
- immovable (fixed) joints.

Figure 7.8 shows some joints that allow different amounts of movement.

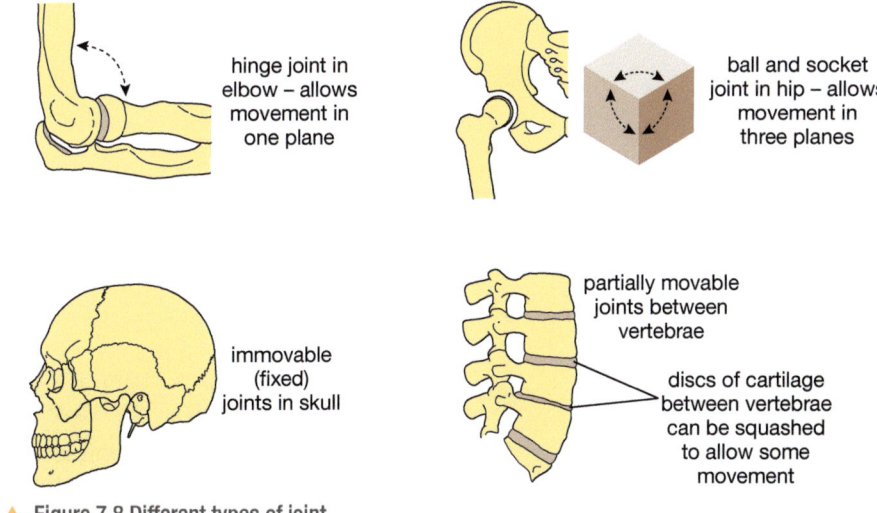

▲ Figure 7.8 Different types of joint.

Freely movable joints include **ball and socket joints**, such as the shoulder and hip, and **hinge joints**, such as the elbow.

A ball and socket joint allows movement in any direction. Imagine your shoulder joint is at the corner of a box. You can move your upper arm in any of the three directions formed by the sides of the box: side, back or top. Compare this with a hinge joint, such as your elbow – due to the shape of the joint, you can only move this in one direction, as when bending your arm.

Partially movable joints allow a little movement. The joints between vertebrae are like this. They allow a small amount of movement when the spinal column bends.

Some joints are immovable or fixed, for example, those between the bones of the skull. The cranium consists of 22 bones but most of them are fused together, allowing no movement.

MUSCLES

Skeletal muscles are organs that are attached to bones and move them by contracting (becoming shorter), pulling on the bone. At the end of a muscle, there are tendons. A **tendon** attaches the muscle to the bone. Tendons have very high tensile strength, like ligaments, but unlike ligaments they are not very elastic. This means that they do not stretch when the muscle contracts.

Muscles cannot push, they can only pull – in other words, they are not able to expand actively. When muscles get longer, it is because they are being stretched by the contraction of another muscle. When a muscle is being stretched, it is relaxed (the opposite of contracted). Because of this, muscles usually work in pairs: one contracts while the other relaxes. These are called **antagonistic pairs**. One of the simplest examples of an antagonistic pair of muscles is the arrangement of the **biceps** and **triceps** muscles in the arm (Figure 7.9).

> **DID YOU KNOW?**
> The word we use to mean 'not very elastic' is inelastic. Both ligaments and tendons have a high tensile strength, but ligaments are fairly elastic while tendons are inelastic. Ligaments join bone to bone across a joint, while tendons join muscle to bone.

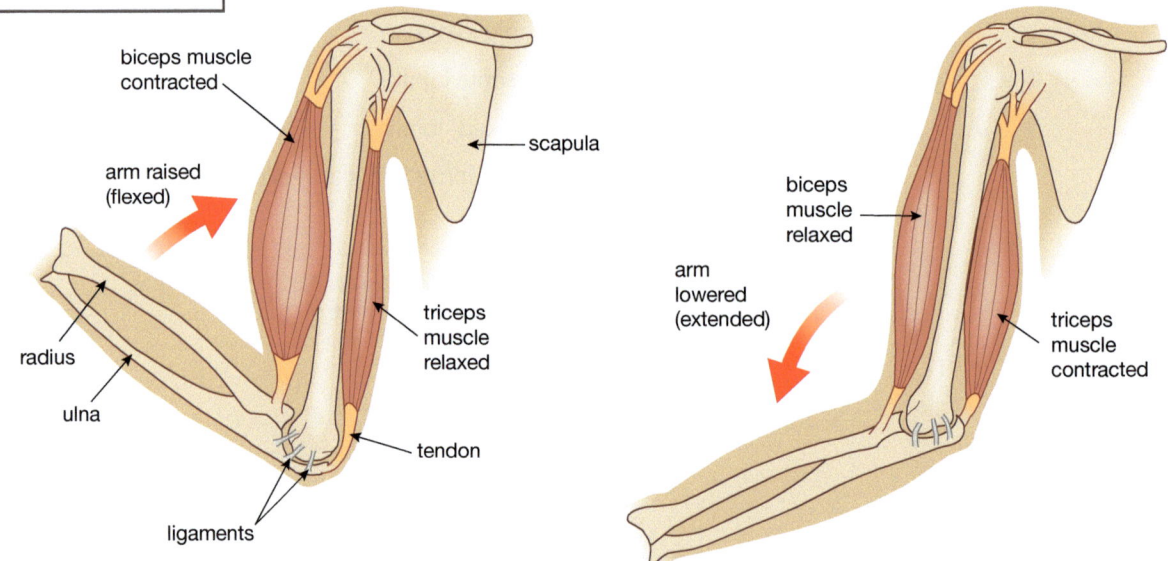

▲ Figure 7.9 The biceps and triceps muscles contract to move the arm at the elbow joint. The biceps flexes the arm, while its antagonistic partner, the triceps, extends the arm.

When the biceps muscle contracts, it bends (or flexes) the arm at the elbow joint (Figures 7.9 and 7.10). Contraction of the triceps straightens (or extends) the arm. Of course, there are other muscles in the arm which produce movement in other directions.

When a muscle contracts, the bone at one end of the muscle moves and the bone at the other end stays still. The place where the muscle is attached to the stationary bone is called the **origin**. The place where it is attached to the moving bone is called the **insertion**. When a muscle contracts, the insertion moves towards the origin.

You can identify other antagonistic pairs of muscles in the body. For example, when we run we use several sets of muscles that cause bending at the hip, knee and ankle joints (Figure 7.11).

UNIT 3 BONES, MUSCLES AND JOINTS

▲ Figure 7.10 A body-builder flexing his arm to lift a heavy weight. You can see clearly the contraction of the biceps muscle.

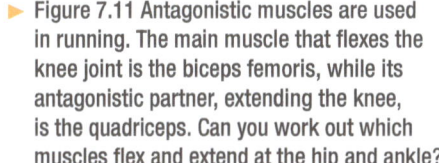

▶ Figure 7.11 Antagonistic muscles are used in running. The main muscle that flexes the knee joint is the biceps femoris, while its antagonistic partner, extending the knee, is the quadriceps. Can you work out which muscles flex and extend at the hip and ankle?

BENDING THE SPINE

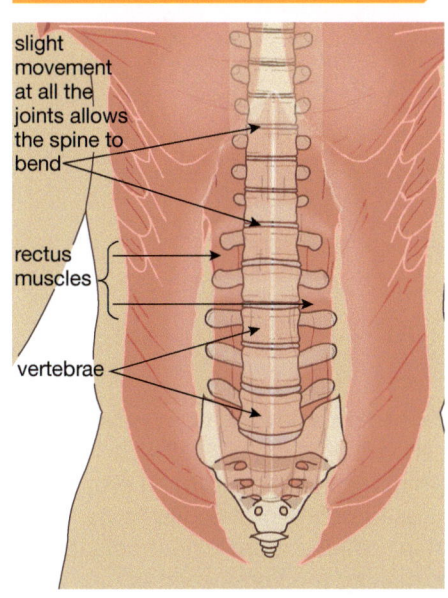

The movements at the elbow, shoulder, knee and hip are easy to see and understand. Other movements are less obvious, but just as necessary. For example, we need to be able to bend our spine from side to side and from front to back. When this happens, the movement takes place at several different points at the same time.

Our spine, or vertebral column, is made from bones called vertebrae. These bones are joined by synovial joints, but the joints cannot bend like the elbow joint. The bones at each joint can be pulled very slightly closer together or further apart to allow a slight bending movement. Because of this slight bending at each joint, the whole spine can bend.

The main muscles responsible for bending the spine from side to side are the *rectus* muscles (Figure 7.12).

◀ Figure 7.12 Contraction and relaxation of the rectus muscles produces limited movement at each joint in the spine.

The vertebrae in different regions of the spine are different. Their shape and size is related to the kind of load they have to take (Figure 7.13).

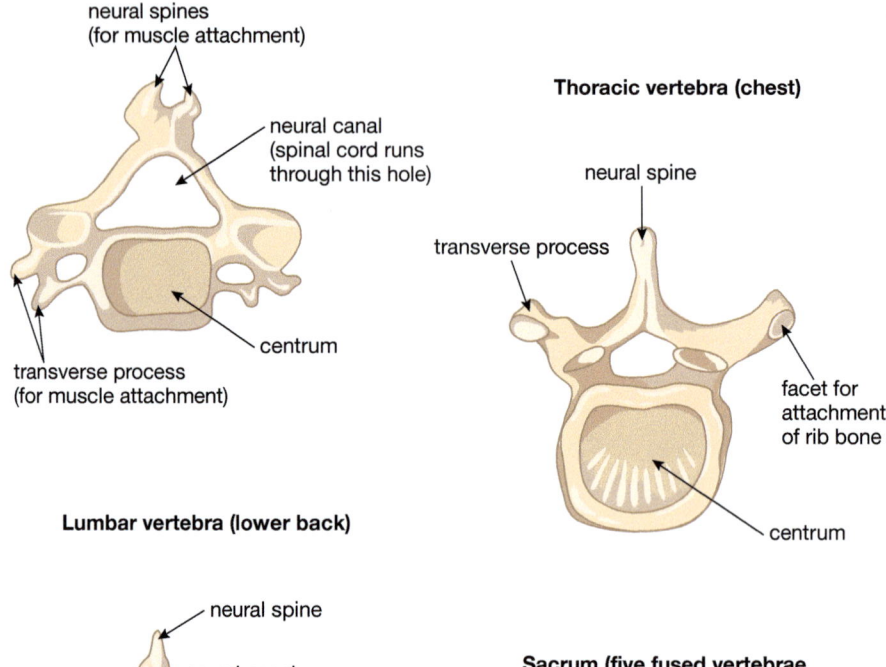

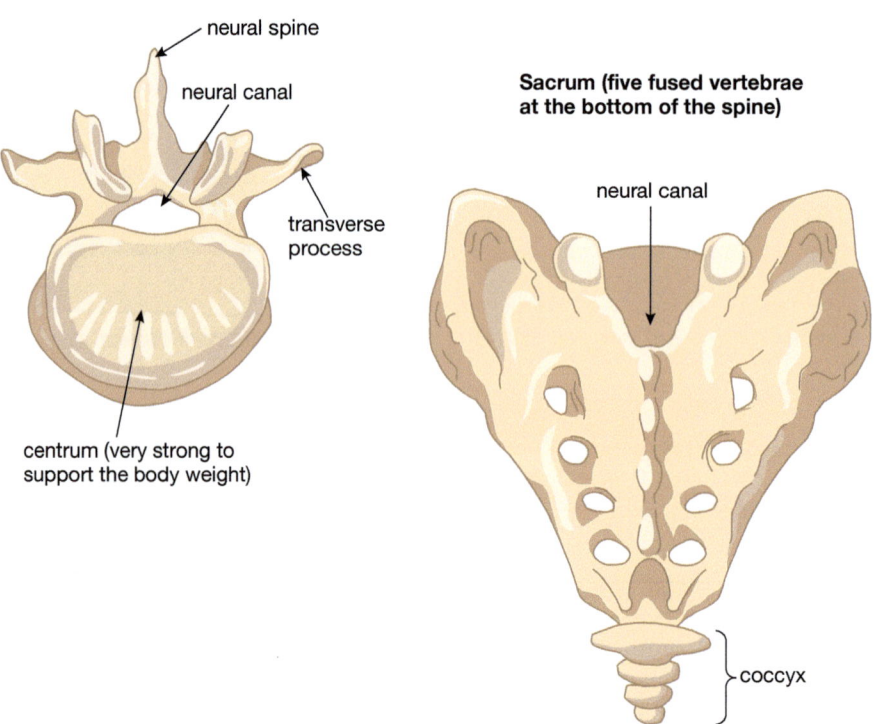

▲ Figure 7.13 Vertebrae from different parts of the spine differ in shape and size according to their position.

THE STRUCTURE OF SKELETAL MUSCLE

Skeletal muscle is made up of highly specialised muscle cells or muscle *fibres*, arranged in bundles in a connective tissue sheath (Figure 7.14). Muscle fibres are adapted for contraction. Under very high magnification, using an electron microscope, we can see that they are composed of fine protein filaments. There are two types of filament, thick and thin. When a muscle contracts, the thin filaments slide past the thick filaments, making the fibres shorter.

REMINDER

Remember that there are three types of muscle in the body – **skeletal**, **cardiac** and **smooth muscle**. The first of these is the muscle attached to the skeleton. It is described as voluntary muscle, because it is under the conscious control of the brain. Cardiac muscle is found only in the heart. It is involuntary, meaning not under conscious control. Smooth muscle is also involuntary. It is found in the wall of the gut, bladder, uterus, sperm ducts, blood vessels and other organs.

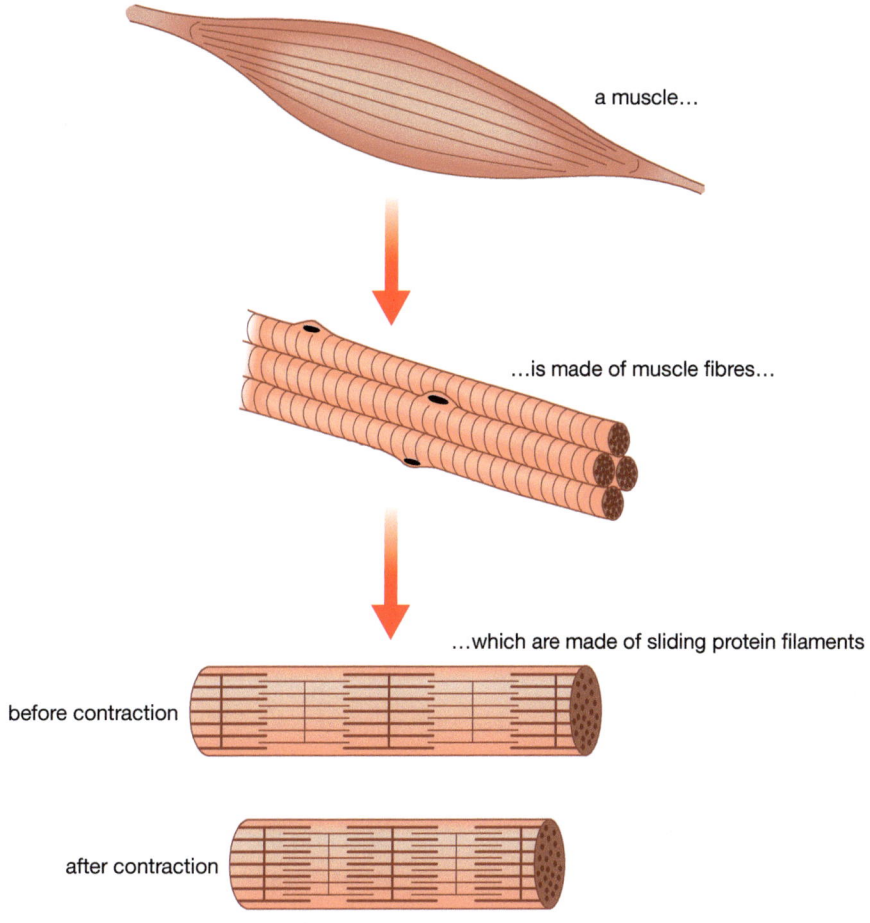

▲ Figure 7.14 Muscles are composed of muscle fibres, made of fine protein filaments. When a muscle contracts, the filaments slide over each other.

DID YOU KNOW?

A healthy balanced diet is not just needed for bone growth. Muscles are largely made of protein, and a balanced diet containing an adequate supply of amino acids is needed for protein synthesis in muscles.

Contraction of muscle fibres needs energy; this comes from respiration (see Chapter 5). Blood vessels supply glucose and oxygen to the muscle fibres. The fibres respire, converting the glucose and oxygen into carbon dioxide and water. Energy is released for the fibres to contract but this process is not 100% efficient and some energy is lost as heat. When we carry out strenuous exercise, our muscles demand a greater supply of glucose and oxygen than usual. In addition, carbon dioxide and heat must be removed at a faster rate. To achieve this, various changes take place in the body.

- The breathing rate increases, so more oxygen is taken into the blood by the lungs and more carbon dioxide is lost. The volume of each breath also increases.
- The heart rate increases, which pumps more oxygenated blood to the muscles.
- Blood is diverted away from places like the gut, and towards the muscles.
- The skin performs processes such as vasodilation and sweating (see Chapter 10) which remove excess heat from the body.

Even when a muscle is relaxed, some of its fibres are contracted. This state of partial contraction is called muscle tone. It keeps our muscles taut, but not enough to cause movement. Muscle tone keeps us upright when we are standing or sitting. Not all the muscle fibres are contracted at once – they take it in turn.

CHAPTER QUESTIONS

SKILLS CRITICAL THINKING

1 a Which parts of the body are protected by:
      i the cranium
      ii the vertebral column
      iii the ribs?
   b Between the vertebrae are discs of cartilage. From your knowledge of the properties of cartilage, suggest what their function is.
   c Name three components of bone.

2 The synovial membrane, synovial fluid and ligaments are parts of a synovial joint. Explain the function of each part.

3 The bar chart shows the percentage of adult men and women diagnosed with osteoporosis.

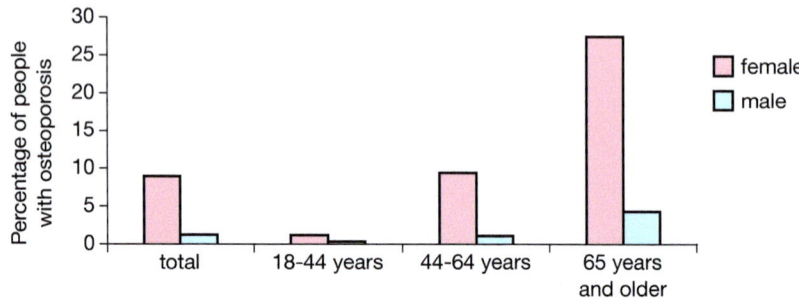

a Describe the symptoms of osteoporosis.

SKILLS ANALYSIS

b What does the bar chart tell us about:
   i the development of osteoporosis in women as they become older
   ii the differences between the development of osteoporosis in women and in men?

SKILLS CRITICAL THINKING

SKILLS ANALYSIS

c State two ways in which osteoporosis in women can be treated.

4 The diagram shows the muscles and bones in a human forearm when the forearm is being raised.

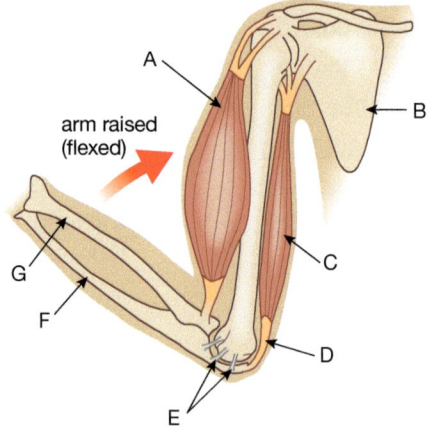

a Name the parts labelled A, B, C, D, E, F and G.

SKILLS CRITICAL THINKING

b A and C are antagonistic muscles. What does this mean?
c i Describe the function of the structures labelled E.
   ii What properties of structures E adapt them to this function?
d What must happen for the forearm to be lowered? Explain your answer.

SKILLS ANALYSIS

5 If a muscle is isolated and stimulated by giving it a brief electrical shock, the muscle responds by 'twitching'. This is shown in the graph.

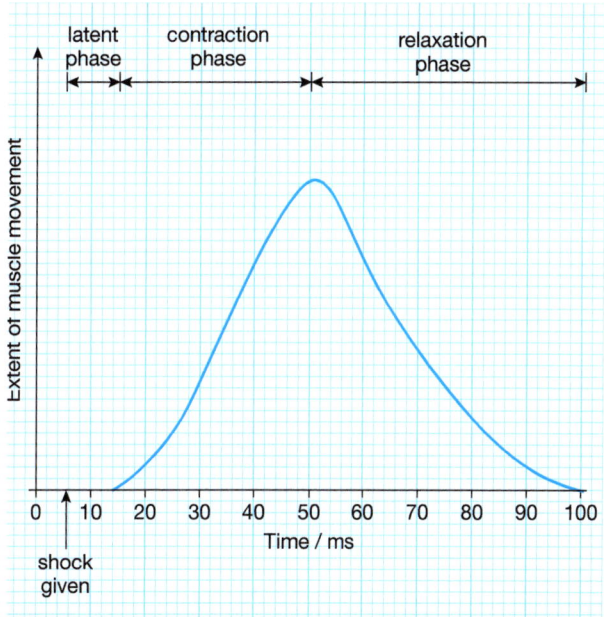

  a In this experiment, how long did each of the three phases of the twitch response last?

SKILLS CRITICAL THINKING

  b For how long could the muscle have been exerting a force? Explain your answer.

  c Muscles often exist in antagonistic pairs. Explain how antagonistic muscles can:
    i bend and straighten the arm
    ii keep the body upright and straight.

SKILLS REASONING

  d Muscles are made from muscle fibres. There are three types of muscle fibre: type I, type IIa and type IIb. The table gives some of the properties of each type.

	Fibre type		
	I	IIa	IIb
Resistance to fatigue	high	moderate	low
Aerobic/anaerobic respiration	aerobic	both	anaerobic
Speed of contraction	slow	intermediate	fast
Tension produced when contracted	low	intermediate	high

  Different athletes have different combinations of the three types of fibre in their muscles. Describe the combination of fibres you would expect to find in:
    i an endurance athlete (such as a marathon runner)
    ii a power athlete (such as a weightlifter).
  e Explain your answers to **d** parts **i** and **ii**.

8 SENSORY RECEPTORS – THE EYE AND THE EAR

Humans need to be able to detect changes in their environment. This is the function of sensory receptors. In this chapter, you will learn about the different kinds of sense organ, in particular the eye and the ear.

LEARNING OBJECTIVES

- Understand that the body has receptors that can detect the stimuli of light, sound, temperature, pressure and taste
- Describe how to investigate the distribution of sensory receptors in the skin
- Explain the structure and function of the eye in:
  - focusing on near and distant objects
  - responding to changes in light intensity
  - stereoscopic vision, allowing better judgement of distance
- Understand defects of the eye and their treatment:
  - long sight and short sight
  - astigmatism
  - cataracts
  - corneal transplants
- Explain the structure and function of the ear in hearing and balance
- Describe how to investigate the range of frequencies audible to the human ear
- Explain how prolonged exposure to loud noise affects the functioning of the ear and hearing

KEY POINT

The surroundings outside the body are called the *external* environment. The inside of the body is known as the *internal* environment. The body also responds to changes in its internal environment, such as temperature and blood glucose levels. You will read about these responses in Chapters 9 and 10.

STIMULUS AND RESPONSE

Imagine you are walking along when you see a football coming at high speed towards your head. If your nerves are working properly, you will probably move quickly to avoid contact. Imagine another situation, where you are very hungry and you smell food cooking. Your mouth might begin to 'water', in other words, secrete saliva.

Each of these situations is an example of a **stimulus** and a **response**. A stimulus is a change in a person's environment (surroundings) and a response is a reaction to that change.

In the first example, the approaching ball was the stimulus, and your movement to avoid it hitting you was the response. The change in your environment was detected by your eyes, which are an example of a **receptor** organ. The response was brought about by contraction of muscles, which are an **effector** organ (they produce an effect). The nervous system links the two and is an example of a coordination system. A summary of the sequence of events is:

stimulus → receptor → coordination → effector → response

In the second example, the receptor for the smell of food was the nose, and the response was the secretion of saliva from glands. Glands secrete (release) chemical substances, and they are the second type of effector organ. Again, the link between the stimulus and the response is the nervous system. The information in the nerve cells is transmitted in the form of tiny electrical signals called **nerve impulses**. The nervous system is described in Chapter 9.

RECEPTORS

The role of any receptor is to detect a stimulus by changing the energy of the stimulus into electrical energy in nerve impulses. For example, the eye converts light energy into nerve impulses, and the ear converts sound energy into nerve impulses (Table 8.1).

▼ Table 8.1 Human receptors and the energy they receive.

Receptor	Type of energy received
eye (retina)	light
ear (organ of hearing)	sound
ear (organ of balance)	mechanical (kinetic)
tongue (tastebuds)	chemical
nose (organ of smell)	chemical
skin (touch/pressure/pain receptors)	mechanical (kinetic)
skin (temperature receptors)	heat
muscle (stretch receptors)	mechanical (kinetic)

Notice how a 'sense' like touch is made up of several components. When we touch a warm surface, we will be stimulating several types of receptor, including touch and temperature receptors, as well as stretch receptors in the muscles. In addition to this, each sense detects different aspects of the energy it receives. For example, the ears do not only detect sounds; they can also detect different loudness and frequencies of sound. The eye not only forms an image; it also detects brightness of light and, in humans, can tell the difference between different light wavelengths (colours). Senses tell us a great deal about changes in our environment.

Safety Note: Use a large paper clip: this can be bent to a suitable shape and has no sharp points. Do not use pins or needles.

ACTIVITY 1

▼ PRACTICAL: INVESTIGATING THE DISTRIBUTION OF TOUCH RECEPTORS IN THE SKIN

Different areas of the skin have a different sensitivity to touch. Some places, such as the fingertips, have a very high density of touch receptors and are more sensitive than other areas, such as the skin on the back of the arm. It is easy to investigate the distribution of touch receptors using a wire paperclip opened up into a U shape. To start with, the ends of the wire are arranged so that they are 2 cm apart.

One person (the experimental 'subject') puts on a blindfold. A second person (the experimenter) takes the U-shaped wire and *gently* touches the subject on the back of their hand, using either one or two points of the wire. The subject is asked to identify whether the experimenter used one or two points. The experimenter records whether the subject was right or wrong, using a tick or cross.

The experimenter repeats this ten times, each time randomly changing the number of points used (one or two). He or she records the number of correct answers out of ten.

The procedure is then repeated three more times, with the points of the paperclip different distances apart, e.g. 1 cm, 0.5 cm and 0.2 cm.

Finally, the whole exercise is repeated on different areas of skin, such as the fingertips, the palm of the hand or the back of the arm.

The results will show that different areas of skin have a different density of touch receptors. One investigation found that the threshold distances (the minimum distance at which a person could distinguish between one and two points) were as shown in Table 8.2.

▼ Table 8.2 Threshold distances for a 'one/two point' test on different areas of skin.

Area of skin	Threshold distance / cm
fingertips	0.2
cheek	0.5
palm of hand	1.0
front of forearm	1.5
back of forearm	3.0

THE EYE

Many animals have eyes, but few animals have eyes as complex as the human eye. Simple animals, such as snails, use their eyes to detect light but cannot form a proper image. Other animals can form images but cannot recognise colours. The human eye does all three. Of course, it is not really the eye that 'sees' anything at all; it is the brain, which interprets the impulses from the eye. To find out how light from an object is converted into impulses representing an image, we need to look at the structure of this complex organ (Figure 8.1).

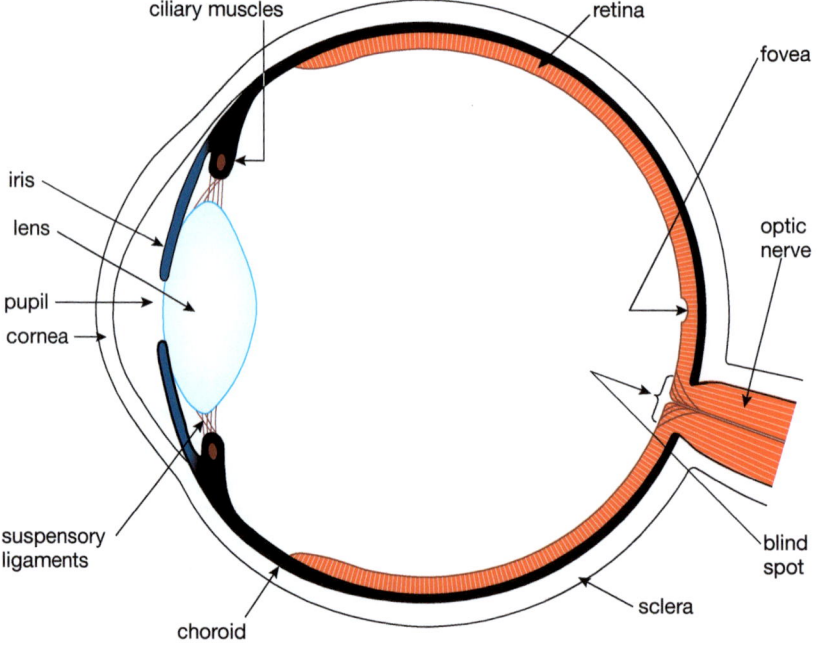

▲ Figure 8.1 A horizontal section through the human eye.

The tough outer coat of the eye is called the **sclera**. This is the visible, white part of the eye. At the front of the eye, the sclera becomes a transparent 'window' called the **cornea**, which lets light into the eye. Behind the cornea is the coloured ring of tissue called the **iris**. In the middle of the iris is a hole called the **pupil**, which lets the light through. It is black because there is no light escaping from the inside of the eye.

Underneath the sclera is a dark layer called the **choroid**. It is dark because it contains many pigment cells, as well as blood vessels. The pigment stops light being reflected around inside the eye, which would prevent a clear image being formed.

The innermost layer of the back of the eye is the **retina**. This is the light-sensitive layer, the place where light energy is converted into the electrical energy of nerve impulses. The retina contains cells called **rods** and **cones**. These cells react to light, producing impulses in sensory neurones. The sensory neurones then pass the impulses to the brain through the **optic nerve**. Rod cells work well in dim light, but they cannot distinguish between different colours, so the brain 'sees' an image produced by the rods in black and white. This is why we cannot see colours very well in dim light: only our rods are working properly. The cones, on the other hand, will only work in bright light. There are three types of cone cell which respond to different wavelengths or colours of light – red, green and blue. We can see all the colours of visible light as a result of these three types of cone being stimulated to different degrees. For example, if red, green and blue are stimulated equally, we see white. Both rods and cones are found throughout the retina, but cones are particularly concentrated at the centre of the retina, in an area called the **fovea**. Cones give a sharper image than rods, which is why we can only see objects clearly if we are looking directly at them, so the image falls on the fovea.

FORMING AN IMAGE

To form an image on the retina, light needs to be bent or *refracted*. Refraction takes place when light passes from one medium to another of a different density. In the eye, this happens first at the air/cornea boundary; it happens again at the lens (Figure 8.2). In fact, the cornea acts as the first lens of the eye.

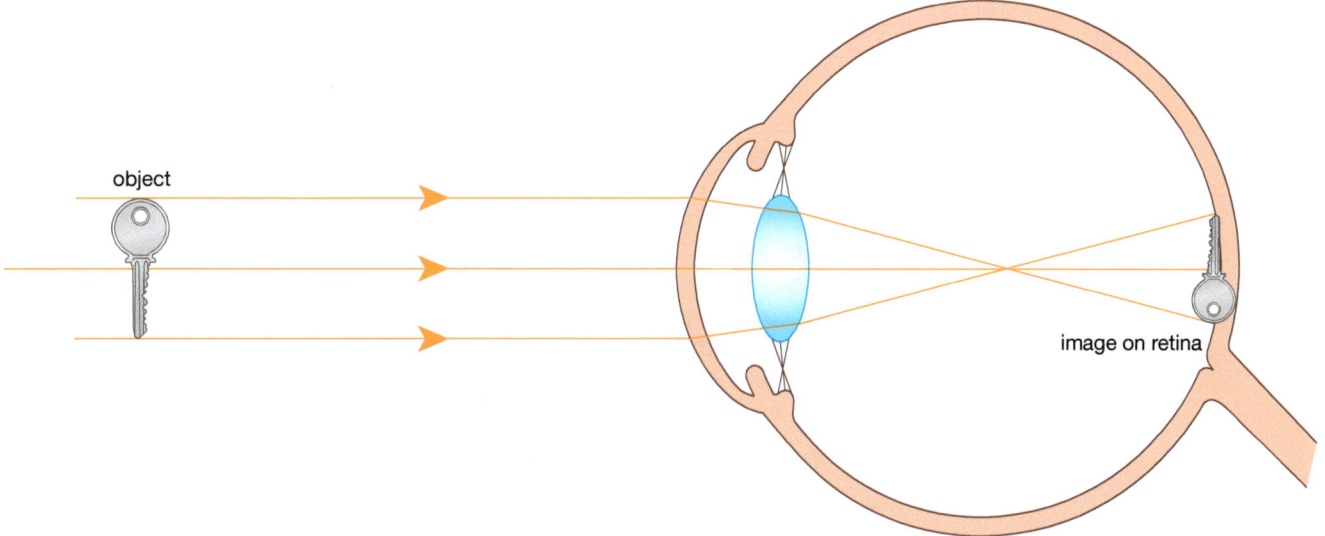

▲ Figure 8.2 How the eye forms an image. Refraction of light occurs at the cornea and lens, producing an inverted image on the retina.

As a result of refraction at the cornea and the lens, the image on the retina is inverted (upside down). The brain interprets the image the right way up.

The role of the iris is to control the amount of light entering the eye, by changing the size of the pupil. The iris contains two types of muscle. Circular muscles form a ring shape in the iris, and radial muscles lie like the spokes of a wheel. In bright light, the pupil is constricted (made smaller). This happens because the circular muscles contract and the radial muscles relax. In dim light, the opposite happens. The radial muscles contract and the circular muscles relax, dilating (widening) the pupil (Figure 8.3).

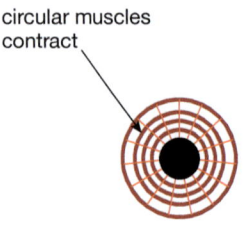
circular muscles contract

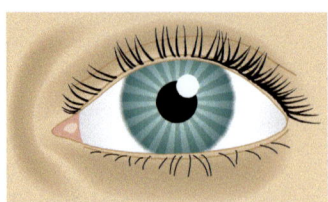
bright light
- circular muscles contract
- radial muscles relax
- pupil constricts

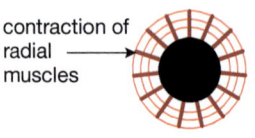
contraction of radial muscles

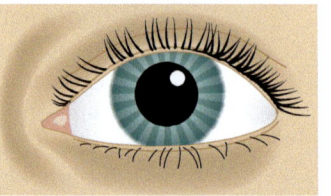
dim light
- circular muscles relax
- radial muscles contract
- pupil dilates

KEY POINT

In the iris reflex, the route from stimulus to response is this:

stimulus (light intensity)
↓
retina (receptor)
↓
sensory neurones in optic nerve
↓
unconscious part of brain
↓
motor neurones in nerve to iris
↓
iris muscles (effector)
↓
response (change in size of pupil)

▲ Figure 8.3 The amount of light entering the eye is controlled by the iris, which alters the diameter of the pupil.

Whenever our eyes look from a dim light to a bright one, the iris rapidly and automatically adjusts the pupil size. This is an example of a reflex action. You will find out more about reflexes in Chapter 9. The purpose of the iris reflex is to make sure the light falling on the retina is at the right intensity. Light that is too bright could damage the rods and cones, and light that is too dim would not form an image. The intensity of light hitting the retina is the stimulus for this reflex. Impulses pass to the brain through the optic nerve, and straight back to the iris muscles, which adjust the diameter of the pupil. This all happens without the need for conscious thought – in fact, we are not even aware of it happening.

THE BLIND SPOT

There is one area of the retina where an image cannot be formed; this is where the optic nerve leaves the eye. At this point, there are no rods or cones, so it is called the **blind spot**. The retina of each eye has a blind spot but they are not a problem, because the brain puts the images from each eye together, cancelling out the blind spots of both eyes. As well as this, the optic nerve leaves the eye towards the edge of the retina, where vision is not very sharp anyway. To 'see' your own blind spot you can do a simple experiment. Cover or close your right eye. Hold this page about 30 cm from your eyes and look at the black dot below. Now, without moving the book or turning your head, read the numbers from left to right by moving your left eye slowly towards the right.

● 1 2 3 4 5 6 7 8 9 10 11 12 13 14 15

You should find that the image of the dot disappears at some point. This is when it is falling on the blind spot. If you try doing this with both eyes open, the image of the dot will not disappear.

UNIT 3 — SENSORY RECEPTORS – THE EYE AND THE EAR

> **DID YOU KNOW?**
> A way to prove to yourself that the eyes form two overlapping images is to try the 'sausage test'. Focus your eyes on a distant object. Place your two index fingers tip to tip, and bring them up in front of your eyes, about 30 cm from your face, while still focusing at a distance. You should see a finger 'sausage' between the two fingers. Now try this with one eye closed. What is the difference?

WHY DO WE HAVE TWO EYES?

There are several advantages to having two eyes, such as cancelling the blind spot and providing a wider field of view. In addition, each eye forms a slightly different image of an object. The brain combines the information from each eye, giving us *stereoscopic* or three-dimensional (3D) vision. This allows us to judge the distance and depth of objects, and to estimate the speed of moving objects more accurately. You can show the benefit of stereoscopic vision by a simple experiment. Close one eye and ask a friend to hold a pencil horizontally in front of you, about 50 cm away. Try to line up your finger with the end of the pencil (without touching it). Now try the same thing with both eyes open. You will find it is harder to do using only one eye.

> **DID YOU KNOW?**
> Do you know which is your 'dominant' eye? Each eye forms a slightly different image of an object, as shown by the 'sausage test'. When your brain combines the images from each eye, it sees the view from one eye as the main or 'dominant' image. You can find out which is your dominant eye by holding up a finger in line with a distant vertical object, such as a window frame. Now close each eye in turn. When you close the dominant eye, the finger will appear to move relative to the distant object. When you close the non-dominant eye, the finger will not move.

ACCOMMODATION

The changes that take place in the eye which allow us to see objects at different distances are called **accommodation**.

You have probably seen the results of a camera or projector not being in focus – a blurred picture. In a camera, we can focus light from objects that are different distances away by moving the lens backwards or forwards, until the picture is sharp. In the eye, a different method is used: the shape of the lens can be changed, rather than its position. A lens that is fatter in the middle (more convex) will refract light rays more than a thinner (less convex) lens. The lens in the eye can change shape because it is made of cells containing an elastic crystalline protein.

Figure 8.1 shows that the lens is held in place by a series of fibres called the **suspensory ligaments**. These are attached like the spokes of a wheel to a ring of muscle, called the **ciliary muscle**. The inside of the eye is filled with a transparent watery fluid which pushes outwards on the eye. In other words, there is a slight positive pressure within the eye. The changes to the eye that take place during accommodation are shown in Figure 8.4.

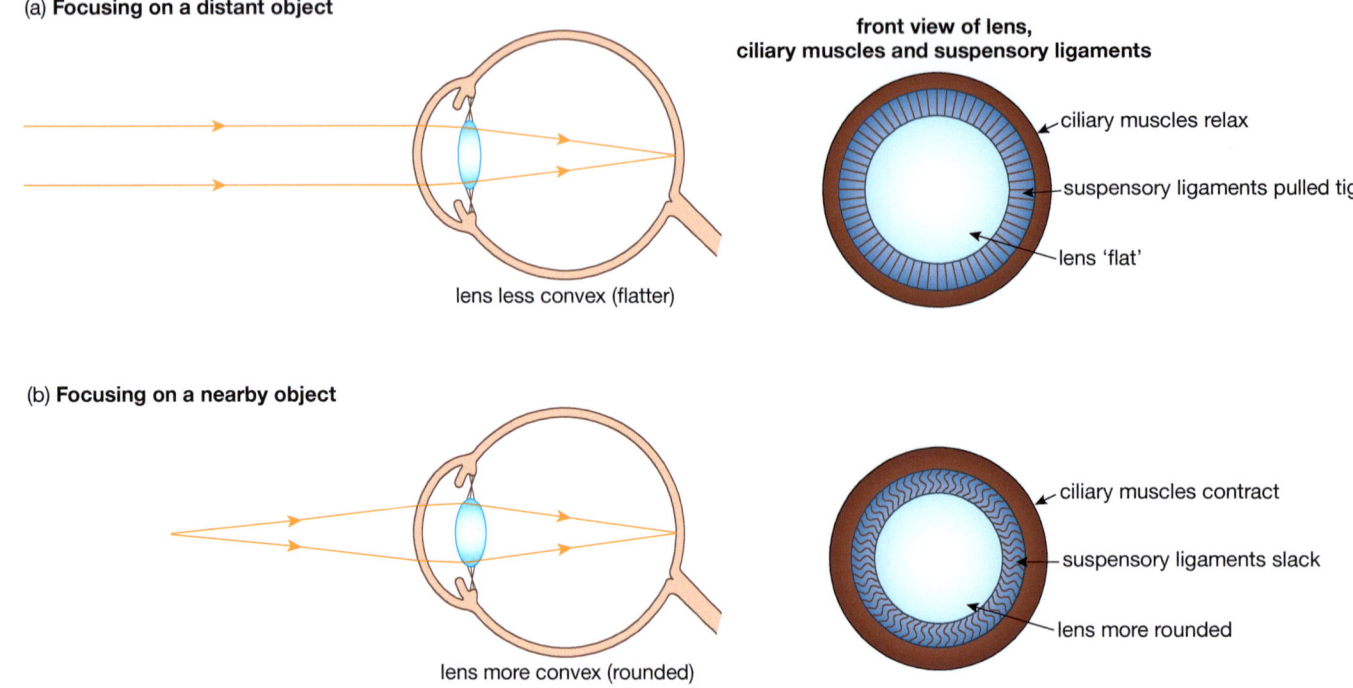

▲ Figure 8.4 Accommodation: how the eye focuses on objects at different distances.

When the eye is focused on a distant object, the rays of light from the object are almost parallel when they reach the cornea (Figure 8.4 (a)). The cornea refracts the rays, but the lens does not need to refract them much more to focus the light on the retina. This means the lens does not need to be very convex. The ciliary muscles relax and the pressure in the eye pushes outwards on the lens, flattening it and stretching the suspensory ligaments. This is the condition when the eye is at rest – our eyes are focused for long distances.

When we focus on a nearby object, for example, when reading a book, the light rays from the object are spreading out (diverging) when they enter the eye (Figure 8.4 (b)). In this situation, the lens has to be more convex in order to refract the rays enough to focus them on the retina. The ciliary muscles now contract; the suspensory ligaments become slack and the elastic lens bulges outwards into a more convex shape.

DEFECTS OF VISION

LONG AND SHORT SIGHT

In some people, the accommodation mechanism does not work properly, and they are unable to see clearly without the help of glasses or contact lenses. There are two main problems, called long sight and short sight.

In the case of **long sight**, either the lens is not convex enough (i.e. it is too flat) or the eyeball is too short from front to back, so that light rays from a nearby object are focused behind the retina (Figure 8.5 (a)). This means that the image falling on the retina will be out of focus. A long-sighted person has difficulty focusing on nearby objects – for example, they will hold a book at arm's length to read it without glasses. Long sight can be corrected by using convex lenses or glasses which converge the light rays before they enter the eye (Figure 8.5 (b)).

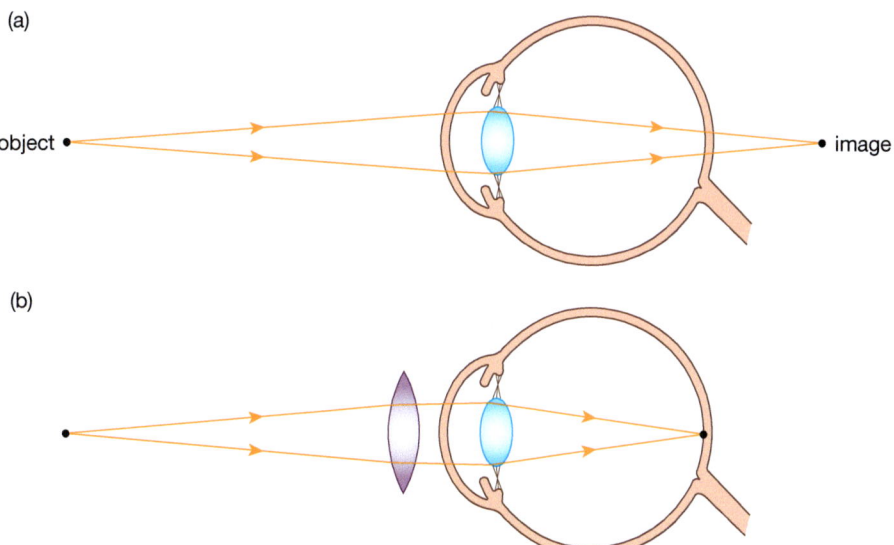

▲ Figure 8.5 (a) With long sight, light is focused behind the retina. (b) Long sight is corrected using a convex lens.

With **short sight**, either the lens is too convex or the eyeball is too long, so that the light rays from a distant object are focused in front of the retina (Figure 8.6 (a)). This produces an out-of-focus image. A short-sighted person has problems focusing on distant objects. This fault can be corrected using concave lenses, which diverge (spread out) the light rays before they enter the eye (Figure 8.6 (b)).

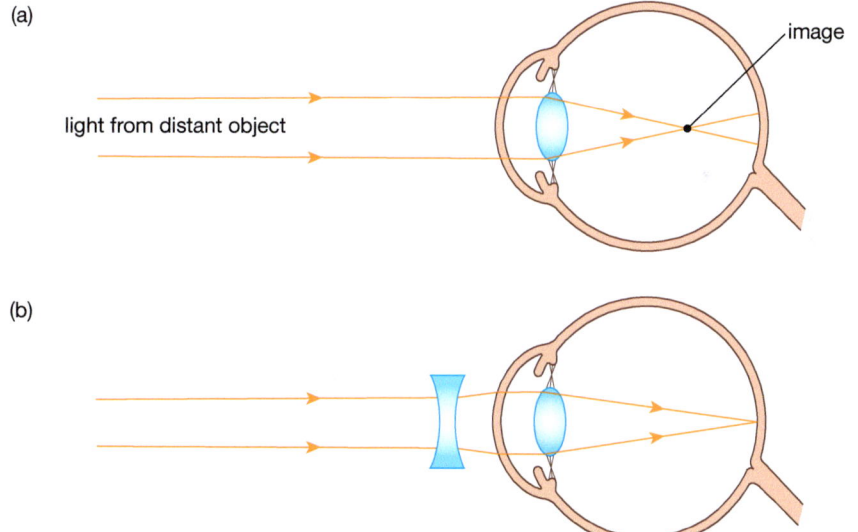

▲ Figure 8.6 (a) With short sight, light is focused in front of the retina. (b) Short sight is corrected using a concave lens.

The causes and effects of long and short sight are summarised in Table 8.3.

▼ Table 8.3 The causes, effects and method of correction of long and short sight.

	Long sight	Short sight
Cause	lens not convex enough, or eyeball too short	lens too convex, or eyeball too long
Result	light focused beyond retina	light focused short of retina
Method of correction	convex lenses to converge light before it enters the eye	concave lenses to diverge light before it enters the eye

ASTIGMATISM

Just like long sight and short sight, **astigmatism** is not a disease or health problem. It is just a minor defect in the structure of the eye that causes blurred vision. There are two types of astigmatism, called regular and irregular.

Regular astigmatism occurs when the surface of the cornea or lens is not a perfectly spherical shape like a football, but is rounder in one direction than the other (Figure 8.7). This shape means that the person's eye will be in focus in one direction (e.g. up and down) but not the other (e.g. side-to-side). This is easily corrected by wearing glasses or contact lenses.

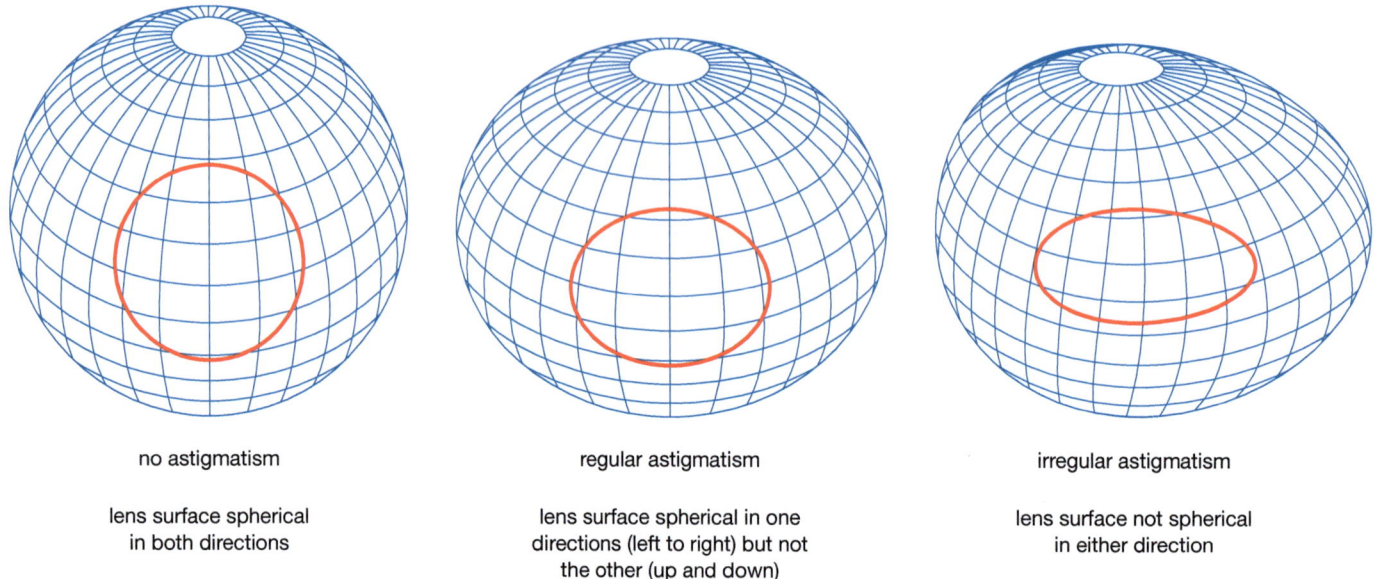

▲ Figure 8.7 The shape of the surface of a normal lens and ones with regular and irregular astigmatism.

Irregular astigmatism is where the curvature of the surface of the cornea or lens is uneven in more than one direction. This cannot be corrected by wearing glasses, but it can be corrected with contact lenses.

CATARACTS

Both long and short sight are more common in older people. However, the elderly can also develop a number of more serious defects of vision, including cataracts. A **cataract** is a condition where the lens of the eye becomes cloudy and opaque, so that the person is unable to see. It cannot be corrected with glasses or contact lenses, and can only be treated by surgery. The surgeon opens up the front of the eye and removes the affected lens, replacing it with an artificial lens. The patient will be able to see again, but they will need to wear glasses.

CORNEAL TRANSPLANTS

When a cornea is diseased or damaged by an injury, it is possible to carry out a corneal transplant. The diseased cornea is removed from the patient and replaced with a cornea taken from a donor. Corneal transplants are generally highly successful. The majority result in restored vision and there is a low rate of rejection of the donated tissue, with the replaced corneas lasting many years or a lifetime. In a small number of cases, where transplant of a living cornea is not possible, artificial corneas made of transparent plastic have been used.

THE EAR

The ear detects sound and is also an organ of balance. The structure of the ear is shown in Figure 8.8.

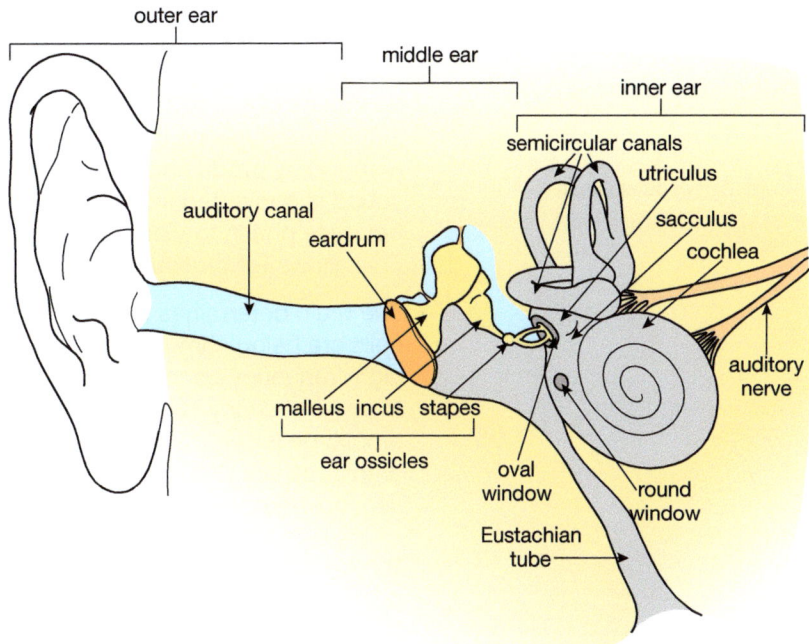

▲ Figure 8.8 The structure of the ear.

DID YOU KNOW?

Why do your ears 'pop' in an aeroplane? As an aeroplane gains height, the air pressure falls, causing the eardrum to bulge outwards. If you swallow, the Eustachian tube opens and allows air to pass into the middle ear from the throat, equalising the pressure. The 'pop' is caused by the eardrum going back to its normal position.

The ear has three parts: the outer, middle and inner ear. The outer ear directs sound waves to the **eardrum** at the end of the auditory canal, causing it to vibrate. The vibrations are passed across the middle ear by three small bones (**ear ossicles**), the **malleus** (hammer), **incus** (anvil) and **stapes** (stirrup). These bones amplify the vibrations as they pass across them. The stapes transmits the vibrations to a membrane-covered opening called the **oval window**, which is at one end of a coiled structure called the **cochlea**. The **Eustachian tube** connects the middle ear with the throat and allows the air pressure to be equalised either side of the eardrum.

A structure called the **organ of Corti** runs along the whole length of the coiled cochlea. The receptor cells that convert the vibrations into nerve impulses are found here. To make this easier to understand, Figure 8.9 shows the cochlea 'uncoiled' and Figure 8.10 shows a cross-section of the cochlea.

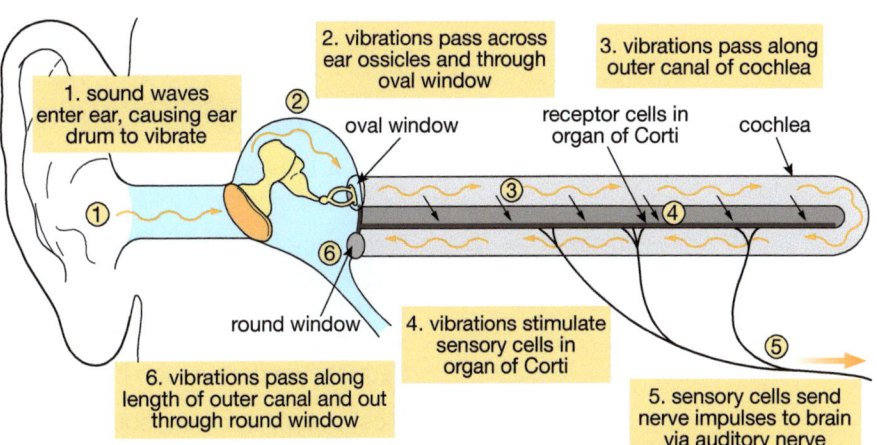

▲ Figure 8.9 Cochlea uncoiled to show how the inner ear detects sounds.

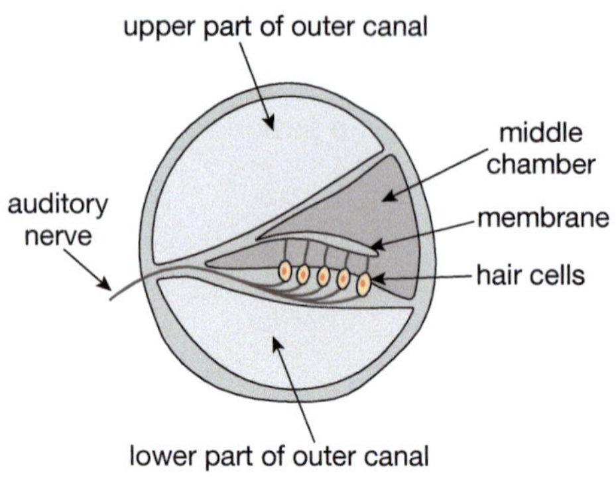

Figure 8.10 Cross-section through the cochlea.

The outer fluid-filled canal runs from the oval window, all the way along the top of the cochlea, around the end of the cochlea and back to a membrane-covered opening called the **round window** at the end of the cochlea. Between the two parts of this canal is a middle chamber, also filled with fluid. The organ of Corti is located in a membrane between the middle and outer canals. The receptor cells in the organ of Corti have sensory 'hairs' embedded in a second membrane (Figure 8.10).

Vibrations of the oval window are transmitted to the fluid in the outer canal (Figure 8.9) causing the sensory hairs to be stretched. The receptor cells respond by producing nerve impulses in the receptor neurones. In this way, sound is converted into nerve impulses. The round window vibrates with the opposite phase to vibrations entering the inner ear through the oval window. This allows for pressure changes between the middle ear and the cochlea.

The brain determines the frequency (pitch) of sounds by detecting which hair cells are being stimulated. Those nearest the oval window are sensitive to high-frequency sounds, while those nearest the round window are sensitive to low-frequency sounds. The loudness of sounds is determined by the amplitude (size) of vibrations of the hair cells. Loud sounds produce high-amplitude vibrations, which result in more nerve impulses per second in the sensory neurones.

ACTIVITY 2

▼ PRACTICAL: INVESTIGATING THE RANGE OF FREQUENCIES THAT CAN BE DETECTED BY THE HUMAN EAR

A signal generator is an electronic device that can produce electrical vibrations over a range of frequencies. If the signal generator is connected to a loudspeaker, it will produce sounds in the same frequency range (Figure 8.11).

Safety Note: Some people are sensitive to particular frequencies and may feel nauseous. Avoid very loud sounds or prolonged periods of high frequency sounds.

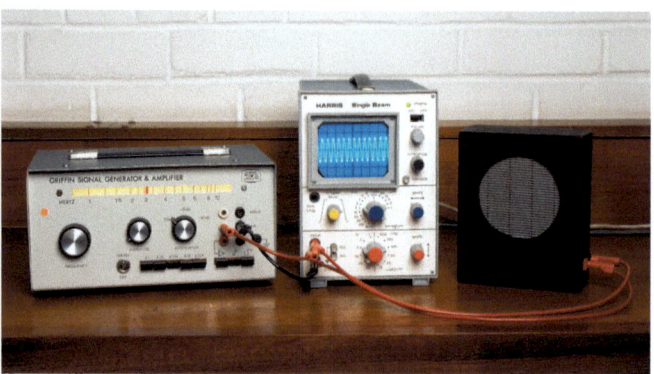

Figure 8.11 A signal generator connected to a loudspeaker and an oscilloscope. (The oscilloscope shows the shape of the sound wave.)

Frequency of sound waves is measured in cycles per second or Hertz (Hz). A signal generator can normally produce waves with frequencies from 1 to 100 000 Hz. The signal generator is turned on and the frequency is changed. The purpose is to measure how low and how high a frequency a person can hear.

Most people can hear sounds in the range 20–20 000 Hz, although 'normal' sounds (such as speech) involve frequencies of a few hundred Hz.

THE EAR AS AN ORGAN OF BALANCE

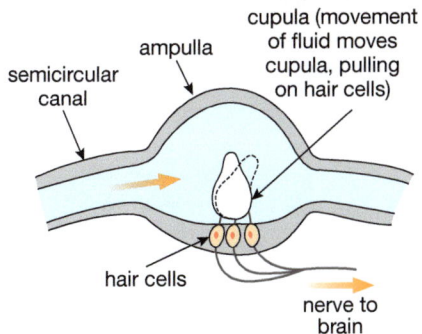

▲ Figure 8.12 Cross-section of an **ampulla** showing how movement is detected.

The **semicircular canals**, as well as the **sacculus** and **utriculus**, are the parts of the ear involved with the sense of balance (Figure 8.8). In the swellings at the ends of the semicircular canals, there are more hair cells, with their hairs embedded in a jelly-like mass called a **cupula** (Figure 8.12).

Movement of fluid in the semicircular canals causes the cupula to pull on the hair cells, which stimulates them to send nerve impulses to the brain. The canals are arranged in three planes at right-angles to each other, so they can detect movement in any direction. The sacculus and utriculus also contain hair cells. Their hairs are embedded in a jelly containing calcium carbonate crystals, called an **otolith**. The weight of the otolith pulls on the hairs, stimulating the hair cells and producing nerve impulses. This gives information to the brain about the position of the head.

LOUD NOISE CAN DAMAGE THE EAR

Every day we experience sound in our environment, such as the noise from traffic, machinery or people talking. Normally, sounds are at safe levels that do not damage our hearing. Loud sounds, however – especially if they are long lasting – can be very harmful to the ears. They can damage the sensitive structures within the ear, producing **noise-induced hearing loss**, or **NIHL**.

A one-off exposure to a very loud noise, such as an explosion, can rupture (tear) the eardrum or damage the delicate bones in the inner ear, although the ear can recover from this. Exposure to loud noise for longer periods of time can cause temporary deafness. It may also cause a ringing or buzzing sound in the ears, called *tinnitus*.

What is less well understood by many people is that *prolonged* exposure to moderately loud or very loud noise can cause damage to the ears that results in permanent hearing loss. The intensity of sound is measured in units called decibels. Here are some examples of the decibel rating of different sounds:

- a humming refrigerator   45 decibels
- normal conversation   60 decibels
- heavy traffic   85 decibels
- an MP3 player at full volume 105 decibels
- loud fireworks   150 decibels

Even after long-term exposure, sounds of less than 75 decibels are unlikely to cause hearing loss. However, long periods of exposure to sounds above 85 decibels can permanently damage the ears.

NIHL is caused by damage to the delicate hair cells in the inner ear. This is a gradual process: when it starts to happen, the person may not notice, or they may ignore the signs of hearing loss until they become worse. Over time, sounds become distorted or muffled, so the person may struggle to follow a conversation, or may have to turn up the volume on their television.

NIHL is completely preventable. All you need to do is avoid exposure to sounds over 85 decibels, and if you are taking part in an activity that involves loud noises, wear earplugs or ear protectors.

CHAPTER QUESTIONS

SKILLS ANALYSIS

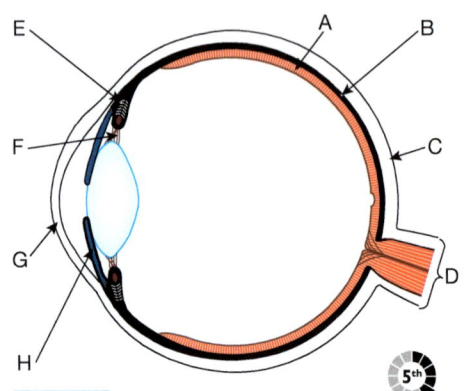

1 The diagram shows a section through a human eye, with parts labelled A to H.

   a The table lists the functions of some of the parts. Copy the table and write the letters of the correct parts in the boxes.

Function	Letter
refracts light rays	
converts light into nerve impulses	
contains pigment to stop internal reflection	
contracts to change the shape of the lens	
takes nerve impulses to the brain	

SKILLS CRITICAL THINKING
SKILLS REASONING

   b i Which label shows the iris?
     ii Explain how the iris controls the amount of light entering the eye.
     iii Why is this important?

SKILLS ANALYSIS

2 The diagram shows the parts of the ear.

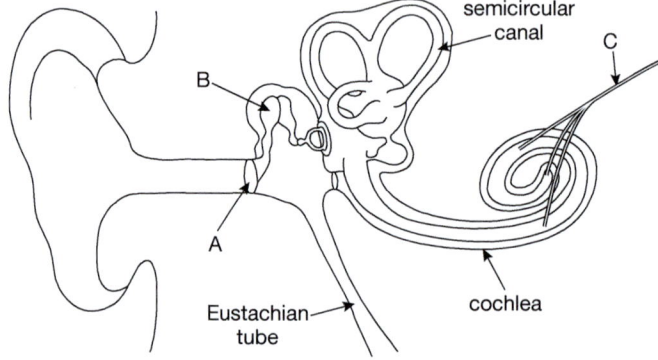

   a Name the parts labelled A, B and C.

SKILLS CRITICAL THINKING

   b Explain the role of the ear bones (ear ossicles).

   c Explain how vibrations in the fluid inside the cochlea are converted into nerve impulses.

   d What is the function of the Eustachian tube?

   e How does the ear allow the brain to distinguish between:
     i sounds of different frequencies (pitch)
     ii sounds of different loudness?

SKILLS EXECUTIVE FUNCTION

3 It is thought that the range of audible frequencies (frequencies of sound that can be detected by the human ear) decreases with age. Describe an investigation you could carry out to test this hypothesis. Make sure you include details of the variables you will need to control.

9 COORDINATION

Coordination means linking up different activities in the body so that things happen at the right time. Coordination also connects a stimulus with a response. Humans have two organ systems that do this. The nervous system uses electrical signals sent through nerves, and the endocrine system uses chemicals called hormones, which are carried in the blood.

LEARNING OBJECTIVES

- Know the basic plan of the central nervous system
- Know the structure of sensory, motor and relay neurones
- Understand how nerve impulses are initiated in receptors and their direction of movement along a neurone
- Understand how impulses are transmitted across a synapse
- Describe the pathway taken by a nerve impulse to cause a response to a stimulus
- Know the structure and function of a reflex arc and the spinal cord
- Know the main areas of the brain and their functions, including the cerebral hemispheres, cerebellum, medulla, pituitary gland and hypothalamus
- Describe the causes, symptoms and treatments of Alzheimer's disease, vascular dementia and Parkinson's disease
- Describe the causes, symptoms and treatments of mental illness, including schizophrenia and depression
- Understand the meaning of the term 'drug' and distinguish between legal and illegal drugs, including:
  - the action of common painkillers such as paracetamol on the nervous system
  - the dangers of heroin, cannabis and cocaine
- Describe the damaging effects of alcohol on the nervous system and liver and the behavioural consequences of excessive and long-term drinking
- Understand the differences between the nervous and hormonal systems
- Understand the action of hormones from:
  - the pituitary gland (ADH* and hormones controlling reproduction*)
  - the adrenal gland (adrenaline)
  - the thyroid gland (thyroxine)
  - the pancreas (insulin and glucagon)
  - the testes and ovaries*
- Understand the role of negative feedback with reference to blood glucose concentration
- Know the role of hormones in growth and development

*In this chapter, the functions of these hormones are described in outline only. They will be covered in more detail in Chapters 10–11.

THE CENTRAL NERVOUS SYSTEM

The biological name for a nerve cell is a **neurone**. The impulses that travel along a neurone are not like an electric current in a wire. They are caused by movements of charged particles (ions) in and out of the neurone. Impulses travel at speeds between about 10 and 100 metres per second. This is much slower than an electric current but it is fast enough to produce a rapid response (see the 'Looking ahead' feature at the end of this chapter).

Impulses from receptors pass along nerves containing **sensory neurones**, until they reach the brain and spinal cord. These two organs are together known as the **central nervous system**, or **CNS** (Figure 9.1).

Other nerves contain **motor neurones**, which transmit impulses to the muscles and glands. Some nerves contain only sensory or motor neurones, while other nerves – 'mixed' nerves – contain both. A typical nerve contains thousands of individual neurones.

> **DID YOU KNOW?**
> The CNS is well protected by the skeleton. The brain is encased in the skull or cranium and the spinal cord runs down the middle of the spinal column, passing through a hole in each vertebra. Nerves connected to the spinal cord are called *spinal* nerves and those connected to the brain are *cranial* nerves.

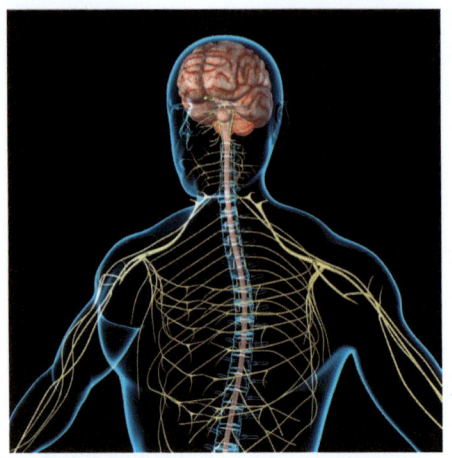

▲ Figure 9.1 The brain and spinal cord form the central nervous system. Cranial and spinal nerves lead to and from the CNS. The CNS sorts out information from the senses and sends messages to muscles.

THE STRUCTURE OF NEURONES

Both sensory and motor neurones can be very long. For instance, a motor neurone leading from the CNS to the muscles in the finger has a fibre about 1 m in length, which is 100 000 times the length of the cell body (Figure 9.2).

The cell body of a motor neurone is found at one end of the fibre, in the CNS. The cell body has fine cytoplasmic extensions, called **dendrons**. These in turn form finer extensions, called **dendrites**. There can be junctions with other neurones on any part of the cell body, dendrons or dendrites. These junctions are called **synapses** (see below). One of the extensions from the motor neurone cell body is much longer than the other dendrons. This fibre is called the **axon** and it carries impulses to the effector organ. The axon divides

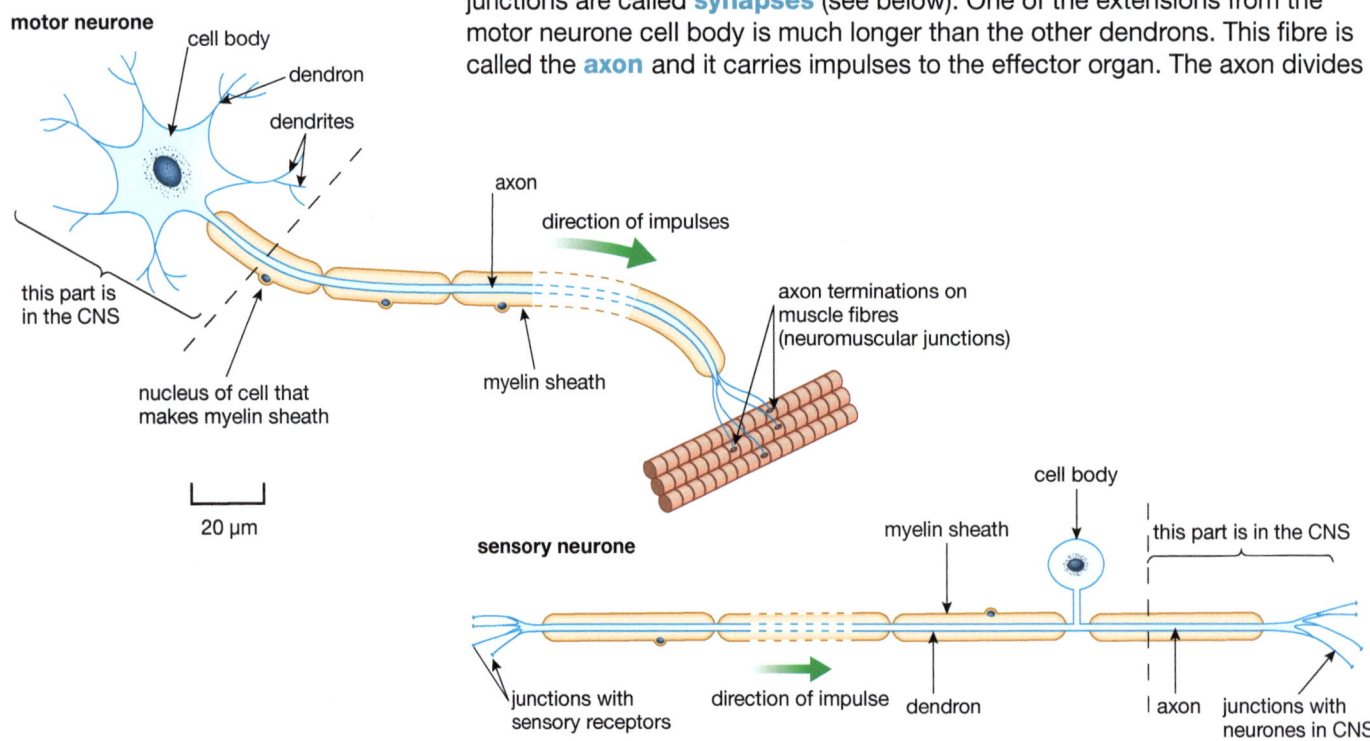

▲ Figure 9.2 The structure of motor and sensory neurones. The cell fibres (axon/dendron) are very long, which is indicated by the dashed sections.

into many nerve endings at the end furthest from the cell body. These fine branches of the axon connect with a muscle at a special sort of synapse called a **neuromuscular junction**. In this way, impulses are carried from the CNS out to the muscle. The signals from nerve impulses are transmitted across the neuromuscular junction, causing the muscle fibres to contract. The axon has an outer covering or 'sheath' made of a fatty material called myelin. This **myelin sheath** insulates the axon, preventing 'short circuits' with other axons, and also speeds up the conduction of the impulses. The sheath is formed by the membranes of special cells that wrap themselves around the axon as it develops. Cells with a myelin sheath are described as myelinated.

A sensory neurone has a similar structure to a motor neurone, but the cell body is located on a side branch of the fibre, just outside the CNS. The fibre from the sensory receptor to the cell body is actually a dendron, while the fibre from the cell body to the CNS is a short axon. As with motor neurones, fibres of sensory neurones are often myelinated.

SYNAPSES

Synapses are essential to the working of the nervous system. The CNS is made of many billions of nerve cells, and these have links with many others, through synapses. In the brain, each neurone may form synapses with thousands of other neurones, so the number of possible pathways through the system is almost unlimited.

A synapse is actually a gap between two nerve cells. The electrical impulses passing through the neurones do not cross this gap. Instead, impulses arriving at a synapse cause the ends of the fine branches of the axon to secrete a chemical called a neurotransmitter. This chemical diffuses across the gap and attaches to the membrane of the second neurone. It then starts off new impulses in the second cell (Figure 9.3). After the neurotransmitter has 'passed on the message', it is broken down by an enzyme.

Remember that many nerve cells, particularly those in the brain, have thousands of synapses with other neurones. The output of one cell may depend on the inputs from many cells adding together. In this way, synapses are important for integrating information in the CNS (Figure 9.4).

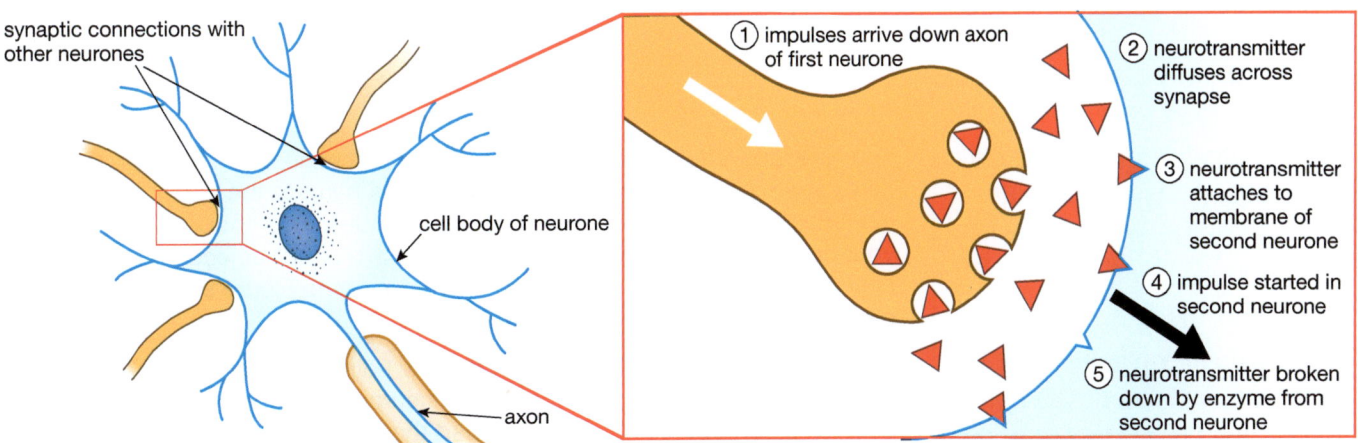

▲ Figure 9.3 The sequence of events happening at a synapse.

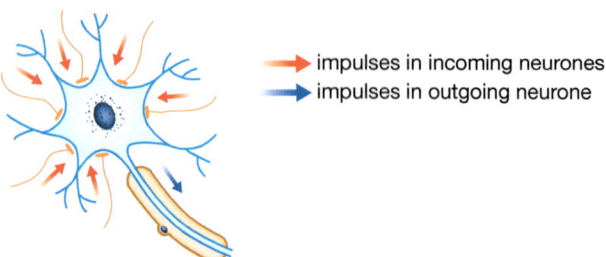

▲ Figure 9.4 Synapses allow the output of one nerve cell to be a result of integration of information from many other cells.

Because synapses rely on the movement of chemicals, it is easy for other chemicals to interfere with the way they work. These chemicals may imitate the neurotransmitter, or prevent it from acting. This is how many well-known drugs, both useful and harmful, work.

REFLEX ACTIONS

> **KEY TERM**
>
> A reflex action is a rapid, automatic (or involuntary) response to a stimulus. The action often (but not always) protects the body. Involuntary means that it is not started by impulses from the brain.

You read in Chapter 8 that the dilation and constriction of the pupil by the iris is an example of a reflex action. You now need to understand more about the nerves involved in a reflex. The nerve pathway of a reflex is called the **reflex arc**. The 'arc' part means that the pathway goes into the CNS and then straight back out again, in a sort of curve or arc (Figure 9.5).

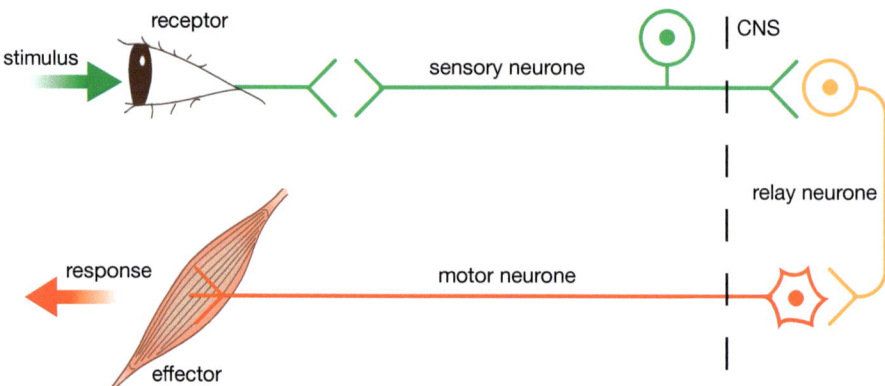

▲ Figure 9.5 Simplified diagram of a reflex arc.

DID YOU KNOW?

'Dorsal' and 'ventral' are words describing the back and front of the body. The dorsal roots of spinal nerves emerge from the spinal cord towards the back of the person, while the ventral roots emerge towards the front. Notice that the cell bodies of the sensory neurones are all located in a swelling in the dorsal root, called the **dorsal root ganglion**.

The iris–pupil reflex protects the eye against damage by bright light. Other reflexes are protective too, preventing serious harm to the body. For example, consider the reflex response to a painful stimulus. This happens when part of your body, such as your hand, touches a sharp or hot object. The reflex results in your hand being moved quickly away from harm. Figure 9.6 shows the nerve pathway of this reflex in more detail.

The stimulus is detected by temperature or pain receptors in the skin. These receptors generate impulses in sensory neurones. The impulses enter the CNS through a part of the spinal nerve called the **dorsal root**. In the spinal cord, the sensory neurones connect by synapses with short **relay neurones**, which in turn connect with motor neurones. The motor neurones emerge from the spinal cord through the **ventral root**, and send impulses back to the muscles of the arm. These muscles then contract, pulling the hand away from the harmful stimulus.

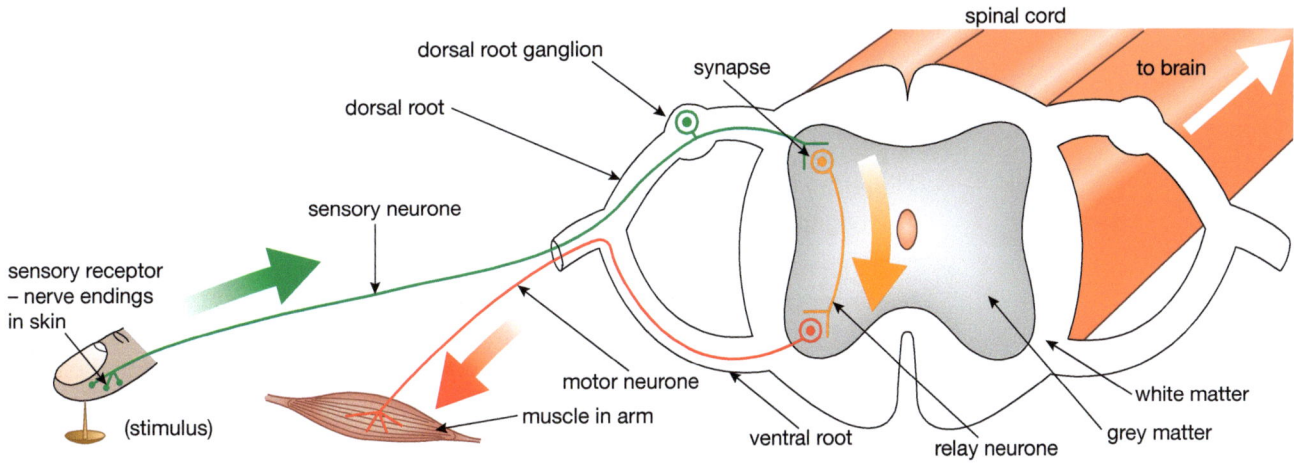

▲ Figure 9.6 A reflex arc in more detail.

The middle part of the spinal cord consists mainly of nerve cell bodies, which give it a grey colour. This is why it is known as **grey matter**. The outer part of the spinal cord is called **white matter**, and has a whiter appearance because it contains many axons with fatty myelin sheaths.

Impulses travel through the reflex arc in a fraction of a second, so the reflex action is very fast and does not need to be started by impulses from the brain. This does not mean, however, that the brain is unaware of what is going on: in the spinal cord, the reflex arc neurones also form synapses with nerve cells leading to and from the brain. The brain therefore receives information about the stimulus. This is how we feel the pain.

THE KNEE-JERK REFLEX

You can demonstrate a spinal reflex on yourself quite easily. It is the well-known **knee-jerk reflex**. Sit down and cross your legs, so that the upper leg hangs freely over the lower one. Grip the muscles of the top of the upper thigh with one hand and gently hit the area below the kneecap with a rubber hammer or the edge of the other hand (Figure 9.7). You may need a little practice but you should eventually see the lower leg jerk forward as the muscles at the front of the thigh contract.

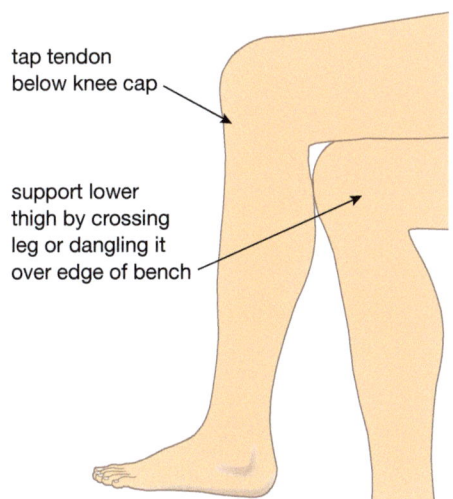

▲ Figure 9.7 Demonstration of the knee-jerk reflex.

The reflex arc that brings this about is very similar to the withdrawal reflex shown in Figure 9.6 but, in this case, the stimulus is detected in stretch receptors in the tendon below the knee (rather than in receptors in the skin). Tapping the tendon causes these receptors to be stretched. They react by sending nervous impulses towards the spinal cord through sensory neurones. The impulses then pass out again to the thigh muscles, through motor neurones, causing contraction of the muscle.

Of course, you would not normally experience a tap on the knee from a rubber hammer in everyday life. This reflex normally acts in situations where the knee joint is unexpectedly flexed. For example, if you trip up, the stretch receptors will be stimulated in the same way, and the contraction of the thigh muscle will help to stop you falling over.

Movements are sometimes a result of reflex actions, but we can also contract our muscles as a **voluntary action**, using nerve cell pathways from the brain linked to the same motor neurones. A voluntary action is under *conscious control*.

KEY POINT

In the knee-jerk reflex, the route from stimulus to response is as follows:

stimulus (the tap below the knee)
↓
stretch receptor
↓
sensory neurones in leg
↓
central nervous system
↓
motor neurones in the leg
↓
thigh muscle
↓
response (knee jerk)

THE BRAIN

We began to understand the functions of different parts of the brain by studying people who had suffered brain damage through accident or disease. Nowadays, we have very advanced electronic equipment that can record the activity in a normal living brain. However, we still do not fully understand the workings of this most complex organ of the body.

Your brain is sometimes called your 'grey matter'. This is because the positions of the grey and white matter are reversed in the brain compared with the spinal cord. The grey matter, mainly made of nerve cell bodies, is on the outside of the brain, while the axons that form the white matter are in the middle of the brain. The brain is made up of different parts, each with a specific function (Figure 9.8).

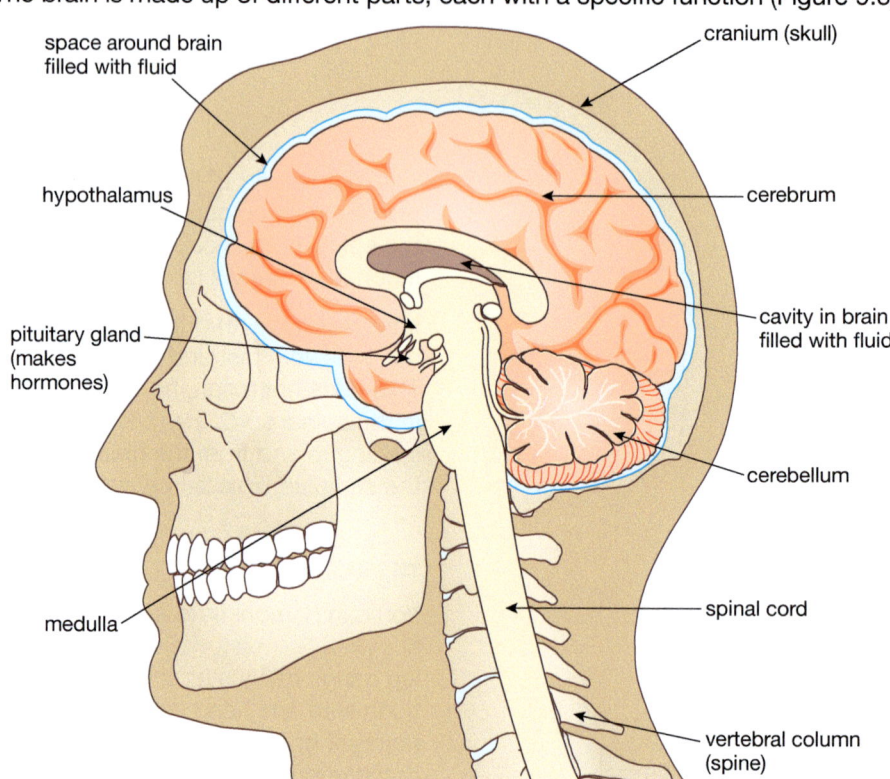

▲ Figure 9.8 Section through the human brain, showing its main parts.

The largest part of the brain is the **cerebrum**, made of two **cerebral hemispheres**. The cerebrum is the source of all our conscious thoughts. It has an outer layer called the **cerebral cortex**, with many folds all over its surface (Figure 9.9).

The cerebrum has three main functions.

- It contains sensory areas that receive and process information from all our sense organs.
- It has motor areas, from which all our voluntary actions originate.
- It is the origin of 'higher' activities, such as memory, reasoning, emotions and personality.

Different parts of the cerebrum carry out particular functions. For example, the sensory and motor areas are always situated in the same place in the cortex (Figure 9.10). Some parts of these areas deal with more information than others. Large parts of the sensory area deal with impulses from the fingers and lips, for example. This is illustrated in Figure 9.11.

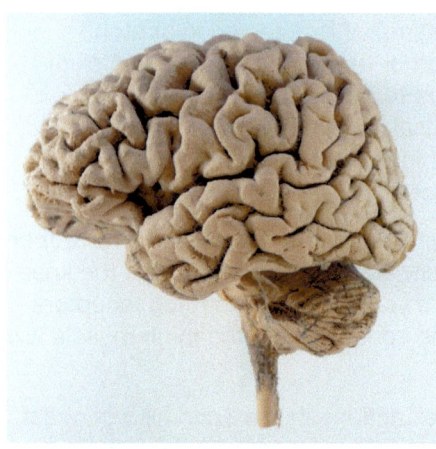

▲ Figure 9.9 A side view of a human brain. Notice the folded surface of the cerebral cortex.

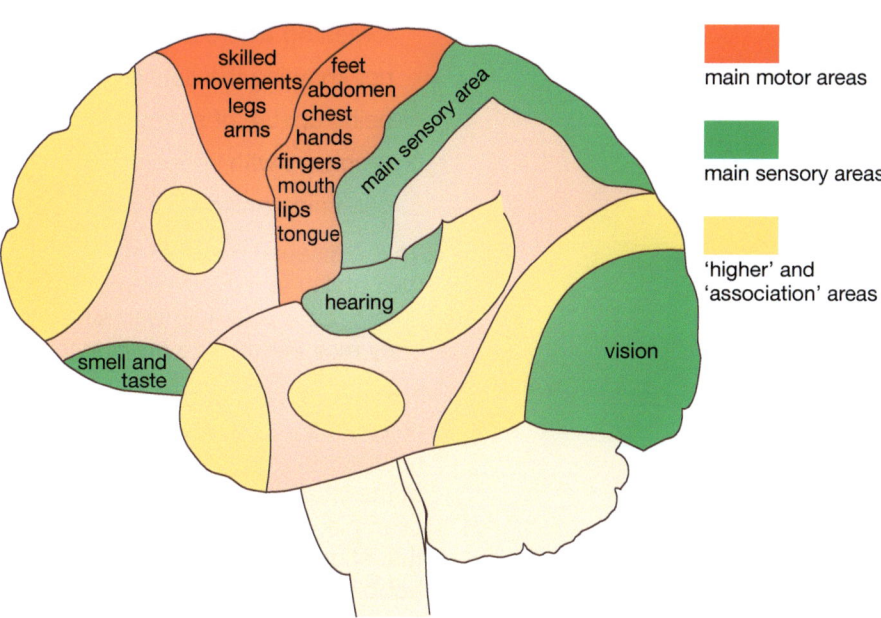

▲ Figure 9.10 Different parts of the cerebrum carry out specific functions.

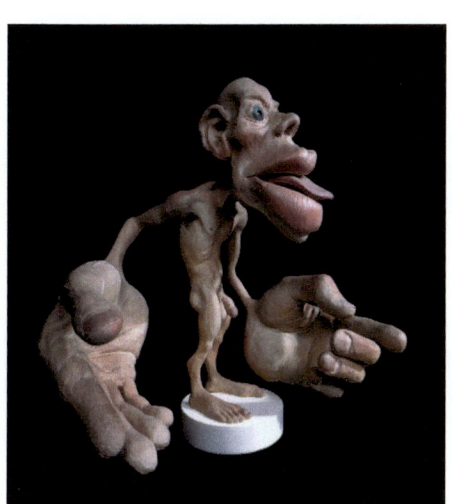

▲ Figure 9.11 A model of a human with its parts drawn in proportion to the amount of sensory information they send to the cortex of the brain. (Note that this does not apply to the eyes, which use more cortex than the rest of the body put together.)

Behind the cerebrum is the **cerebellum**. This region is concerned with coordinating the contraction of sets of muscles, as well as maintaining balance. This is important when you are carrying out complicated muscular activities, such as running or riding a bike. Underneath the cerebrum, connecting the spinal cord with the rest of the brain, is the brain stem or medulla. This controls basic body activities such as heartbeat and breathing rate.

The **pituitary gland** is located at the base of the brain, just below a part of the brain called the **hypothalamus**. The pituitary gland secretes a number of chemical 'messengers', called **hormones**, into the blood. Hormones are described later in this chapter.

DISEASES OF THE BRAIN

There are many different brain diseases. Some only develop in people as they grow old. We will look at three of these age-related brain diseases: **Alzheimer's disease**, **vascular dementia** and **Parkinson's disease**. You should know about the cause, symptoms and treatments for each disease.

ALZHEIMER'S DISEASE

Alzheimer's disease (usually just called Alzheimer's) is a disease that causes *dementia*. People with dementia show a group of symptoms that affect normal activities such as eating or getting dressed. Their personalities may change, so that they cannot control their emotions, become agitated or see things that are not there.

Cause: The cause is not well understood, but it seems to be due to a build-up of two proteins in brain cells, called *amyloid* and *tau*. As these proteins build up, they damage and kill brain cells. As the disease progresses, more and more brain cells are damaged, leading to the symptoms of Alzheimer's.

Symptoms: Early symptoms include:

- forgetting recent events, names and faces
- becoming increasingly repetitive, e.g. asking the same question after a short period of time

- forgetting where things are or putting them in strange places
- not knowing the date or time of day
- being unsure of surroundings or getting lost
- problems with speech and finding the right words
- becoming anxious or irritable, or showing little interest in what is going on.

As the disease progresses, a person with Alzheimer's gets worse. They lose their ability to think, remember or make decisions. They need more and more nursing care and help with simple everyday tasks such as dressing and eating. They may experience *hallucinations*, seeing people or things that are not there.

Treatment: As yet, there is no cure for Alzheimer's. In the early stages of the disease, some people are helped by a drug that increases the amount of a neurotransmitter called acetylcholine. Other drugs can help to relieve some of the later symptoms of the disease. Non-drug treatments are also used; for example, patients with Alzheimer's may be encouraged to take part in activities that stimulate thinking skills, such as games, discussion groups and simple hobbies.

VASCULAR DEMENTIA

Cause: Vascular dementia is the second most common type of dementia after Alzheimer's. It is caused by diseased and damaged blood vessels in the brain (the 'vascular system'). The blood vessels become blocked or leak, reducing the supply of oxygen and nutrients to the brain cells. This causes the brain cells to die.

Symptoms: Like Alzheimer's, vascular dementia causes problems with memory loss, thinking and reasoning. There are several different types of vascular dementia, affecting different areas of the brain. Each type can produce a range of different symptoms.

Treatment: As with Alzheimer's, there is no cure, and drugs used to treat Alzheimer's do not work with vascular dementia. It is possible to slow down the development of the disease by treating the cause. For example, drugs may be used to reduce high blood pressure or lower blood cholesterol and reduce the risk of strokes. Non-drug treatments aim to help patients to cope with the loss of abilities and encourage them to take part in activities that stimulate the brain.

PARKINSON'S DISEASE

Cause: Parkinson's disease affects about 1 in 500 people and is more common in men than in women. It normally begins in middle age or later life. It is caused by the death of certain neurones in the brain. These neurones produce a neurotransmitter called **dopamine**. A decrease in dopamine levels produces abnormal brain activity, leading to the symptoms of Parkinson's disease. The underlying cause is unknown, although research has shown that the brain cells of people with the disease contain structures called *Lewy bodies*. Lewy bodies contain several substances, in particular a naturally-occurring protein called *alpha-synuclein*. Research is concentrating on these Lewy bodies to try to establish the cause of Parkinson's.

Symptoms: The symptoms of Parkinson's disease vary from person to person, but they generally start with a *tremor* (shaking) in the hand or fingers. This gradually gets worse, developing over the months and years. Over time, the person's movements become slow, steps shorten and it becomes difficult for them to walk. Muscle stiffness can occur, which causes pain and slows movement. Their posture (the way they hold their body) becomes bent over, and the person may have problems with balance. There is a loss of automatic movements, such as blinking, smiling and the normal swinging of the arms during walking. The person may have difficulty speaking clearly.

Treatment: Although there is no cure, medication can control the symptoms of the disease. A wide range of drugs can be used. The most effective is called *levodopa*. This is a natural chemical that passes into the patient's brain and is converted into dopamine. Other drugs include:

- dopamine *agonists*, which mimic the action of dopamine in the brain
- chemicals that inhibit the enzymes that break down dopamine in nerve cells
- chemicals that block the transmission of impulses to the muscles, to control tremors.

Another treatment that is used is a type of surgery called *deep brain stimulation* (DBS). Usually, this is only used for patients with advanced Parkinson's disease. In DBS, surgeons implant electrodes into a part of the brain. The electrodes are connected to a generator in the patient's chest, which sends electrical pulses to the brain. This helps to reduce tremors.

As well as drug and surgical treatment, people with Parkinson's disease can benefit from lifestyle remedies such as healthy eating and aerobic exercise, as well as physiotherapy.

MENTAL ILLNESS

Mental illnesses are health conditions that involve abnormal changes in a person's thoughts, emotions or behaviour. These changes cause distress or problems in functioning in family, work or social situations. Mental illness is very common. For example, research in the United States shows that 19% of adults experience some form of mental illness, and 4% have a serious mental illness. There are many different types of mental illness. We will look at just two examples: schizophrenia and depression.

SCHIZOPHRENIA

About one in a hundred people have schizophrenia. It affects men and women equally. Most people who are diagnosed with the condition are aged between 18 and 35.

Symptoms: A person with schizophrenia may show a number of common symptoms, including:

- a lack of interest in things
- becoming anxious and confused
- feeling disconnected from their surroundings
- difficulty in concentrating
- wanting to avoid people
- having hallucinations – seeing things that are not there
- hearing voices
- having delusions – believing things to be true when they are not (refusing to accept reality)
- suffering from *paranoia* – being suspicious of other people and thinking that people want to harm them.

Causes: Psychiatrists (doctors who treat mental illness) generally agree that schizophrenia is caused by a combination of factors:

- **Dopamine:** The same neurotransmitter that is lower than normal in patients with Parkinson's disease is *higher* than normal in people with schizophrenia. It is thought that high levels of dopamine may help to start the development of schizophrenia, but no one knows how.

- **Stress:** Stressful or life-changing events may trigger schizophrenia, for example, being out of work, living in poverty, being homeless or the death of someone close to you.

- **Heredity:** In some families, the condition occurs more often than one would expect, suggesting that there may be a genetic link. It is thought that having some genes may make people more likely to develop schizophrenia.

- **Drug abuse:** Some people develop schizophrenia as a result of the use of drugs such as cannabis and cocaine (see below).

There are probably other factors involved. Brain injuries, viruses, hormones (particularly in women), diet and allergies have all been suggested as possibilities.

Treatment: Different things work for different people. Antipsychotic drugs (tranquilisers) are used to treat symptoms such as hallucinations and hearing voices. However, they do not work for all patients, and can cause unpleasant side effects. Non-drug treatments can be very successful. These include:

- **talking treatments:** where the patient has a regular time and place to talk about their troubles with a trained health professional

- **cognitive behavioural therapy (CBT):** a particular type of talking treatment, where the patient is encouraged to identify connections between their thoughts, feelings and behaviour, and to develop practical ways to manage any negative patterns of thinking or behaviour

- **family intervention therapy:** which aims to help families develop communication, problem solving, information sharing and skills to deal with different situations.

DEPRESSION

Most of us have felt sad or 'low' at some point in our lives and may have described this as being depressed. This is a normal reaction to the problems in life. However, when a person has intense feelings of sadness that last for days or weeks, they may be suffering from a mental condition called **depression**. People with depression often feel that nothing is worthwhile and that life is pointless.

Symptoms: Apart from feeling sad and upset, people with depression can experience several other symptoms. They may feel:

- restless or irritable
- guilty about their condition, worthless, having no 'self-esteem'
- unable to relate to other people
- unable to enjoy life or find pleasure in anything
- a sense of unreality.

Severe depression can be life threatening, because it can make people lose the desire to continue living. This can lead to suicidal thoughts or actions.

Causes: There is no single cause of depression, and it does not seem to be caused by any change in the brain's chemistry. It can be caused by various factors, including other mental health problems, poor physical health, bad childhood experiences, life-changing events (such as a death in the family) and drug or alcohol abuse. As with schizophrenia, there seems to be a genetic component – the tendency to develop depression runs in families.

Treatment: As with schizophrenia, talking treatments such as cognitive behavioural therapy can be very effective in relieving the symptoms of depression. There are also drugs called *antidepressants*, which can be used on their own or in combination with talking therapies.

LEGAL AND ILLEGAL DRUGS

Drugs are chemicals that affect the normal chemical reactions taking place in a person's body. Many drugs are useful. For example, aspirin is an effective painkiller. A number of drugs, however, act by interfering with the nervous system, and some of these can have very harmful side effects. This is one reason why many drugs are illegal. We will look at some legal and illegal drugs.

PARACETAMOL

One of the most widely used legal drugs is **paracetamol**. This is a medication that is taken by mouth and used to treat common 'aches and pains' such as headache and toothache. It also reduces a high temperature (fever) caused by a cold or flu. Despite its widespread use, we do not know for sure how paracetamol works! It may inhibit an enzyme that controls the production of chemicals called *prostaglandins*. These naturally-occurring substances have various protective roles in the body, but one of their side effects is that they produce pain and fever. Paracetamol is an effective painkiller and is safe if you take no more than the recommended dose. An overdose of paracetamol, however, can cause permanent liver damage and death.

ALCOHOL

Alcohol is a legal 'recreational' drug in many countries around the world. The alcohol in beer, wine and spirits slows down the nervous system, even when drunk in small quantities, and increases the time a person takes to react to a stimulus. This is why driving after drinking alcohol is so dangerous. The driver will not be able to react quickly to sudden danger, such as a person walking into the road (Figure 9.12).

▲ Figure 9.12 Alcohol in the bloodstream increases reaction times and is one cause of car accidents.

Larger amounts of alcohol in the body interfere with the drinker's balance and muscular control, and lead to blurred vision and slurred speech. High concentrations of alcohol in the blood can even cause coma and death.

Many people drink moderate amounts of alcohol to relax. To some people, however, alcohol is an addictive drug. Long-term alcohol abuse leads to serious medical problems. Alcohol is quickly absorbed into the blood through the stomach and intestines, and is taken around the body. The liver breaks the alcohol down (a process called detoxification) but if a person drinks large amounts regularly, the liver may not be able to cope. The person can develop a disease called **cirrhosis**, where the liver does not perform its usual functions properly and toxins in the blood build up to high levels. This disease is usually fatal. Alcohol also damages the brain and stomach lining.

ILLEGAL DRUGS

Most countries of the world have laws against the use of certain drugs that cause serious harm to the drug-user's body or to society as a whole, for example, by causing the break-up of families or an increase in crime. We will look at three examples of illegal drugs – cannabis, heroin and cocaine. All three are *psychoactive* drugs – they are 'mind-altering', affecting brain function and changing a person's mood or consciousness.

CANNABIS

Cannabis is obtained from a plant called *Cannabis sativa* (Figure 9.13). It is one of the most widely used illegal drugs in the world. The dried leaves of the cannabis plant are smoked as *marijuana* or *hashish*. It contains a psychoactive substance called THC or tetrahydrocannabinol.

The effect of this drug on the body depends on its strength and the amount of cannabis used. The most common effects are a sense of relaxation, cheerfulness and an increased awareness of sounds and colours. However, cannabis has many dangerous effects. The user may become confused or disorientated and suffer hallucinations. They can become anxious, depressed and even suicidal. Cannabis is also dangerous to the lungs, since it is usually smoked with tobacco.

▲ Figure 9.13 A cannabis plant.

HEROIN

Heroin is both a legal and an illegal drug. It is a very strong painkiller – a modified form of morphine. Under its medical name, 'diamorphine', it is used by doctors to treat people who are in severe pain (for example, cancer patients). However, most people will have heard of heroin as an illegal drug. A heroin user normally injects the drug into a vein although it may be smoked or inhaled.

As well as being a painkiller, heroin is a narcotic, producing a powerful feeling of pleasure and contentment known as a 'high'. Heroin is also a powerful depressant drug, slowing down the nervous system and producing deep drowsiness.

When a person starts to use heroin, they rapidly develop a tolerance to the drug. This means that they will need to inject more and more of the drug to produce the same effects. Use of heroin rapidly leads to *addiction*, where the person becomes psychologically dependent on taking the drug. Addicts are unable to live without regularly injecting heroin, and their lives become centred on obtaining and using the drug. People who are addicted to heroin often turn to crime to fund their addiction. Their family and social lives suffer, and they are less likely to be able to maintain a job.

If a heroin addict tries to stop or reduce their use of the drug, within hours they will suffer from *withdrawal symptoms*. These include sweats, chills, severe muscle and bone aches, vomiting, cramps and diarrhoea, along with psychological symptoms such as anxiety and depression.

There are a number of other medical problems linked with heroin use. Repeated injections can cause skin infections and abscesses. The illegally-bought heroin may be contaminated with other substances, which are added to dilute the drug (making more profit for the drug dealer). These substances are sometimes toxic. Addicts frequently die from an accidental overdose of the drug, because they do not know how strong their supply is.

Addicts who inject the drug may share needles with other users. This greatly increases their risk of catching a number of infectious diseases, including hepatitis and HIV.

COCAINE

Cocaine is a drug made from the leaves of the coca plant, *Erythroxylum coca*. It has some medicinal uses as a local anaesthetic, but is much more widely known as an illegal drug. Cocaine is a white powder that is usually inhaled ('snorted') through the nose. It can also be rubbed onto the gums or dissolved in water and injected into the blood. Cocaine that has been processed to make a solid 'rock' is heated to produce vapours that are inhaled. In this form, it is called 'crack' cocaine.

Cocaine is a powerful stimulant drug. It affects synapses in the brain and produces a short-term 'high' in the user. This involves unusual behaviour patterns, such as:

- extreme happiness and energy
- mental alertness
- unusual sensitivity to sounds, touch and sights
- irritability
- extreme distrust of other people (paranoia).

Large amounts of cocaine can trigger bizarre, unpredictable and violent behaviour in the user.

There are many health problems associated with cocaine use. It increases the heart rate and blood pressure to dangerously high levels. It can cause nosebleeds, dizziness, nausea (feeling sick), fever, seizures (fits), breathing difficulties, tremors, and muscle twitches. In addition, cocaine is highly addictive and long-term use can result in a number of severe medical problems, such as a breakdown of the large intestine due to reduced blood flow.

CHEMICAL COORDINATION – THE ENDOCRINE SYSTEM

The nervous system is a coordination system forming a link between stimulus and response. The body has a second coordination system, which does not involve nerves. This is the endocrine system. It consists of organs called **endocrine glands**, which make chemical messenger substances called hormones. Hormones are carried in the bloodstream.

GLANDS AND HORMONES

A gland is an organ that releases or *secretes* a substance. **Secretion** means that cells in the gland make a chemical which passes out of the cells. The chemical then travels somewhere else in the body, where it carries out its function. There are two types of gland: exocrine and endocrine glands. **Exocrine glands** secrete their products through a tube or **duct**. For example, salivary glands in your mouth secrete saliva down salivary ducts, and tear glands secrete tears through ducts that lead to the surface of the eye. Endocrine glands have no duct and so are called ductless glands. Instead, their products, the hormones, are secreted into the blood vessels that pass through the gland (Figure 9.14).

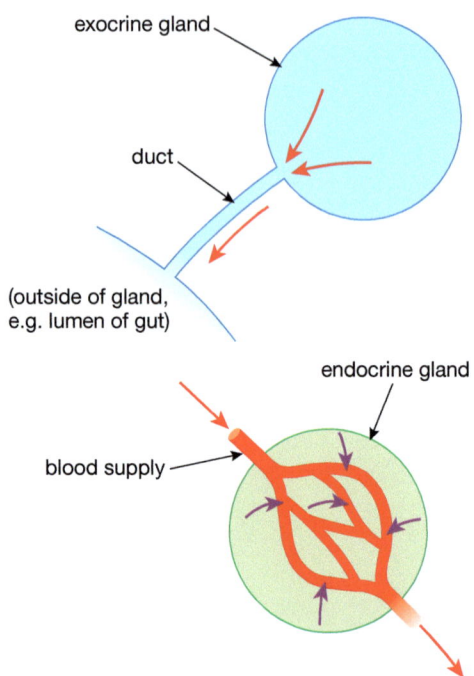

▲ Figure 9.14 Exocrine glands secrete their products though a duct, while endocrine glands secrete hormones into the blood.

Because hormones are carried in the blood, they can travel to all areas of the body. They usually only affect certain tissues or organs, called target organs. The target organs can be a long distance from the gland that made the hormone. A hormone only affects a tissue or organ if the cells of that tissue or organ have special chemical receptors for the hormone. For example, the hormone insulin affects the cells of the liver, which have insulin receptors.

> **DID YOU KNOW?**
> The receptors for some hormones are located in the cell membrane of the target cell. Other hormones will target receptors in the cytoplasm or in the nucleus. Without specific receptors, a cell will not respond to a hormone at all.

THE DIFFERENCES BETWEEN NERVOUS AND ENDOCRINE CONTROL

Although the nervous and endocrine systems both act to coordinate body functions, they do this in different ways. These differences are summarised in Table 9.1.

Table 9.1 The nervous and endocrine systems compared.

Nervous system	Endocrine system
works by nerve impulses transmitted through nerve cells (although chemicals are used at synapses)	works by hormones transmitted through the bloodstream
nerve impulses travel fast and usually have an 'instant' effect	hormones travel more slowly and generally take longer to act
response is usually short-lived	response usually lasts longer
impulses act on individual cells, such as muscle fibres, and so have a very localised effect	hormones can have widespread effects on different organs (although they only act on particular tissues or organs if the cells have the correct receptors)

THE POSITIONS OF THE ENDOCRINE GLANDS

The main endocrine glands are shown in Figure 9.15. Table 9.2 lists some of the hormones that they make, and their main functions. Several of these hormones will be covered in more detail in other chapters of this book.

Table 9.2 Some of the main endocrine glands, the hormones they produce and their functions.

Gland	Hormone	Some functions of the hormones
pituitary	follicle stimulating hormone (FSH)	stimulates egg development and oestrogen secretion in females and sperm production in males
	luteinising hormone (LH)	stimulates egg release (ovulation) in females and testosterone production in males
	antidiuretic hormone (ADH)	controls the water content of the blood
	growth hormone	speeds up the rate of growth and development in children
thyroid	thyroxine	controls the body's metabolic rate (how fast chemical reactions take place in cells)
pancreas	insulin	lowers blood glucose
	glucagon	raises blood glucose
adrenals	adrenaline	prepares the body for physical activity
testes	testosterone	controls the development of male secondary sexual characteristics
ovaries	oestrogen	controls the development of female secondary sexual characteristics
	progesterone	regulates the menstrual cycle

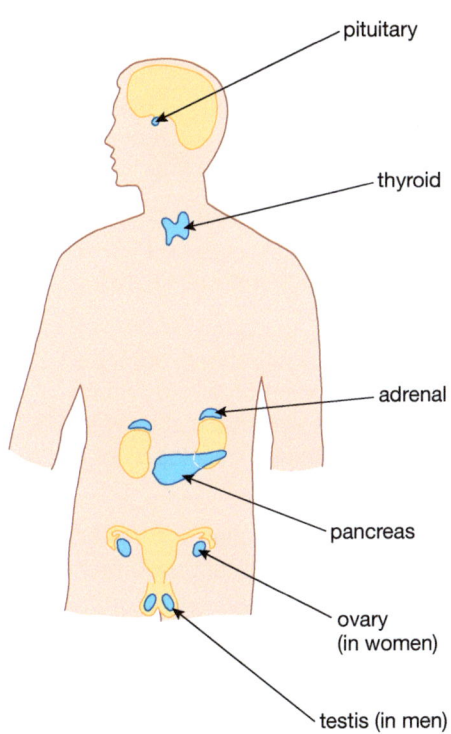

Figure 9.15 The main endocrine glands of the body.

The pituitary gland is found at the base of the brain. It secretes a number of hormones, including several that regulate reproduction. These functions are described in Chapter 11. The pituitary contains neurones linking it to a part of the brain called the hypothalamus, and some of its hormones are produced under the control of the brain. For example, **antidiuretic hormone** (**ADH**) is made in the cell bodies of neurones in the hypothalamus. It is secreted from the ends of fibres of these neurones and stored in the pituitary gland. The target organs of ADH are the kidneys, where the hormone is involved in controlling the water content of the blood. This process is described in more detail in Chapter 10, but a key point is that receptors in the hypothalamus monitor the water potential of the blood and control ADH release by the pituitary. This shows how the nervous system and endocrine system can sometimes act together in coordinating a response.

The **thyroid** is a gland in the neck, shaped a little like a butterfly. It secretes several hormones, including **thyroxine**, which speeds up the metabolic rate of the body. The metabolic rate is the rate at which the cells of the body carry out all their chemical reactions. The release of thyroxine is also under the control of the hypothalamus of the brain (see below).

The pancreas is both an endocrine and an exocrine organ. It secretes two hormones called **insulin** and **glucagon**, both of which are involved in the regulation of blood glucose. The pancreas is also a gland of the digestive system, secreting enzymes through the pancreatic duct into the small intestine (see Chapter 4). The endocrine functions of the pancreas are described below.

The sex organs of males and females produce sex cells (gametes) and are also endocrine organs. Both the **testes** and the **ovaries** make hormones involved in the control of reproduction. This topic is covered in Chapter 11.

We will now look at the functions of three endocrine glands in more detail – the thyroid, adrenal glands and pancreas.

THE THYROID AND CONTROL OF METABOLIC RATE

The thyroid makes a number of hormones, including thyroxine. The thyroid gland is the only endocrine gland that stores large amounts of a hormone. All other glands make their hormones as required, but the thyroid contains about a 100-day supply of thyroxine. Thyroxine contains the element iodine. Lack of iodine in the diet is one cause of an enlarged thyroid gland, called a goitre (Figure 9.16).

If a person fasts (eats nothing) overnight, and their rate of oxygen consumption at rest is then measured, this is called their **basal metabolic rate** (**BMR**). Thyroxine increases the BMR, stimulating cells to respire aerobically to produce more ATP (see Chapter 5).

The body needs to increase its metabolic rate in various situations, such as during exercise and in cold surroundings when ATP can be broken down to produce heat to maintain body temperature. Thyroxine also increases protein synthesis in the body, so it is needed for normal growth and development in children.

▲ Figure 9.16 An enlarged thyroid gland, called a goitre.

However, when changes are not needed, the BMR is controlled within very exact limits. The hypothalamus in the brain detects the slightest drop in the BMR and responds by making a hormone called **thyrotropin releasing hormone** (**TRH**). TRH travels the short distance from the hypothalamus to the pituitary, where it stimulates the release of another hormone, called **thyroid-stimulating hormone** (**TSH**). TSH passes in the bloodstream to the thyroid, where it stimulates synthesis and release of thyroxine.

Thyroxine causes the metabolic rate to return to normal. In turn, this 'switches off' production of TRH and TSH and stops the release of thyroxine.

NEGATIVE FEEDBACK

Control of metabolic rate is a good example of a principle called **negative feedback**. Negative feedback is a control mechanism in the body. During negative feedback, a change in the body is detected and this starts a process to return conditions to normal. Many conditions are monitored and maintained by negative feedback, including metabolic rate and body temperature (see Chapter 10). Many negative feedback processes involve hormones as 'messengers', such as ADH in the control of the water content of the blood (also described in Chapter 10). The information pathway forms a loop, so it is also known as a negative feedback loop. The negative feedback loop which controls metabolic rate is shown in Figure 9.17.

Negative feedback depends on there being a normal, 'correct' value for the condition that is being monitored. For example, our normal body temperature is 37 °C. Biologists call the normal value the 'set point'. The set point is monitored by the body, and any change from the set point is corrected by negative feedback.

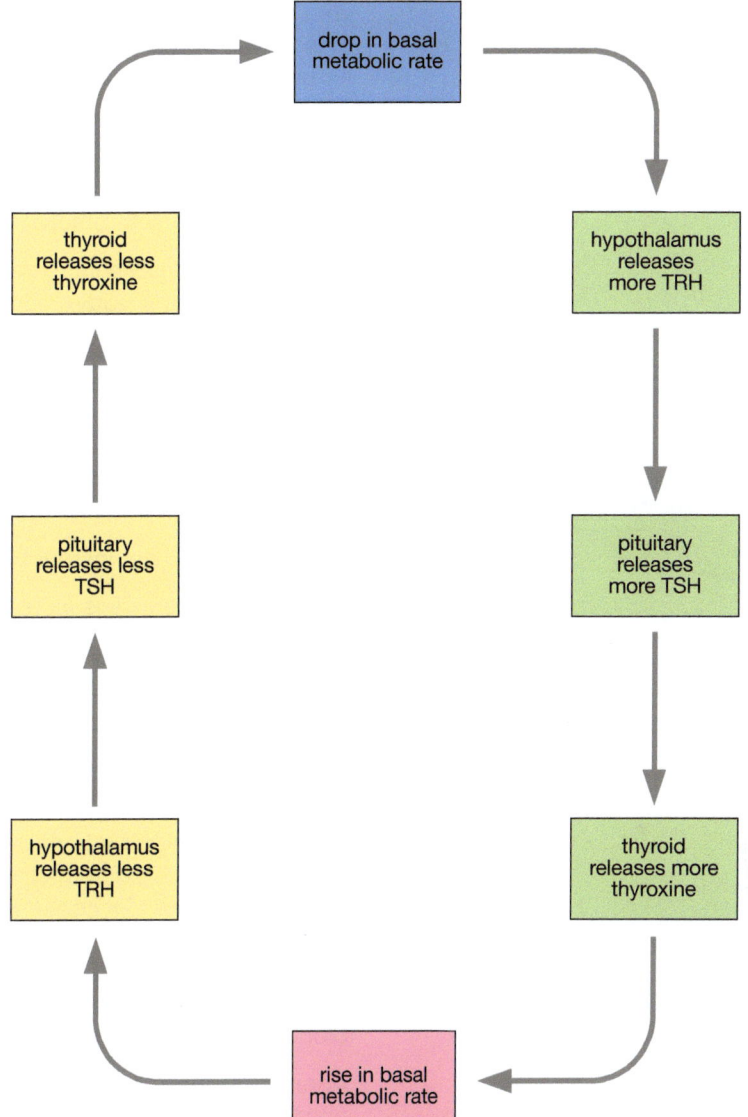

▲ Figure 9.17 Thyroxine controls metabolic rate by a negative feedback loop.

THE EFFECTS OF HORMONES ON GROWTH AND DEVELOPMENT

Much of the growth and development of the body is under the control of hormones. Growth hormone from the pituitary gland stimulates growth, cell division (mitosis) and regeneration of cells. It is particularly important in childhood, and is used to treat children with growth disorders. As well as causing an increase in metabolic rate (which is needed for growth), thyroxine increases the production of growth hormone. When humans become sexually mature (at **puberty**), the sex hormones **testosterone** (in males) and **oestrogen** (in females) have several effects on the body. These effects include an increase in the production of growth hormone, which brings about a 'growth spurt' (sudden growth) in teenage children. This will be discussed in more detail in Chapter 11.

ADRENALINE – THE 'FIGHT OR FLIGHT' HORMONE

> **DID YOU KNOW?**
> 'Adrenal' means 'next to the kidneys' which describes where the adrenal glands are located, on top of these organs (see Figure 9.15).

When you are frightened, excited or angry, your **adrenal glands** secrete the hormone adrenaline.

Adrenaline acts at a number of target organs and tissues, preparing the body for action. In animals other than humans, this action usually means dealing with an attack by an enemy, where the animal can stay and fight or run away. This is why adrenaline is known as the 'fight or flight' hormone. This is not a situation humans encounter often, but there are plenty of other times when adrenaline is released (Figure 9.18).

When the body prepares for action, its muscles need a good supply of oxygen and glucose for respiration. Adrenaline triggers several changes to ensure this happens, along with other changes to prepare for 'fight or flight'.

- The breathing rate increases and breaths become deeper, taking more oxygen into the body.
- The heart beats faster, sending more blood to the muscles, so that they receive more glucose and oxygen for respiration.
- Blood is diverted away from the intestine and into the muscles.
- In the liver, stored carbohydrate is changed into glucose and released into the blood. The muscle cells absorb more glucose and use it for respiration.
- The pupils dilate, increasing visual sensitivity to movement.
- Mental awareness is increased, so reactions are faster.

In humans, adrenaline is not only released in 'fight or flight' situations. It is released in many other stressful activities too, such as preparing for a race, going for a job interview or taking an exam.

▲ Figure 9.18 Many human activities cause adrenaline to be produced, not just 'fight or flight' situations!

CONTROLLING BLOOD GLUCOSE

You saw earlier that adrenaline can raise blood glucose levels by releasing carbohydrate stored in the liver. The liver cells contain carbohydrate in the form of glycogen. Glycogen is made from long chains of glucose units joined together to form larger molecules (see Chapter 3). Glycogen is a good storage product because it is insoluble. When the body is short of glucose, the glycogen can be broken down into glucose, which then passes into the bloodstream.

Adrenaline raises blood glucose concentration in an emergency, but other hormones act all the time to control blood glucose levels, keeping them fairly constant. The main hormone controlling glucose is insulin. Insulin is made by special cells in the pancreas. It stimulates the liver cells to take up glucose and convert it into glycogen, lowering the level of glucose in the blood.

> **KEY POINT**
> Insulin does not convert glucose into glycogen. The liver cells do this. The insulin just stimulates the liver cells to take up glucose from the blood.

The other hormone from the pancreas that affects blood glucose levels is glucagon. This stimulates the liver cells to break down glycogen into glucose, raising the concentration of glucose in the blood if it is too low.

Together, these two hormones work to keep the blood glucose level approximately constant, at a little less than 1 gram of glucose in every dm^3 (cubic decimetre) of blood. Both hormones are released by special cells in the pancreas, in direct response to the level of glucose in the blood passing through this organ. In other words:

$$\text{glucose} \underset{\text{glucagon}}{\overset{\text{insulin}}{\rightleftharpoons}} \text{glycogen}$$

> **DID YOU KNOW?**
> Most of the cells of the pancreas are concerned with making digestive enzymes. However, in the pancreas tissue, there are small groups of cells called the Islets of Langerhans (Figure 9.19). These contain two types of cell. Larger α (alpha) cells secrete glucagon, and smaller β (beta) cells secrete insulin.

The concentration of glucose in your blood will start to rise after you have had a meal. Sugars from digested carbohydrate pass into the blood and are carried to the liver in the hepatic portal vein (see Chapters 4 and 6). Here the glucose is converted to glycogen, so the blood leaving the liver in the hepatic vein will have a lower concentration of glucose.

Control of blood glucose levels by insulin and glucagon operates through negative feedback. If the glucose concentration of the blood rises, the release of insulin causes glucose levels to fall until the normal level is achieved again. Glucagon works the other way: if glucose levels fall, glucagon stimulates responses that raise the glucose level. Having two hormones with opposite effects allows for better control.

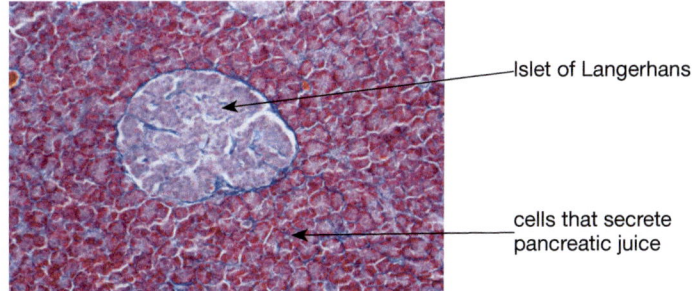

▲ Figure 9.19 Pancreas tissue including an Islet of Langerhans.

DIABETES

Some people have a disease where their pancreas cannot make enough insulin to keep their blood glucose level constant, so it rises to very high concentrations. This disease is called diabetes. One symptom of diabetes can be detected by a chemical test on urine. Normally, people have no glucose at all in their urine. Someone suffering from diabetes may have such a high concentration of glucose in their blood that it is excreted in their urine.

> **DID YOU KNOW?**
> We should really call this disease 'type 1' diabetes. There is also a 'type 2' diabetes, where the pancreas produces insulin but the body shows *insulin resistance*; the insulin has less effect than it should do. In people with type 2 diabetes, the pancreas tries to makes extra insulin at first. Eventually, however, it cannot make enough insulin to maintain blood glucose at a normal level. Type 2 diabetes is common in people who are overweight and eat a poor diet that is high in sugar and other carbohydrates. It is also more common in middle-aged or older people, whereas type 1 diabetes can develop at any age and is common in children. Type 2 diabetes can be prevented and controlled by a good diet and regular exercise.

Another symptom of this kind of diabetes is a constant thirst. This is because the high blood glucose concentration stimulates receptors in the hypothalamus of the brain. These 'thirst centres' are stimulated so that, by drinking, the person will dilute their blood.

Severe diabetes is very serious. If it is untreated, the sufferer loses weight and becomes weak and eventually falls into a coma and dies.

Carbohydrates in the diet, such as starch and sugars, are the source of glucose in the blood, so a person with diabetes can help to control their blood

> **DID YOU KNOW?**
> Insulin is a protein. If it was taken by mouth, in tablet form, it would be broken down by protease enzymes in the gut. Instead it is injected into muscle tissue, where it is slowly absorbed into the bloodstream.

sugar if they limit the amount of carbohydrate they eat. However, a person with type 1 diabetes also needs to receive daily injections of insulin to keep the glucose in their blood at the right level.

People with diabetes can check their blood glucose using test strips or special sensors. How these work is described in Chapter 3.

> **DID YOU KNOW?**
> Insulin for the treatment of diabetes has been available since 1921, and has kept millions of people alive. It was originally extracted from the pancreases of animals such as pigs and cows, and much insulin is still obtained in this way. However, since the 1970s, human insulin has been produced commercially, from genetically modified (GM) bacteria. The bacteria have their DNA 'engineered' to contain the gene for human insulin (see Chapter 1).

LOOKING AHEAD – WHAT IS A NERVE IMPULSE?

A nerve impulse is an electrical signal that travels along the axon of a nerve cell. When the cell is not transmitting an impulse, there is a small potential difference (voltage) across the nerve cell membrane. The potential inside the axon is about −70 mV compared to the outside. This 'resting potential' is caused by differences in concentrations of various ions inside and outside the cell.

When the cell is stimulated, sodium ions (Na^+) rush into the axon through the membrane. The inflow of positively charged ions causes the potential to become positive – we say that it is *depolarised*. This sudden switch in voltage is called an 'action potential'. It only lasts for a few milliseconds (Figure 9.20). After a brief 'overshoot' below −70 mV it returns to normal (*repolarised*), as other positive ions (potassium, K^+) pass out through the membrane.

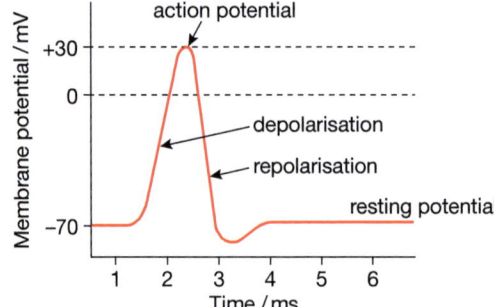

▲ Figure 9.20 An action potential.

A nerve impulse is a propagated action potential. The action potential stimulates the next part of the cell membrane, so that the depolarisation spreads along the axon. After the action potential has passed, ion exchange pumps in the membrane sort out the imbalance of Na^+ and K^+ ions. The pumps use ATP for active transport – this is one reason why nerve cells need a lot of metabolic energy from respiration.

Nerve cells are called 'excitable cells' because they can change their membrane potential in this way. Other excitable cells include muscle and receptor cells. If you continue to study biology beyond International GCSE level you will probably learn more about this interesting topic.

UNIT 3 — COORDINATION

CHAPTER QUESTIONS

SKILLS ANALYSIS

1 The diagram shows some parts of the nervous system involved in a simple reflex action that happens when a finger touches a hot object.

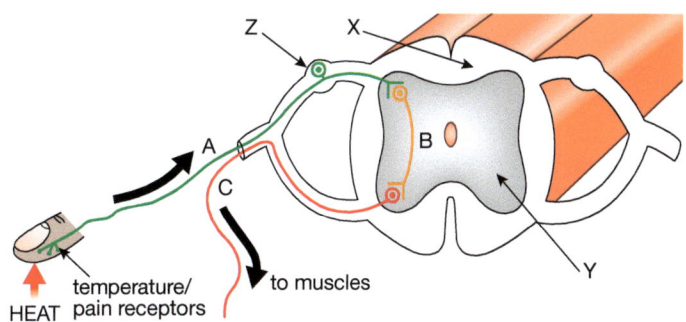

  a What type of neurone is:
   i neurone A
   ii neurone B
   iii neurone C?

SKILLS CRITICAL THINKING

  b Describe the function of each of these types of neurone.

SKILLS ANALYSIS

  c Which parts of the nervous system are shown by the labels X, Y and Z?
  d In what form is information passed along neurones?

SKILLS CRITICAL THINKING

  e Explain how information passes from one neurone to another.

2 a Which part of the human brain is responsible for controlling each of the following actions?
   i keeping your balance when you walk
   ii maintaining your breathing when you are asleep
   iii making your leg muscles contract when you kick a ball
  b A 'stroke' is caused by a blood clot blocking the blood supply to part of the brain.
   i One patient, after suffering a stroke, was unable to move his left arm. Which part of his brain was affected?
   ii Another patient lost her sense of smell following a stroke. Which part of her brain was affected?
  c Strokes can also cause vascular dementia.
   i What are the symptoms of vascular dementia?
   ii Describe two treatments that slow down the development of vascular dementia.

3 Hormones are secreted by endocrine glands.
  a Explain the meaning of the three terms underlined in this sentence.

SKILLS ANALYSIS

  b Identify the hormones A to D in the table.

Hormone	One function of this hormone
A	stimulates the liver to convert glucose to glycogen
B	controls the 'fight or flight' responses
C	brings about a sudden increase in growth in teenage boys
D	regulates the menstrual cycle

SKILLS REASONING

4 Explain these observations:
   a Mice that had their thyroid gland removed were less able to survive in freezing conditions than mice with intact thyroid glands.
   b People living in areas of the world where there is no iodine in the drinking water can develop a condition called a goitre.

5 The graph shows the changes in blood glucose in a healthy woman over a 12-hour period.

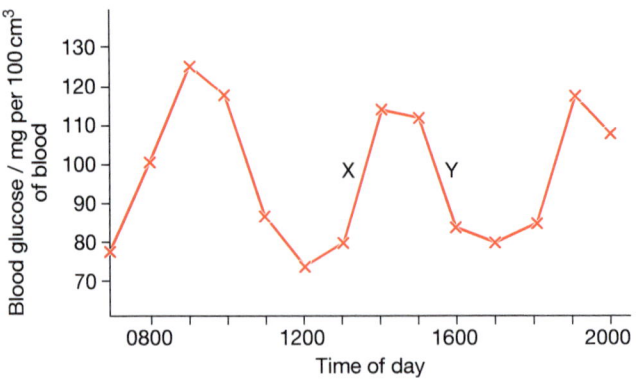

SKILLS ANALYSIS

a Explain why there was a rise in blood glucose at X.

b How does the body bring about a decrease in blood glucose at Y? Your answer should include the words insulin, liver and pancreas.

SKILLS REASONING

c Diabetes is a disease where the body cannot control the concentration of glucose in the blood.
   i Why is this dangerous?

SKILLS CRITICAL THINKING

   ii Describe two ways in which a person with diabetes can monitor their blood glucose level.
   iii Explain two ways in which a person with diabetes can help to control their blood glucose level.

SKILLS ANALYSIS

6 a Explain the meaning of the term 'negative feedback'.
   b Describe the negative feedback loop involving the hormone glucagon that takes place in response to a fall in blood glucose.

7 The graph shows the effect of different amounts of alcohol in the blood on the average reaction time of a group of adult men.

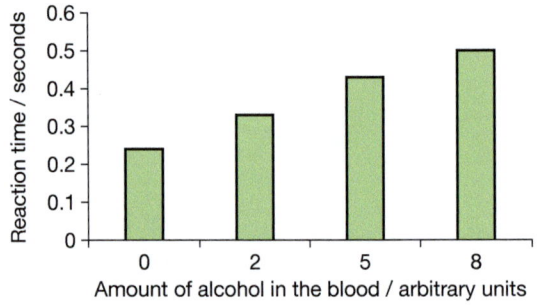

SKILLS ANALYSIS

a Describe the results shown by the graph.

SKILLS PROBLEM SOLVING

b Calculate the change in reaction time when the concentration of alcohol in the blood increases from 0 to 8 units.

SKILLS REASONING

c Use these results to explain why it is dangerous to drink alcohol and drive.

SKILLS CRITICAL THINKING

d What are the effects on the liver of regularly drinking large amounts of alcohol?

10 HOMEOSTASIS AND EXCRETION

Homeostasis maintains a balance of substances in the body and excretion removes the waste products of metabolism. The kidneys are involved in both these processes. This chapter is mainly concerned with the structure and function of the kidneys. You will also learn about the skin and how it is involved in regulating body temperature.

LEARNING OBJECTIVES

- Understand the concept of homeostasis
- Know the definition of excretion – the removal of metabolic waste, including urea, carbon dioxide and water
- Describe the functions of the liver in bile production, regulation of blood sugar, and urea formation; and detoxification, including the breakdown of alcohol
- Know the structure and functions of the renal system
- Explain why the composition of urine may vary
- Describe the role of the hypothalamus and pituitary gland in osmoregulation
- Explain the role of ADH in regulating the water content of the blood
- Describe the advantages and disadvantages of renal dialysis and kidney transplants
- Describe how to investigate diffusion through a partially permeable membrane such as Visking tubing
- Know the structure and functions of the skin and explain the role of sweat glands, vasoconstriction, vasodilation and shivering in temperature regulation
- Understand the role of negative feedback in temperature control

Inside our bodies, conditions are kept relatively constant. This is called **homeostasis**. The kidneys are organs which have a major role to play, both in homeostasis and in the removal of waste products (**excretion**). They filter the blood, removing substances and controlling the concentration of water and dissolved substances (solutes) in the blood and other body fluids.

HOMEOSTASIS

If you drink a litre of water and wait for half an hour, your body will soon respond to this change by producing about the same volume of urine. In other words, it will automatically balance your water input and water loss. Drinking is the main way that our bodies gain water, but there are other sources (Figure 10.1). Some water is present in the food we eat, and a small amount is formed by cell respiration. The body also loses water, mostly in urine; smaller volumes of water are lost in sweat, faeces and exhaled air. Every day, we gain and lose about the same volume of water, so the total water content of our bodies stays more or less the same. This is an example of homeostasis. The word 'homeostasis' means 'steady state', and refers to keeping conditions inside the body relatively constant. Homeostasis is very dependent on negative feedback mechanisms. You read about the role of negative feedback in Chapter 9, and will see more examples of it in this chapter.

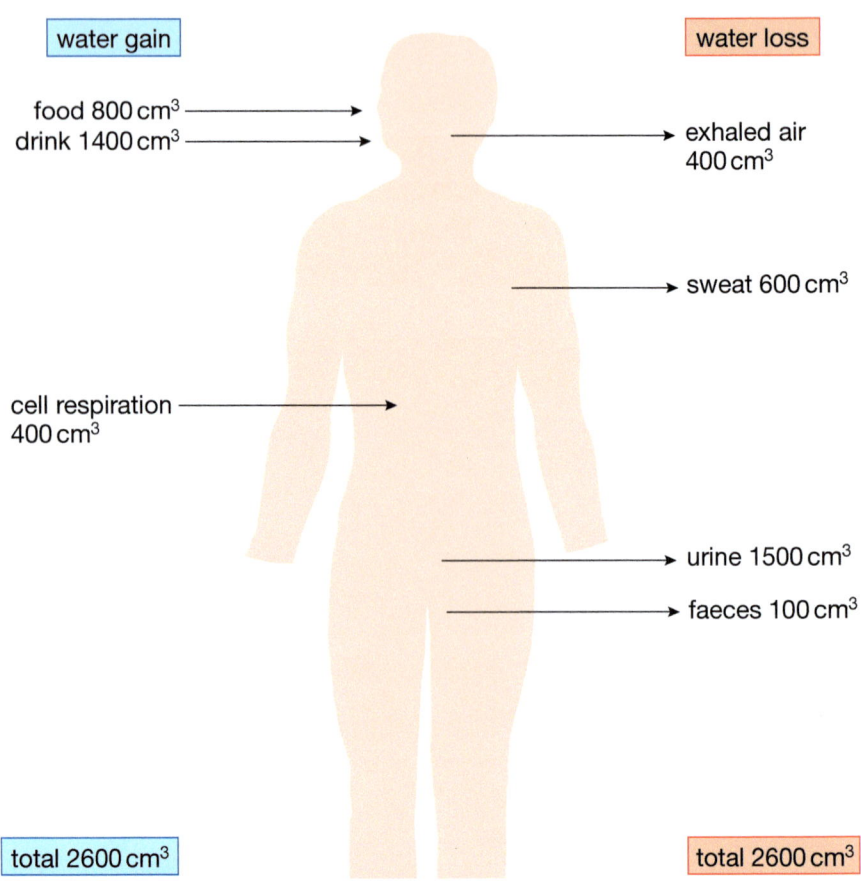

▲ Figure 10.1 The daily water balance of an adult.

KEY TERM

The meaning of 'homeostasis' is 'keeping the conditions in the internal environment of the body relatively constant'. One aspect of homeostasis is the maintenance of the water and salt content of the internal environment. This is called **osmoregulation**.

The inside of the body is known as the **internal environment**. You have probably heard of the 'environment', which means the 'surroundings' of an organism. The internal environment is the surroundings of the cells inside the body. It particularly means the blood, together with another liquid called tissue fluid (see Chapter 6, page 99).

Tissue fluid is a watery solution of salts, glucose and other solutes. It surrounds all the cells of the body, forming a pathway for the transfer of nutrients between the blood and the cells. Tissue fluid is formed by leakage from blood capillaries. It is similar in composition to blood plasma, but does not contain the plasma proteins.

It is not just water and salts that are kept constant in the body. Many other components of the internal environment are maintained. For example, the level of carbon dioxide in the blood is regulated, along with the blood pH, the concentration of dissolved glucose (see Chapter 9) and the body temperature.

Homeostasis is important because cells need to be bathed in a fluid that allows them to function correctly. For example, if the tissue fluid contains too many solutes, the cells will lose water by osmosis and become dehydrated. If the tissue fluid is too dilute, the cells will swell up with water. Both conditions will prevent the cells working efficiently and might cause permanent damage. If the pH of the tissue fluid is not correct, it will affect the activity of the cell enzymes, as will a body temperature much different from 37 °C. It is also important that excretory products are removed. Substances such as urea must be prevented from building up in the blood and tissue fluid, where they would be toxic to cells.

EXCRETION

> **KEY POINT**
>
> Excretion can be defined as 'the process by which waste products of metabolism are removed from the body'. In humans, the main nitrogenous excretory substance is urea. Another excretory product is carbon dioxide, which is produced by cell respiration and excreted by the lungs (Chapter 5). Water is also a product of respiration, and lost from the body in urine, sweat and breath. Skin is an excretory organ, secreting small amounts of urea and salts along with the water.

An adult human produces about 1.5 dm³ of urine every day, although this volume depends very much on the amount of water drunk and the volume of water lost in other forms, such as sweat. Every litre of urine contains about 40 g of waste products and salts (Table 10.1).

▼ Table 10.1 Some of the main solutes in urine.

Substance	Amount / g per dm³
urea	23.3
ammonia	0.4
other nitrogenous waste	1.6
sodium chloride (salt)	10.0
potassium	1.3
phosphate	2.3

> **KEY TERM**
>
> 'Salts' in urine or in the blood are present as ions. For example, the sodium chloride in Table 10.1 will be in solution as sodium ions (Na^+) and chloride ions (Cl^-). Urine contains many other ions, such as potassium (K^+), phosphate (HPO_4^{2-}) and ammonium (NH_4^+), and removes excess ions from the blood.

Notice the words *nitrogenous waste*. Urea and ammonia are two examples of nitrogenous waste. This means that they contain the element nitrogen. All animals have to excrete a nitrogenous waste product; in humans, this is urea.

> **KEY POINT**
>
> The liver is a 'chemical factory', with over two hundred different metabolic functions. These include:
> - breaking down proteins to form urea
> - making bile for the digestion of lipids (Chapter 4)
> - regulation of blood glucose level (Chapter 9)
> - converting alcohol to harmless products (*detoxification*, see Chapter 9)

The reason for this is quite complicated. Carbohydrates and fats only contain the elements carbon, hydrogen and oxygen. However, proteins also contain nitrogen. If the body has too much carbohydrate or fat, these substances can be stored, for example, as glycogen in the liver or as fat under the skin and around other organs. Excess proteins, or their building blocks (amino acids) *cannot* be stored. The amino acids are first broken down in the liver. They are converted into carbohydrate (which is stored as glycogen) and urea, the main nitrogen-containing waste product. The urea passes into the blood, to be filtered out by the kidneys during the formation of urine.

Note that urea is made by chemical reactions in the cells of the body (the body's *metabolism*). 'Excretion' means removing waste of this kind. When the body gets rid of solid waste from the digestive system (faeces), this is *not* excretion: faeces contain few products of metabolism, just the remains of undigested food, along with bacteria and dead cells.

So, the kidney is really carrying out two functions. It is a *homeostatic* organ, controlling the water and salt (ion) concentration in the body and it is an *excretory* organ, removing nitrogenous waste.

THE URINARY SYSTEM

The human urinary system is shown in Figure 10.2.

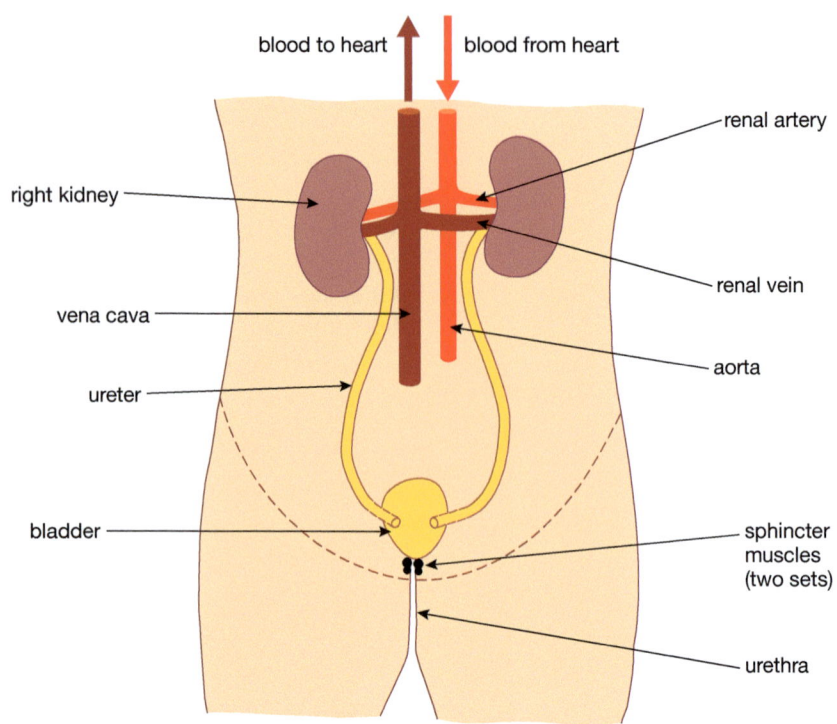

▲ Figure 10.2 The human urinary system.

Each kidney is supplied with blood through a short **renal artery**. This leads straight from the body's main artery, the aorta, so the blood entering the kidney is at a high pressure. Inside each kidney, the blood is filtered. The 'cleaned' blood passes out through each **renal vein** to the main vein, the vena cava. The urine passes out of the kidneys through two tubes, the **ureters**, and is stored in a muscular bag called the **bladder**.

The bladder has a tube leading out of the body, called the **urethra**. The wall of the urethra contains two ring-shaped muscles, called sphincter muscles. They can contract to close the urethra and hold back the urine. The lower sphincter muscle is voluntary (under conscious control), while the upper sphincter muscle is involuntary – it automatically relaxes when the bladder is full.

DID YOU KNOW?
A baby cannot control its voluntary sphincter. When the bladder is full, the baby's involuntary sphincter relaxes and releases the urine. A toddler learns to control this muscle and hold back the urine.

THE KIDNEYS

If you cut a kidney lengthwise, you would be able to see the structures shown in Figure 10.3.

There is not much that you can see without the help of a microscope. The darker outer region is called the **cortex**. This contains many tiny blood vessels that branch from the renal artery. It also contains microscopic tubes that are not blood vessels. They are the filtering units, called kidney tubules or **nephrons** (from the Greek word *nephros*, meaning kidney). The tubules then run down through the middle layer of the kidney, called the **medulla**. The medulla has bulges called pyramids pointing inwards towards the concave side of the kidney. The tubules in the medulla eventually join up and lead to the tips of these pyramids, where they empty urine into a space called the **pelvis**. The pelvis connects with the ureter, carrying the urine to the bladder.

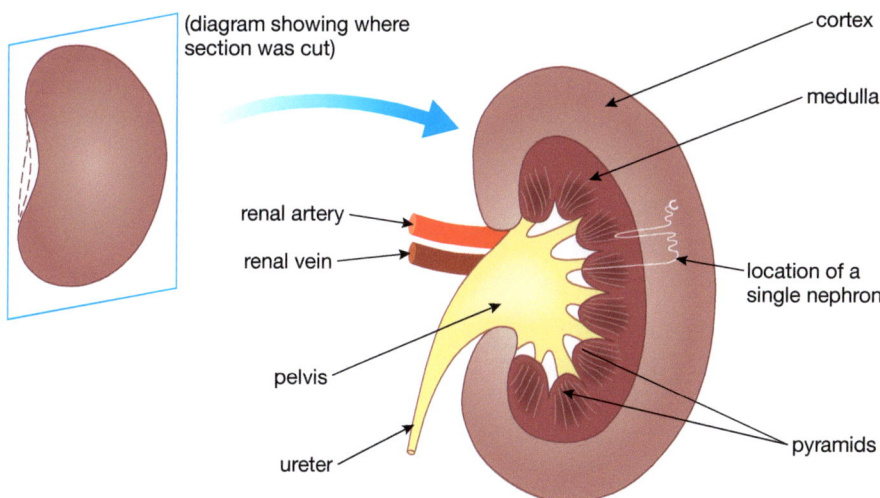

▲ Figure 10.3 Section through a kidney cut along the plane shown.

THE STRUCTURE OF THE NEPHRON

By careful dissection, biologists have been able to find out the structure of a single tubule and its blood supply (Figure 10.4). There are about a million of these tubules in each kidney.

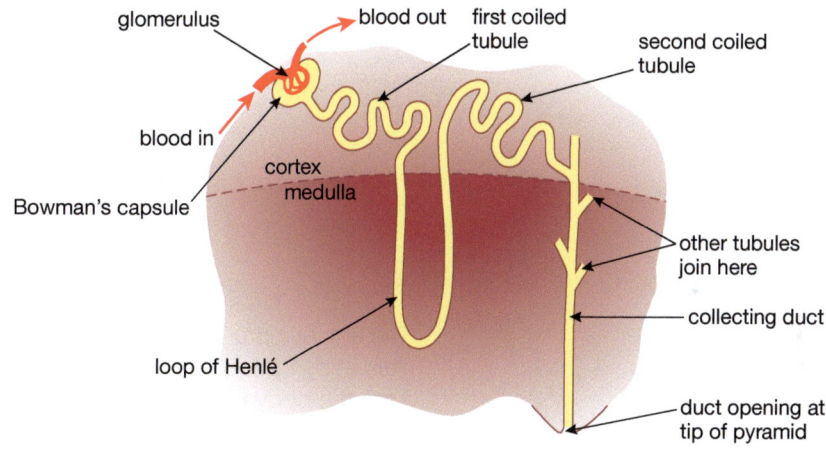

▲ Figure 10.4 A single nephron, and its position in the kidney. Each kidney contains about a million of these filtering units.

KEY TERM

In some biology books, the first and second coiled tubules are called the proximal and distal convoluted tubules. 'Proximal' means near the start of the tubule, and 'distal' means near the end.

ULTRAFILTRATION IN THE BOWMAN'S CAPSULE

At the start of the nephron is a hollow cup of cells called the **Bowman's capsule**. This surrounds a ball of blood capillaries called a **glomerulus** (plural glomeruli). It is here that the blood is filtered. Blood enters the kidney through the renal artery, which divides into smaller and smaller arteries. The smallest arteries (arterioles) supply the capillaries of the glomerulus (Figure 10.5).

A blood vessel with a smaller diameter carries blood away from the glomerulus, leading to capillary networks which surround the other parts of the nephron. Because of the resistance to flow caused by the glomerulus, the pressure of the blood in the arteriole leading to the glomerulus is very high. This pressure forces fluid from the blood through the walls of the capillaries and the Bowman's capsule, into the space in the middle of the capsule. Blood in the glomerulus is separated from the space in the capsule by two layers of cells, the capillary wall and the wall of the capsule. Between the two cell layers is a third layer, called the **basement membrane**, which is not made of cells.

HOMEOSTASIS AND EXCRETION

> **DID YOU KNOW?**
> The cells of the glomerulus capillaries do not fit together very tightly. There are spaces between them, making the capillary walls much more permeable than others in the body. The cells of the Bowman's capsule also have gaps between them, so they only act as a coarse filter. It is the basement membrane which acts as the fine molecular filter.

These layers act like a filter, allowing water, ions and small molecules to pass through, but holding back blood cells and large molecules such as proteins. The fluid that enters the capsule space is called the **glomerular filtrate**. This process, where the filter separates different sized molecules under pressure, is called **ultrafiltration**.

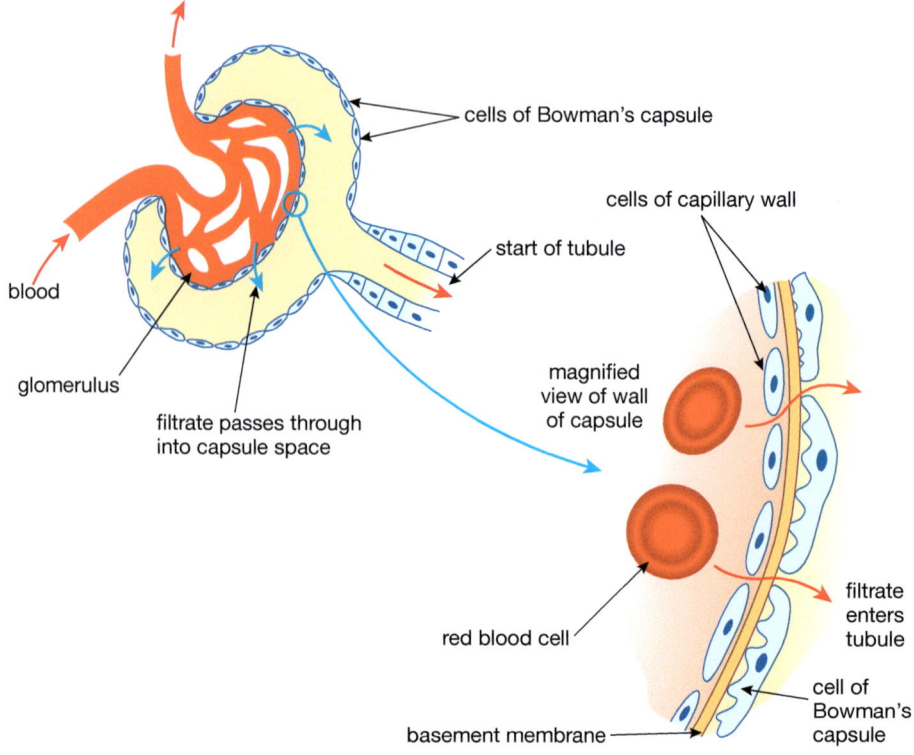

▲ Figure 10.5 A Bowman's capsule and glomerulus.

CHANGES TO THE FILTRATE IN THE REST OF THE NEPHRON

The kidneys produce about 125 cm³ (0.125 dm³) of glomerular filtrate per minute. This adds up to 180 dm³ per day. Remember though, only 1.5 dm³ of urine is lost from the body each day; this is less than 1% of the volume filtered through the capsules. The other 99% of the glomerular filtrate is *reabsorbed* into the blood.

We know this because scientists have analysed samples of fluid from the space in the middle of the nephron. The diameter of the space is only 20 μm (0.02 mm) but it is possible to pierce the tubule with a microscopic glass pipette and extract the fluid for analysis. Figure 10.6 shows the structure of the nephron and the surrounding blood vessels in more detail.

There are two coiled regions of the tubule in the cortex, separated by a U-shaped loop that runs down into the medulla of the kidney. This is called the **loop of Henlé**. After the second coiled tubule, several nephrons join up to form a **collecting duct**, where the final urine passes out into the pelvis.

Sample 1 shows the results of analysing the blood before it enters the glomerulus. Samples 2–4 show the results of analysing the fluid at three points inside the tubule. The flow rate is a measure of how much water is in the tubule. If the flow rate falls from 100% to 50%, this is because 50% of the water in the tubule has gone back into the blood. To make the explanation easier, the concentrations of dissolved protein, glucose, urea and sodium are shown by different letters (a to d). These letters show the relative concentration of each substance at different points along the tubule. For example, urea at a concentration '3c' is three times more concentrated than when it is 'c'.

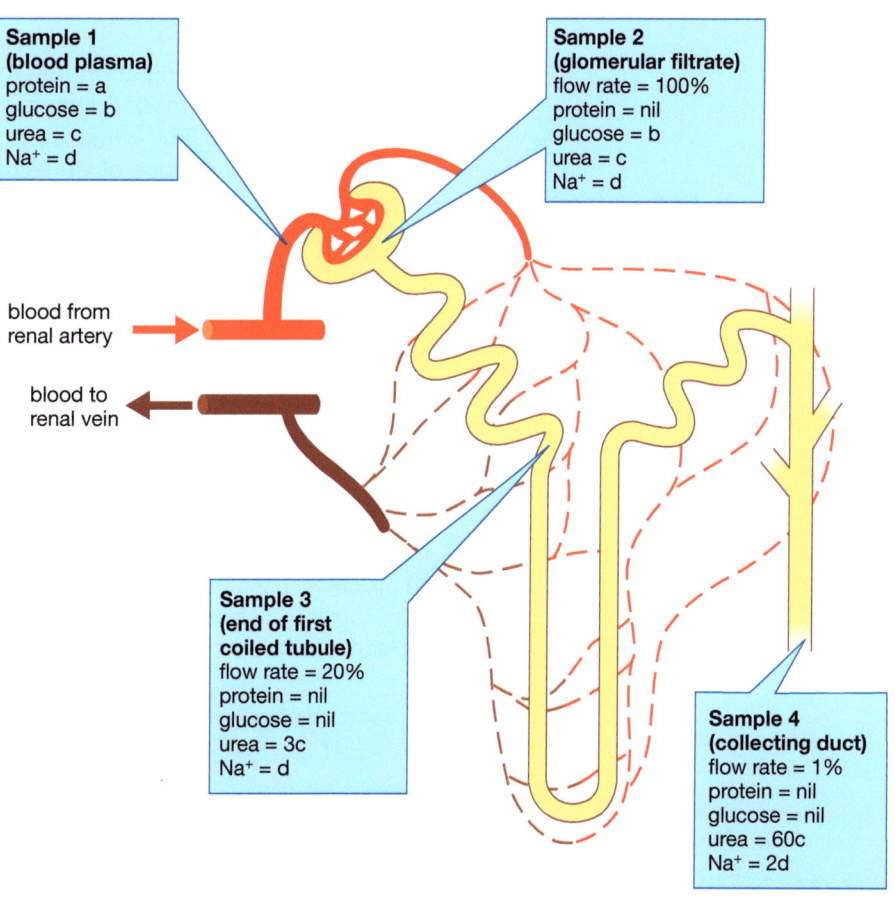

▲ Figure 10.6 A nephron and its blood supply. Samples 1–4 show what is happening to the fluid as it travels along the nephron.

KEY TERM

The kidney tubule reabsorbs different amounts of various substances. This is called **selective reabsorption**. Sodium ions and glucose are both reabsorbed into the blood against a concentration gradient, by active transport (see Chapter 2). Water follows by osmosis.

In the blood (sample 1), the plasma contains many dissolved solutes, including protein, glucose, urea and salts (only sodium ions, Na^+, are shown here). As we saw above, protein molecules are too big to pass through into the tubule, so the protein concentration in sample 2 is zero. The other substances are at the same concentration as in the blood.

Now look at sample 3, taken at the end of the first coiled part of the tubule. The flow rate that was 100% is now 20%. This must mean that 80% of the water in the tubule has been reabsorbed back into the blood. If no solutes have been reabsorbed along with the water, their concentrations should be five times what they were in sample 2. The concentration of sodium has not changed; therefore, 80% of this substance must have been reabsorbed (and some of the urea too). The glucose concentration is now zero, so all of the glucose has been taken back into the blood in the first coiled tubule. This is necessary because glucose is a useful substance that is needed by the body.

Finally, look at sample 4. By the time the fluid passes through the collecting duct, its flow rate is only 1%. This is because 99% of the water has been reabsorbed. Protein and glucose are still zero, but most of the urea is still in the fluid. The level of sodium is only 2d: not all of it has been reabsorbed, but it is still twice as concentrated as in the blood.

This description has only looked at a few of the more important substances. Other solutes are concentrated in the urine by different amounts. Some, like ammonium ions, are secreted into the fluid as it passes along the tubule. The concentration of ammonium ions in the urine is about 150 times what it is in the blood.

THE LOOP OF HENLÉ

You might be wondering about the role of the loop of Henlé. The full answer to this is rather complicated. A simple answer is that it is involved in concentrating the fluid in the tubule by causing more water to be reabsorbed into the blood. Mammals with long loops of Henlé can make more concentrated urine than ones with short loops. Desert animals have many long loops of Henlé, so they can produce very concentrated urine, conserving water in their bodies. Animals which have easy access to water, such as otters or beavers, have short loops of Henlé. Humans have a mixture of long and short loops.

> **KEY POINT**
>
> Here is a summary of what happens in the kidney nephron.
>
> Part of the plasma leaves the blood in the Bowman's capsule and enters the nephron. The filtrate consists of water and small molecules. As the fluid passes along the nephron, all the glucose – along with most of the sodium and chloride ions – is absorbed back into the blood by active transport in the first coiled part of the tubule. Water is reabsorbed by osmosis. In the rest of the tubule, more water and ions are reabsorbed, and some solutes, such as ammonium ions, are secreted into the tubule. The final urine contains urea at a much higher concentration than in the blood. It also contains controlled quantities of water and ions.

CONTROL OF THE BODY'S WATER CONTENT

Not only can the kidney produce urine that is more concentrated than the blood, it can also *control* the concentration of the urine, and so *regulate* the water content of the blood (osmoregulation). This chapter began by asking you to think about what would happen if you drank a litre of water. The kidneys respond to this 'upset' to the body's water balance by making a larger volume of more dilute urine. Conversely, if the blood becomes too concentrated, the kidneys produce a smaller volume of urine. These changes are controlled by a hormone produced by the pituitary gland, at the base of the brain. The hormone is called antidiuretic hormone, or ADH.

'Diuresis' means the flow of urine from the body, so 'antidiuresis' means producing less urine. ADH starts to work when your body loses too much water. This would happen, for example, if you were sweating heavily and not drinking any water.

The loss of water means that the concentration of the blood starts to increase. This is detected by receptor cells in a region of the brain called the hypothalamus, located above the pituitary gland (see Figure 9.8, page 154). These cells are sensitive to the solute concentration of the blood, and cause the pituitary gland to release more ADH. The ADH travels in the bloodstream to the kidney. At the kidney tubules, it causes the collecting ducts to become more permeable to water, so more water is reabsorbed into the blood. This makes the urine more concentrated, so the body loses less water and the blood becomes more dilute. When the water content of the blood returns to normal, this acts as a signal to 'switch off' the release of ADH. The kidney tubules then reabsorb less water.

> **DID YOU KNOW?**
>
> As well as causing the pituitary gland to release ADH, the receptor cells in the hypothalamus also stimulate a 'thirst centre' in the brain. This makes the person feel thirsty, so they will drink water, diluting the blood.

Similarly, if someone drinks a large volume of water, the blood will become too dilute. This leads to lower levels of ADH secretion, so the kidney tubules become less permeable to water and more water passes out of the body in the urine. In this way, through the action of ADH, the level of water in the internal environment is kept constant.

> **KEY POINT**
>
> The action of ADH is another example of negative feedback (see Chapter 9). A change in conditions in the body is detected, and starts a process which works to return conditions to normal. When the conditions are returned to normal, the corrective process is switched off (Figure 10.7).
>
> In the situation described in the text, where the body loses too much water, the blood becomes too concentrated. This switches on ADH release, which acts at the kidneys to correct the problem. The word 'negative' means that the process works to eliminate the change. When the blood returns to normal, ADH release is switched off. The feedback pathway forms a 'closed loop'. Many conditions in the body are regulated by negative feedback loops like this.

▶ Figure 10.7 In homeostasis, the extent of a correction is monitored by negative feedback.

KIDNEY FAILURE – WHAT HAPPENS WHEN THINGS GO WRONG

A kidney can stop working as a result of disease or an accident. We can live perfectly happily with only one kidney, but if both stop working we will die within a week or so, because poisonous waste will build up in the blood. If a person's kidneys fail, there are two ways in which they can be kept alive:

- a **kidney transplant** can be carried out
- their blood can be filtered through an artificial kidney machine, in a process called **renal (kidney) dialysis**.

KIDNEY TRANSPLANTS

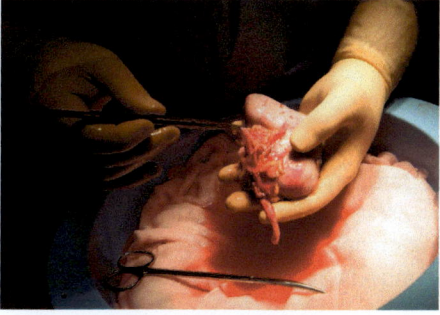

▲ Figure 10.8 A kidney being made ready for a transplant operation.

A kidney transplant is an operation in which a patient receives another person's kidney. The kidney is often donated by a close relative, or it may come from a person who has had a fatal accident (Figure 10.8).

A kidney transplant is a straightforward operation, and kidneys were one of the first human organs to be transplanted successfully. The main problem comes soon after the operation, when the patient's immune system may reject the new kidney. Organ rejection is described in the section on heart transplants (Chapter 6, page 108). There are a number of ways in which doctors can try to stop the kidney being rejected:

- Tissue typing can be used to match the donor kidney with the tissues of the patient. A kidney from a close relative is often used – both donor and recipient can live perfectly well with only one functioning kidney.

- The patient can be given immunosuppressant drugs until the kidney is 'accepted' by the body.
- The bone marrow of the patient can be treated with radiation. This reduces the production of white blood cells, which are responsible for the *immune response* (see Chapter 6).

Kidney transplants have a very high success rate, with 80% of transplant patients surviving for longer than five years after the operation. They are the most simple and effective way of treating a patient whose kidneys have failed permanently. Unfortunately, there are not enough donor organs available, and many people are on waiting lists for a kidney transplant. While on the list, they can be treated using renal dialysis.

KIDNEY DIALYSIS MACHINES

The artificial kidney, or renal dialysis machine, filters the patient's blood, removing urea and other waste, as well as excess water and salts. The filter is a special *dialysis membrane* called Visking tubing, which looks rather like cellophane. You read about Visking tubing in Chapter 2, where it was used as a partially permeable membrane to demonstrate osmosis. Visking tubing is a thin membrane with millions of tiny holes in it. The holes will let small molecules like water, ions and urea pass through, but larger molecules, such as proteins or blood cells, will not be able to pass through (see Activity 3 below).

Blood from the patient flows on one side of the membrane, and a watery liquid called *dialysis solution* flows past the other side, in the opposite direction (Figure 10.9). The dialysis solution is a solution of salts and glucose in exactly the concentrations that the body needs.

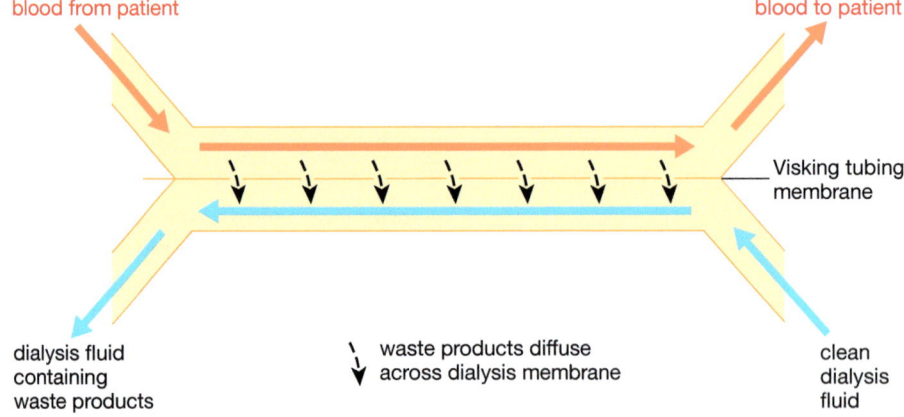

▲ Figure 10.9 The principle of renal dialysis. The dialysis membrane filters the blood, removing toxic waste.

As the blood flows past the membrane, urea and unwanted water and salts diffuse through the holes in the membrane into the dialysis fluid. Cells and large molecules such as proteins are kept back in the blood. The dialysis fluid is replaced with fresh solution all the time. After several hours, the patient's blood has been 'cleaned' of toxic waste and the correct balance of water and salts has been established.

The surface area of the dialysis membrane separating the blood and dialysis fluid must be large to filter the blood effectively. To achieve this, the membrane can be arranged in different ways. In some dialysis machines, it is in the form of many long narrow tubes; in other machines, it is arranged as a stack of flat sheets.

In order to carry out dialysis, it is easier to take blood from a vein than an artery, because veins are closer to the skin and have a wider diameter than arteries (see Chapter 6). However, the blood pressure in veins is too low, so an operation is first carried out to join an artery to a vein. This raises the blood pressure. A tube is then permanently connected to the vein, so that the patient can be linked to the machine without having to use a needle each time. The purified blood returns to the patient through a second tube joined to the vein, and the 'used' dialysis fluid is discarded.

The kidney dialysis machine is a complex and expensive piece of equipment (Figure 10.10). It has pumps to keep the blood and dialysis fluid flowing, traps to prevent air bubbles getting into the blood, and the oxygenation and temperature of the blood are controlled. Although dialysis will keep a person alive if their kidneys have failed, it is a time-consuming and unpleasant process. The patient has to have their blood 'cleaned' for many hours, two or three times a week. A transplant, if one becomes available, is a much better option.

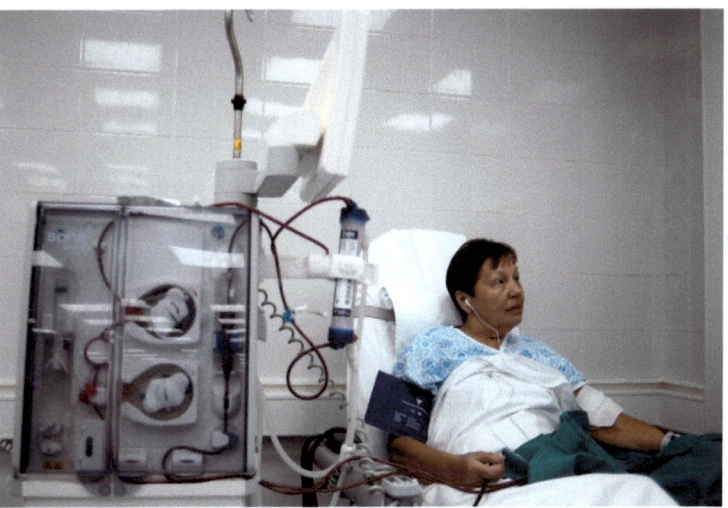

▲ Figure 10.10 A patient connected to a kidney dialysis machine.

Safety Note: Wash hands after filling and tying off the Visking 'sausage'. Wear eye protection when testing and use a water bath to heat the test tube during the glucose test.

ACTIVITY 3

▼ PRACTICAL: INVESTIGATING DIFFUSION THROUGH A VISKING TUBING MEMBRANE

Visking tubing is a partially permeable membrane with very small holes in it. The holes are large enough to allow water and other small molecules through, but too small to allow large molecules through.

A Visking tubing 'sausage' is filled with a solution of starch, glucose and protein. The tubing is rinsed with distilled water and placed in a test tube containing water (Figure 10.11). After 30 minutes, the water in the test tube is tested for each of the three substances:

- **Starch test:** Iodine solution is added. If the test is positive, the mixture turns blue-black.
- **Glucose test:** A sample is heated with Benedict's solution. A positive result is shown by a brick-red precipitate.
- **Protein test:** A sample is taken and an equal volume of 5% potassium hydroxide is added to it, followed by two drops of 1% copper sulfate solution. A purple colour indicates the presence of protein.

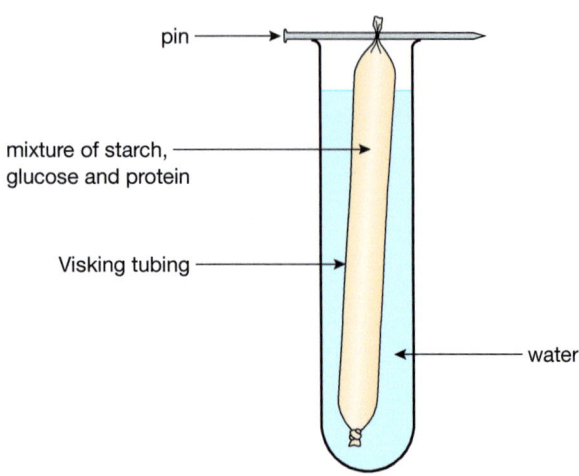

▲ Figure 10.11 Investigating whether Visking tubing is permeable to starch, glucose or protein.

A student carried out this investigation and found that only glucose was present in the water. Can you explain this result?

(Note: In exam questions, you may also see this apparatus used as a model of the gut to demonstrate which molecules can pass through the wall of the ileum into the blood.)

CONTROL OF BODY TEMPERATURE

You may have heard mammals described as 'warm blooded'. A better word for this is that they are **homeotherms**. This means that they keep their body temperature constant, despite changes in the temperature of their surroundings. For example, the body temperature of humans is kept steady at about 37 °C, give or take a few tenths of a degree. This is another example of homeostasis. Apart from mammals and birds, all other animals are 'cold blooded'. For example, if a lizard is kept in an aquarium at 20 °C, its body temperature will be 20 °C too. If the temperature of the aquarium is raised to 25 °C, the lizard's body temperature will rise to 25 °C as well. We can show the difference between homeotherms and other animals as a graph (Figure 10.12).

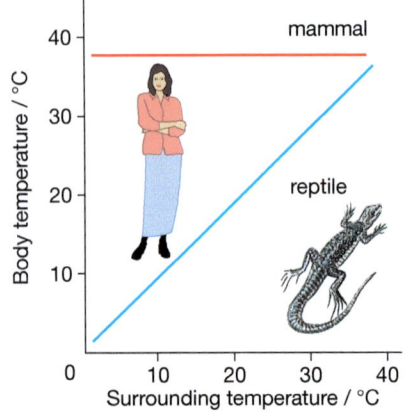

▲ Figure 10.12 The temperature of a homeotherm, such as a mammal, is kept constant at different external temperatures, whereas the lizard's body temperature changes.

In the wild, lizards actually keep their temperature more constant than in Figure 10.12 by adapting their behaviour. For example, in the morning they may lie in the sun to warm their bodies; at midday, if the sun is too hot, they may retreat to holes in the ground to cool down.

The real difference between homeotherms and all other animals is that homeotherms can keep their temperature constant by using physiological changes to generate or lose heat. For this reason, mammals and birds are also called endotherms, meaning 'heat from inside'.

As endotherms, we use heat from the chemical reactions in our cells to warm our bodies. The body then controls its heat loss by regulating processes such as sweating and blood flow through the skin. We use behavioural ways to control our temperature too. For example, we put on extra clothes in winter.

What is the advantage of a human maintaining a body temperature of 37 °C? It means that all the chemical reactions taking place in the cells of the body can go on at a steady, predictable rate. A human's metabolism does not slow down in cold environments. If you watch goldfish in a garden pond, you will

> **DID YOU KNOW?**
> **Physiology** is a branch of biology that deals with how the body works, for example, how muscles contract, or how nerves send impulses. In this chapter, you have learnt about the physiology of the kidney.

notice that in summer, when the pond water is warm, they are very active, swimming about quickly. In winter, when the temperature drops, the fish slow down and become quite inactive. This would happen to humans too, if our body temperature varied.

It is also important that the body does not become too hot. The cells' enzymes work best at 37 °C. At higher temperatures, enzymes – like all proteins – are destroyed by denaturing (see Chapter 3).

MONITORING BODY TEMPERATURE

Our core body temperature is monitored by a part of the brain called the **thermoregulatory centre**. This is located in the hypothalamus of the brain (see Figure 9.8, page 154). It acts as the body's *thermostat*.

If we go into a warm or cold environment, the first thing that happens is that temperature receptors in the skin send electrical impulses to the hypothalamus. This stimulates the brain to change our behaviour. We start to feel hot or cold, and usually do something about it, such as finding shade, having a cold drink, or putting on more clothes.

If changes to our behaviour are not enough to keep our body temperature constant, the thermoregulatory centre in the hypothalamus detects a change in the temperature of the blood flowing through it. It then sends signals via nerves to other organs of the body, which regulate the temperature by physiological means.

Control of body temperature is another example of a negative feedback system. The thermoregulatory centre detects a change in body temperature and brings about mechanisms to correct this change and bring the temperature back to normal.

> **DID YOU KNOW?**
> A thermostat is a switch that is turned on or off by a change in temperature. It is used in electrical appliances to keep their temperature steady. For example, a thermostat in an iron can be set to 'hot' or 'cool' to keep the temperature of the iron set for ironing different materials.

THE SKIN AND TEMPERATURE CONTROL

The human skin is the outer surface of the body, and has a number of functions including:

- forming a tough outer layer able to resist mechanical damage
- acting as a barrier to the entry of disease-causing microorganisms
- forming an impermeable surface, preventing loss of water
- acting as a sense organ for touch and temperature changes
- controlling the loss of heat through the body surface.

Figure 10.13 shows the structure of human skin. It is made up of three layers, the epidermis, dermis and hypodermis.

The outer **epidermis** consists of dead cells that stop water loss and protect the body against invasion by microorganisms such as bacteria. The **hypodermis** contains fatty tissue, which insulates the body against heat loss and is a store of energy. The middle layer, the **dermis**, contains many sensory receptors. It is also the location of sweat glands and many small blood vessels, as well as hair follicles. These last three structures are involved in temperature control.

Imagine that the hypothalamus detects a rise in the central (core) body temperature. Immediately, it sends nerve impulses to the skin. These bring about changes to correct the rise in temperature.

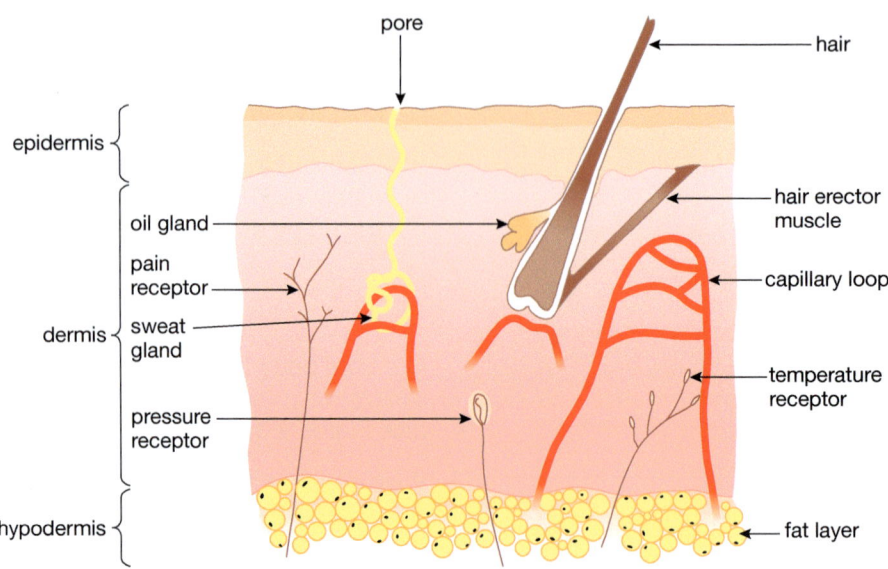

▲ Figure 10.13 A section through human skin.

First of all, the **sweat glands** produce greater amounts of sweat. This liquid is secreted onto the surface of the skin. When a liquid evaporates, it turns into a gas. This change needs energy, called the latent heat of vaporisation. When sweat evaporates, the energy is supplied by the body's heat, which cools the body down. It is not that the sweat is cool – it is secreted at body temperature. It only has a cooling action when it evaporates. In very humid atmospheres (e.g. a tropical rainforest) the sweat stays on the skin and does not evaporate. It then has very little cooling effect.

Secondly, hairs on the surface of the skin lie flat against the skin's surface. This happens because of the relaxation of tiny muscles called **hair erector muscles** attached to the base of each hair. In cold conditions, these muscles contract and the hairs are pulled upright. The hairs trap a layer of air next to the skin and, since air is a poor conductor of heat, this acts as insulation. In warm conditions, the thinner layer of trapped air means that more heat will be lost. This is not very effective in humans, because the hairs over most of our body do not grow very large. It is very effective in hairy mammals, such as cats or dogs.

Lastly, there are tiny blood vessels called capillary loops in the dermis. Blood flows through these loops, radiating heat to the outside and cooling the body down. If the body is too hot, arterioles (small arteries) leading to the capillary loops dilate (widen). This increases the blood flow to the skin's surface (Figure 10.14) and is called **vasodilation**.

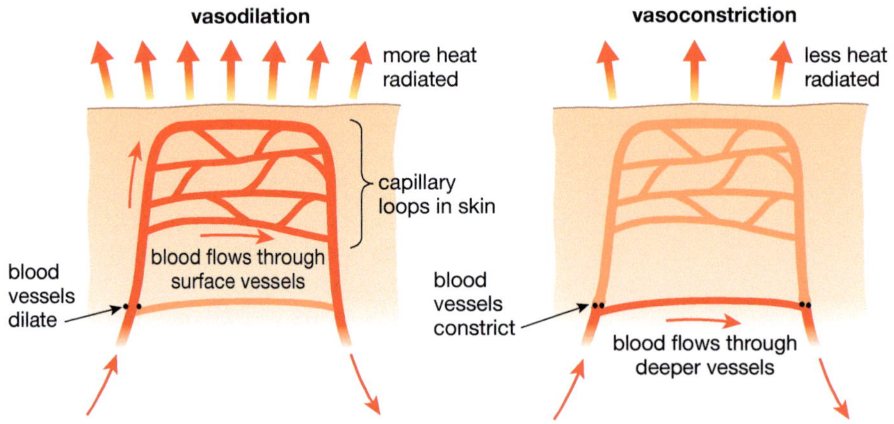

> **HINT**
>
> Students often describe vasodilation incorrectly. They talk about the blood vessels 'moving nearer to the surface of the skin'. They do not move at all, it is just that more blood flows through the surface vessels.

▲ Figure 10.14 Blood flow through the surface of the skin is controlled by vasodilation and vasoconstriction.

In cold conditions, the opposite happens. The arterioles leading to the surface capillary loops constrict (become narrower) and blood flow to the surface of the skin is reduced, so less heat is lost. This is called **vasoconstriction**. Vasoconstriction and vasodilation are brought about by tiny rings of muscle in the walls of the arterioles, called sphincter muscles. These are like the sphincters you read about earlier in this chapter, at the outlet of the bladder.

The body can also control heat loss and heat gain in other ways. In cold conditions, the body's metabolism speeds up, generating more heat. The liver, a large organ, can produce a lot of metabolic heat in this way. The hormone adrenaline stimulates the increase in metabolism (see Chapter 9). **Shivering** also takes place; this is where the muscles contract and relax rapidly, and can generate a large amount of heat.

Sweating, vasodilation and vasoconstriction, hair erection, shivering and changes to the metabolism, along with behavioural actions, work together to keep the body temperature within a few tenths of a degree of the 'normal' 37 °C. If the difference is any bigger than this, it shows that something is wrong. For instance, a temperature of 39 °C might be due to an illness.

CHAPTER QUESTIONS

SKILLS CRITICAL THINKING

1 Explain the meaning of the following terms:
   a homeostasis
   b excretion
   c ultrafiltration
   d selective reabsorption
   e endotherm.

SKILLS ANALYSIS

2 Look at the simple diagram of a nephron (kidney tubule) below.

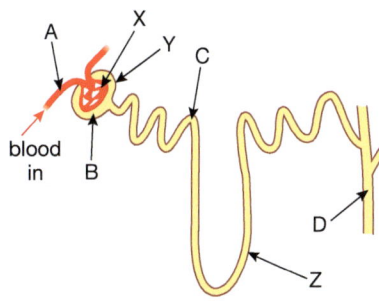

   a What are the names of the parts labelled X, Y and Z?

SKILLS CRITICAL THINKING

   b Four places in the nephron and its blood supply are labelled A, B, C and D. Which of the following substances are found at each of these four places?

   water   urea   protein   glucose   salt

3 The hormone ADH controls the amount of water removed from the blood by the kidneys. Write a short description of the action of ADH in a person who has lost a lot of water by sweating, but has been unable to replace this water by drinking. Explain how this is an example of negative feedback. Your answer should be about half a page in length.

SKILLS CRITICAL THINKING

4 Construct a table with three columns to show the changes that take place when a person is put in a hot or cold environment. The table has been started for you.

Changes taking place	Hot environment	Cold environment
sweating		
blood flow through capillary loops		vasoconstriction decreases blood flow through surface capillaries so less heat is radiated from the skin
hairs in skin		
shivering		
metabolism		

SKILLS REASONING

5 Humans are able to maintain a constant body temperature, which is usually higher than that of their surroundings.

   a Explain the advantage of maintaining a constant high body temperature.

   b The temperature of the blood is monitored constantly by the brain. If the brain detects a drop in blood temperature, the following things happen: the arterioles leading to the skin capillaries constrict, less sweat is formed and shivering begins.
      i Explain how each response helps the body to keep warm.
      ii Explain how the structure of arterioles allows them to constrict.

   c When the weather is hot, we produce less urine.
      i What is the name of the hormone that controls the amount of urine produced by the body?
      ii Explain why the body produces less urine in hot weather.
      iii Explain how the hormone named in (i) works in the kidney to reduce the amount of urine produced.

SKILLS CRITICAL THINKING

SKILLS REASONING

SKILLS CRITICAL THINKING

6 A student carried out an investigation into the diffusion of substances across a partially permeable membrane. He placed 5 cm^3 of starch suspension with 5 cm^3 of amylase solution in a Visking tubing bag. He then placed the bag in a test tube of water (see diagram). After 60 minutes, he tested the water around the Visking tubing for the presence of starch, reducing sugar and protein.

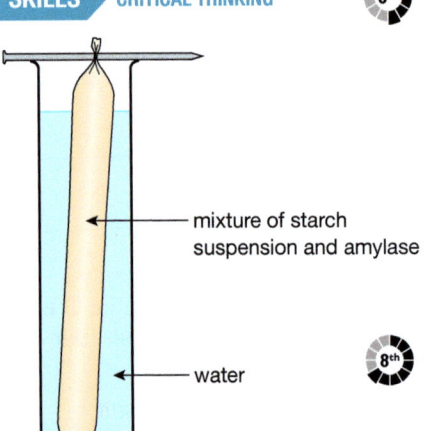

   a What is meant by a 'partially permeable membrane'?
   b Describe the tests for starch, reducing sugar and protein, including the results if the test is positive.
   c Explain why the student tested the water for protein.
   d Only the test for reducing sugar was positive. Explain these results as fully as possible.

UNIT QUESTIONS

SKILLS REASONING

1 Which of the following is *not* part of the appendicular skeleton?

A scapula

B sternum

C femur

D humerus

(Total 1 mark)

2 What do the Haversian canals in compact bone contain?

A cartilage

B calcium salts

C blood vessels and nerves

D stem cells

(Total 1 mark)

3 Which phrase best describes a tendon?

A contractile and inelastic

B tough and elastic

C elastic and contractile

D tough and inelastic

(Total 1 mark)

4 Which statement about synovial joints is true?

A They always allow movement in three planes

B They do not allow movement

C They always allow some movement

D They always allow movement in one plane

(Total 1 mark)

SKILLS ANALYSIS

5 The diagram below shows a section through the eye.
Which row in the table shows the cornea and the choroid?

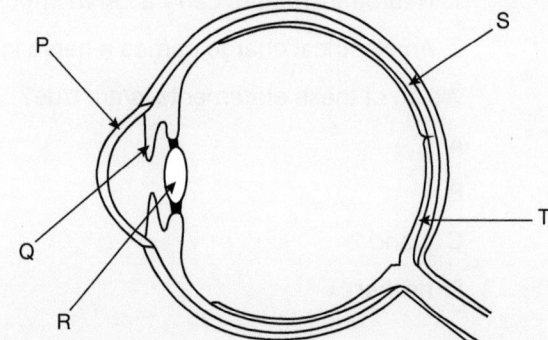

	cornea	choroid
A	R	T
B	P	S
C	R	P
D	Q	T

(Total 1 mark)

SKILLS REASONING

6 A boy sits in the shade under a tree, reading a book. He looks up at an aeroplane in the sunny sky. Which of the following changes will take place in his eyes?

A The pupils dilate and the lens becomes less convex

B The pupils dilate and the lens becomes more convex

C The pupils constrict and the lens becomes less convex

D The pupils constrict and the lens becomes more convex

(Total 1 mark)

SKILLS CRITICAL THINKING

7 What structure forms the boundary between the outer ear and the middle ear?

A eardrum

B cochlea

C oval window

D stapes

(Total 1 mark)

8 Which of the following structures is *not* found in the inner ear?

A utriculus

B semicircular canals

C Eustachian tube

D organ of Corti

(Total 1 mark)

9 Which row in the table correctly describes the three types of neurone?

	Sensory neurone	Relay neurone	Motor neurone
A	Connects neurones within the CNS	Conducts impulses to the effector from the CNS	Conducts impulses from the receptor to the CNS
B	Conducts impulses from the effector to the CNS	Connects neurones within the CNS	Conducts impulses from the CNS to the receptor
C	Connects neurones within the CNS	Conducts impulses from the receptor to the CNS	Conducts impulses to the effector from the CNS
D	Conducts impulses from the receptor to the CNS	Connects neurones within the CNS	Conducts impulses to the effector from the CNS

(Total 1 mark)

10 Below are two statements about how nerve cells work.

1. Neurotransmitters carry a nerve impulse along a neurone.

2. An electrical charge carries a nerve impulse across a synapse.

Which of these statements is/are true?

A 1

B 2

C 1 and 2

D neither

(Total 1 mark)

SKILLS CRITICAL THINKING

11 Read the following statements.

1. Insulin converts glucose to glycogen.
2. Insulin causes blood glucose levels to fall.
3. Glucose is stored as glucagon in the liver.
4. Glycogen can be broken down to release glucose into the blood.

Which of these statements is/are true?

A 2 only

B 2 and 4

C 3 and 4

D 1 and 2

(Total 1 mark)

12 Which of the following will *not* happen when the hormone adrenaline is released?

A An increase in heart rate

C An increase in blood flow to the gut

C Dilation of the pupils

D An increase in breathing rate

(Total 1 mark)

13 Which of the following is *not* an example of excretion?

A Loss of water in urine and sweat

B Loss of carbon dioxide from the lungs

C Removal of urea in urine

D Elimination of faeces from the alimentary canal

(Total 1 mark)

14 The following structures are parts of the human urinary system.

1. ureter
2. kidney
3. urethra
4. bladder

In which order does a molecule of urea pass through these structures?

A $2 \rightarrow 1 \rightarrow 4 \rightarrow 3$

B $2 \rightarrow 3 \rightarrow 4 \rightarrow 1$

C $1 \rightarrow 3 \rightarrow 2 \rightarrow 4$

D $3 \rightarrow 4 \rightarrow 1 \rightarrow 2$

(Total 1 mark)

SKILLS CRITICAL THINKING

15 Below are three statements about the action of antidiuretic hormone (ADH).

1. More ADH is released when the water content of the blood rises.
2. ADH increases the permeability of the collecting duct.
3. When ADH is released, more water is reabsorbed.

Which of these statements are true?

A 1 and 2

B 1 and 3

C 2 and 3

D 1, 2 and 3

(Total 1 mark)

SKILLS REASONING

16 If the human body temperature starts to rise, which of the following happens?

A Vasoconstriction of arterioles in the skin

B Contraction of hair erector muscles

C Decrease in the rate of metabolism

D Decrease in production of sweat by glands in the skin

(Total 1 mark)

SKILLS CRITICAL THINKING

17 Copy and complete the following paragraph, putting the most suitable word or words in the spaces:

Muscles are normally found in _____ pairs, such as the triceps and biceps of the arm. When the _____ contracts, it flexes the arm, whereas when the _____ contracts, it straightens or extends the arm. When muscles are exercised, they need an increased supply of energy. This is supplied by the process of cell _____ , which uses _____ and _____ and makes _____ and water. Exercise also produces heat.

(Total 7 marks)

18 Cataracts can be treated by a simple eye operation, where a surgeon removes the lens of the eye and replaces it with an artificial lens. The surgeon selects a lens that will allow the eye to see distant objects in focus. After the operation, the patient is able to see again, but the eye is unable to carry out accommodation, and the patient will need to wear glasses for close-up work, such as reading.

a What is a cataract? (1)

b What is meant by 'accommodation'? (2)

c Why is accommodation not possible after a cataract operation? (2)

SKILLS REASONING

d What sort of glasses will the patient need to wear to read a book? Explain your answer. (2)

(Total 7 marks)

19 The diagram shows a motor neurone.

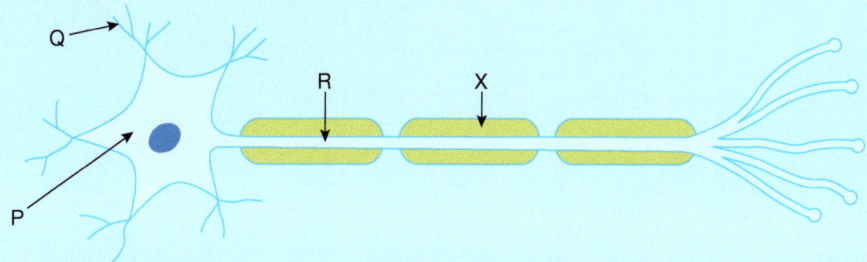

a Name the parts of the neurone labelled P, Q and R. (3)

b This motor neurone is 1.2 metres in length. It takes 0.016 seconds for an impulse to pass along the neurone. Calculate the speed of conduction of the impulse. (1)

c Neurones need energy from ATP to conduct impulses. Name the organelle in the cell that provides most of this ATP. (1)

d Structure R is surrounded by a sheath, labelled X in the diagram.

   i What is the function of this sheath? (1)

   ii Some diseases can cause damage to this sheath. Suggest what would happen to a person's nervous responses if this sheath was damaged. (1)

(Total 7 marks)

20 The bar chart shows the volume of urine collected from a person before and after drinking 1000 cm³ (1 dm³) of distilled water. The person's urine was collected immediately before the water was drunk and then at 30-minute intervals for four hours.

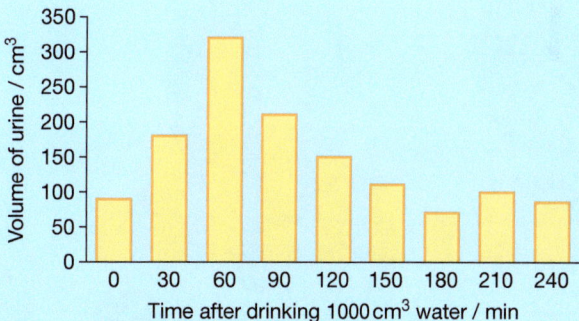

a Describe how the output of urine changed during the course of the experiment. (3)

b Explain the difference in the urine produced at 60 minutes and at 90 minutes. (4)

c The same experiment was repeated with the person sitting in a very hot room. How would you expect the volume of urine collected to differ from the first experiment? Explain your answer. (2)

d Between 90 and 120 minutes, the person produced 150 cm³ of urine. If the rate of filtration at the glomeruli during this time was 125 cm³ per minute, calculate the percentage of filtrate reabsorbed by the kidney tubules. (3)

(Total 12 marks)

| REPRODUCTION 193 | HEREDITY 212 | MICROORGANISMS 228 |

UNIT 4

One of the unique features of living organisms is their ability to reproduce and form new individuals. Chapter 11 looks at the process of sexual reproduction in humans and how it brings about genetic variation in offspring. In Chapter 11, you will learn how reproduction in humans is controlled by the action of hormones during the menstrual cycle. This chapter also covers methods of contraception to prevent pregnancy. Chapter 12 deals with the topic of heredity. This is the science of genetics and how genes are passed on from one generation to the next. Some diseases are caused by faulty genes and some of these non-infectious diseases are discussed in Chapter 12. In Chapter 13, you will learn about microorganisms and some important infectious diseases that are caused by pathogenic microorganisms.

11 REPRODUCTION

Reproduction allows humans to pass on their genes. This happens at fertilisation, through the fusion of special sex cells or gametes – the sperm and egg cells. In this chapter, you will learn about human reproduction, from production of gametes through to the birth of the baby, as well as growth and development to adulthood.

LEARNING OBJECTIVES

- Know the terms diploid and haploid
- Know the stages of meiosis, which result in the production of haploid gametes, and the significance of meiosis in bringing about variation in a species
- Describe the structure of the gametes (egg and sperm) and relate their structure to their function
- Know that the process of fertilisation involves the fusion of a male and female gamete to produce a zygote
- Understand that random fertilisation produces genetic variation in the offspring
- Describe how a zygote divides to form an embryo
- Know the structure and function of the male and female reproductive systems
- Explain the roles during pregnancy of:
  - the placenta, umbilical cord and amniotic fluid
  - the hormone progesterone
- Describe the birth process
- Explain the advantages of breastfeeding
- Know the roles of oestrogen and testosterone in the development of secondary sexual characteristics
- Understand the roles of FSH, LH, oestrogen and progesterone in the menstrual cycle
- Describe natural, barrier and hormonal methods of contraception, intrauterine devices and sterilisation
- Explain the advantages and disadvantages of different methods of contraception
- Describe the process of IVF and how it can improve the chances of pregnancy
- Describe in outline the growth and development of the body to maturity, including human growth curves

Humans reproduce sexually, producing specialised sex cells called gametes. The male gamete is a mobile cell called a **sperm**. The female gamete is a stationary egg cell (also called an **ovum**).

The sperm must move to the egg and fuse (join) with it. This is called fertilisation (Figure 11.1). The single cell formed by fertilisation is called a zygote. This cell will divide many times by mitosis to form all the cells of the offspring.

There are four main stages in sexual reproduction.

- Gametes (sperm and eggs) are produced.
- The male gamete is transferred to the female gamete.
- Fertilisation occurs – the sperm fuses with the egg.
- A zygote is formed, which develops into a new individual.

The offspring produced by sexual reproduction show genetic variation, as a result of both gamete production and fertilisation.

▲ Figure 11.1 A sperm fertilising an egg cell.

KEY POINT

Remember that most cells of the body have two sets of identical chromosomes called homologous pairs. Body cells are called diploid cells. Eggs and sperm have half the normal number of chromosomes (one from each homologous pair) and are called haploid cells (see Chapter 1).

MEIOSIS AND GAMETE PRODUCTION

Sperm are produced in the male sex organs – the testes. Eggs are produced in the female sex organs – the ovaries. Both are formed when cells inside these organs undergo a special sort of division called meiosis. Meiosis produces gametes that are genetically different from each other and have only half as many chromosomes as the original cell.

Meiosis is a more complex process than mitosis and takes place in two stages called meiosis I and meiosis II, resulting in four haploid cells. Each daughter cell is genetically different from the other three and from the original parent cell.

The process of meiosis is shown in Figure 11.2.

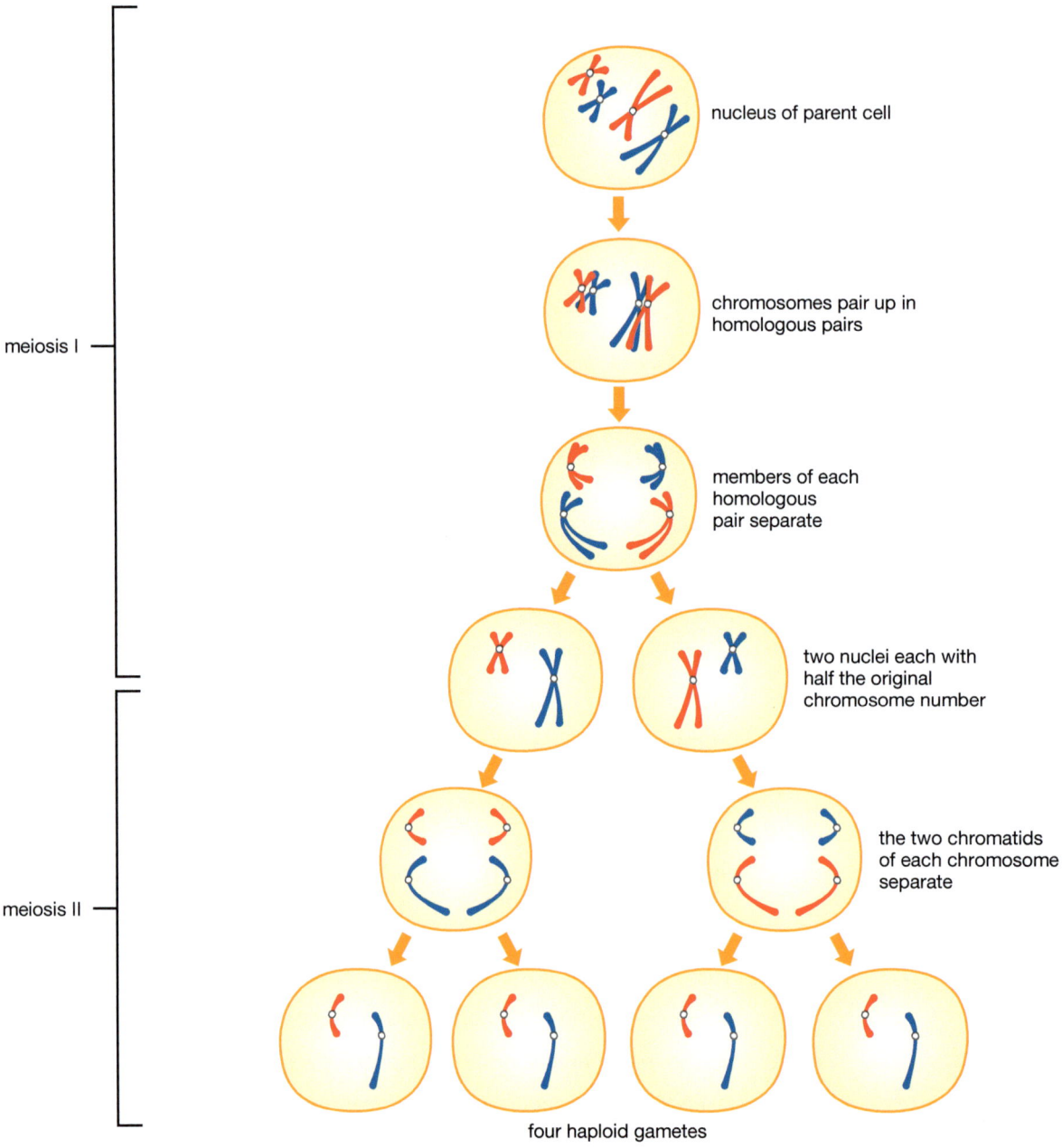

▲ Figure 11.2 The stages of meiosis. For simplicity, the parent cell contains only two homologous pairs of chromosomes (one long pair, one short). To help you to see what happens, one member of each pair is coloured red and one blue. The cell membrane is shown, but the nuclear membrane has been omitted. A spindle forms during each division, but these have also been omitted for clarity.

Before a cell divides, each chromosome makes a copy of itself, called a **chromatid**. This involves DNA replication (see Chapter 1). This is needed so that there is enough genetic material to be shared between the four daughter cells. The cell then divides twice.

- During the first division, one chromosome from each homologous pair goes into each of two daughter cells.

- During the second division, the chromosome separates into the two chromatids. One chromatid goes into each of the daughter cells from this division.

Remember that a chromatid is a copy of a chromosome. After the chromatids separate we call them chromosomes again. The end result is that the final four daughter cells each contain one chromosome from each homologous pair – they form four haploid gametes.

MEIOSIS PRODUCES GENETIC VARIATION

In body cells, each member of a pair of homologous chromosomes carries genes for the same features, arranged in the same order along the chromosome. However, each chromosome may have a different form of the gene. These different forms are called **alleles**. For example, one chromosome may have the allele for brown eyes and the other chromosome may have the allele for blue eyes. (Genes and alleles are explained in more detail in Chapter 12.)

During meiosis I, the members of each homologous pair are shared between the two daughter cells independently of each of the other homologous pairs. This means that there is much possible *genetic variation* in the daughter cells. Figure 11.3 shows how three pairs of homologous chromosomes can form eight possible gametes. In humans there are 23 pairs of chromosomes, which can produce an enormous amount of genetic variation.

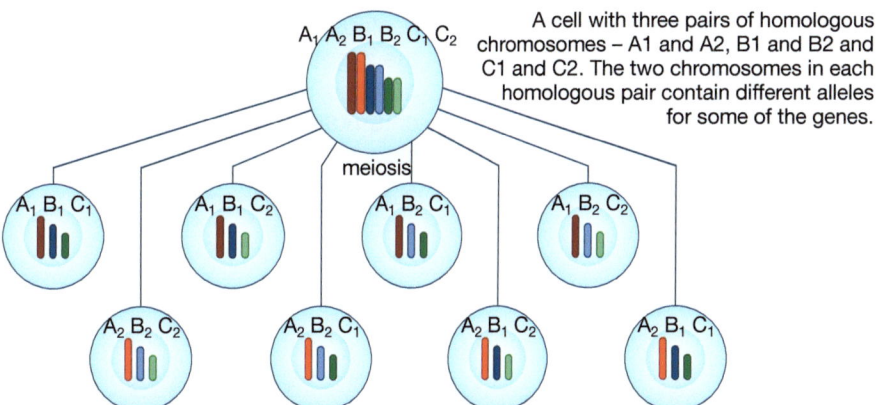

A cell with three pairs of homologous chromosomes – A1 and A2, B1 and B2 and C1 and C2. The two chromosomes in each homologous pair contain different alleles for some of the genes.

As a result of the two divisions of meiosis, each sex cell formed contains one chromosome from each homologous pair. This gives eight combinations. As A1 and A2 contain different alleles (as do B1 and B2, and C1 and C2) the eight possible sex cells will be genetically different.

▲ Figure 11.3 Meiosis produces genetic variation in gametes.

EXTENSION WORK

There is a mathematical rule for predicting how many combinations of chromosomes there can be. The rule is:
number of possible combinations = 2^n
where n = number of *pairs* of chromosomes.
With two pairs of chromosomes, the number of possible combinations = 2^2 = 4.
With three pairs of chromosomes, the number of possible combinations = 2^3 = 8.
With the 23 pairs of chromosomes in human cells, the number of possible combinations = 2^{23} = 8 388 608!

You should take some time to compare meiosis with mitosis (see Chapter 1, page 11-12). Table 11.1 summarises the similarities and differences between the two types of cell division.

▼ Table 11.1 Comparison of meiosis and mitosis.

Feature of the process	Mitosis	Meiosis
Chromosomes are copied before division begins	yes	yes
Number of cell divisions	one	two
Number of daughter cells produced	two	four
Daughter cells are haploid or diploid	diploid	haploid
Genetic variation in the daughter cells	no	yes

GAMETES AND FERTILISATION

The role of a sperm is to swim to the egg and fertilise it, and sperm are highly specialised for this (Figure 11.4). They have a tail-like flagellum that allows them to swim, and mitochondria in the mid-piece to supply ATP for energy. The head of the sperm contains the nucleus with the father's chromosomes. A sperm is much smaller than an egg (see Figure 11.1). The egg is about 0.1 millimetres in diameter, which is approximately the size of a full stop on this page. It is one of the biggest cells in the human body. It has a large amount of cytoplasm, and a nucleus containing the mother's chromosomes. The cytoplasm contains stores of energy and materials, so that after the egg is fertilised, the zygote can start dividing and developing into an embryo.

Once a sperm reaches an egg, its nucleus enters the egg and fuses with its nucleus. As each gamete has only half the normal number of chromosomes, the zygote formed by fertilisation will have the full number of chromosomes. In humans, sperm and egg each have only 23 chromosomes. The zygote has 46 chromosomes, like most other cells in the body. Figure 11.5 shows the main stages in fertilisation.

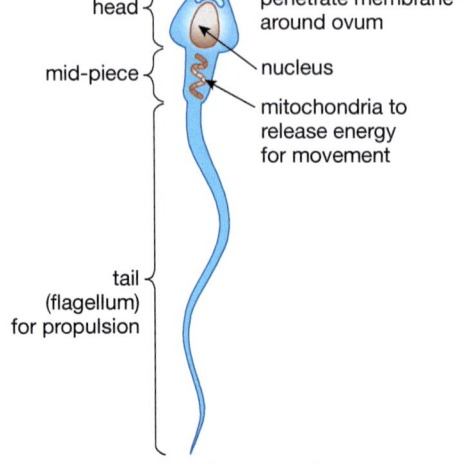

▲ Figure 11.4 The adaptations of a sperm.

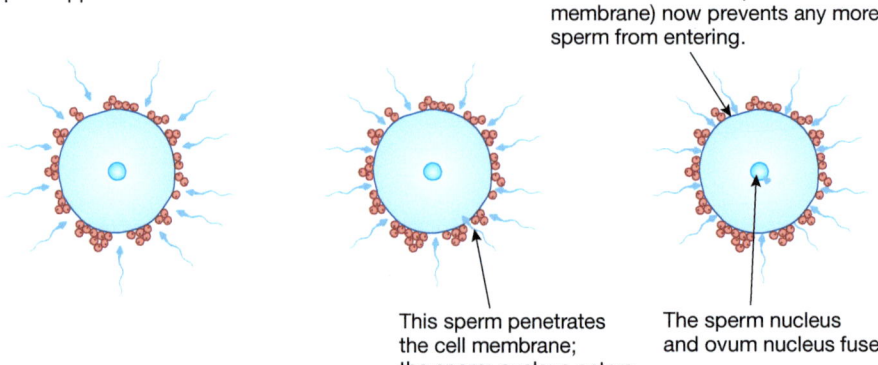

▲ Figure 11.5 The main stages of fertilisation.

FERTILISATION PRODUCES MORE GENETIC VARIATION

Fertilisation does more than just restore the diploid chromosome number – it provides an additional source of genetic variation. The sperm and eggs are all genetically different because they are formed by meiosis. Each time fertilisation takes place, it brings together a different combination of genes. In humans, any one of the billions of sperm formed by a male during his life could, potentially, fertilise any one of the thousands of eggs formed by a female.

This variation applies to both male and female gametes. So, using our estimate of about 8.5 million different types of human gamete (see the Extension box on page 195), there can be 8.5 million different types of sperm and 8.5 million different types of egg. When fertilisation takes place, any sperm could fertilise any egg. The number of possible combinations of chromosomes (and genes) in the zygote is 8.5 million × 8.5 million = 7.2×10^{13}, or 72 trillion!

DEVELOPMENT OF THE ZYGOTE

A zygote divides to produce all the cells that make up the adult. These cells must have the full number of chromosomes, so the zygote divides repeatedly by mitosis. Figure 11.6 shows the involvement of meiosis, mitosis and fertilisation in the human life cycle.

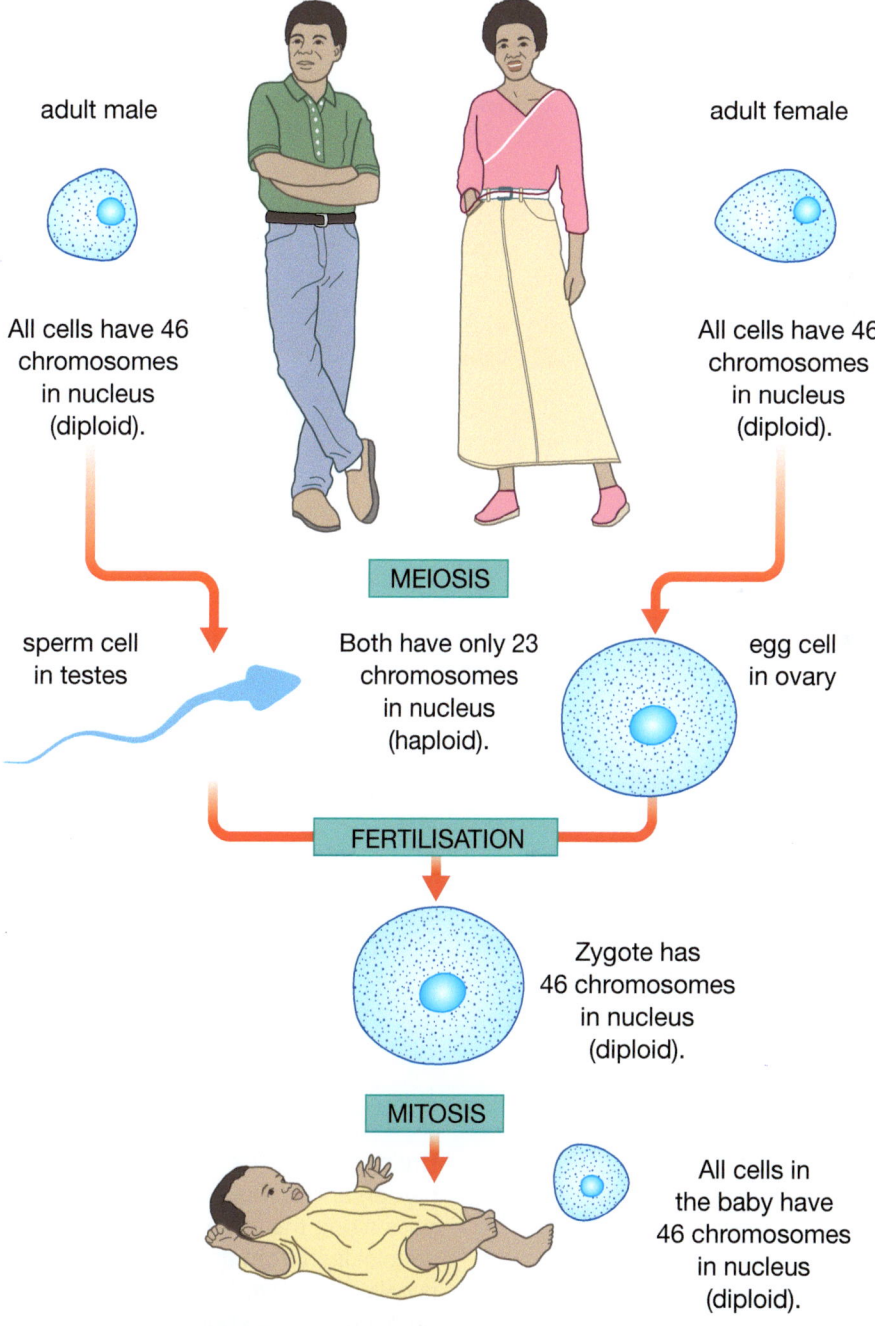

▲ Figure 11.6 The involvement of meiosis, mitosis and fertilisation in the human life cycle.

Mitosis is not the only process involved in development, otherwise all that would be produced would be a ball of cells. During the process, cells move around and different shaped structures are formed. Also, different cells specialise to become bone cells, nerve cells, muscle cells, and so on. This process is called differentiation – see Chapter 1.

THE HUMAN REPRODUCTIVE SYSTEMS

Figures 11.7 and 11.8 show the structure of the human female and male reproductive systems.

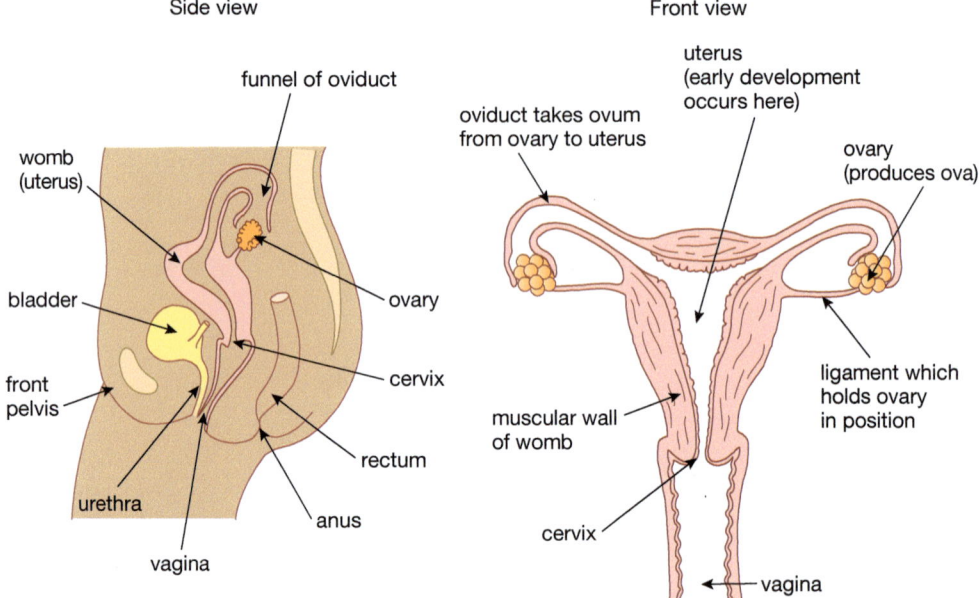

▲ Figure 11.7 The human female reproductive system.

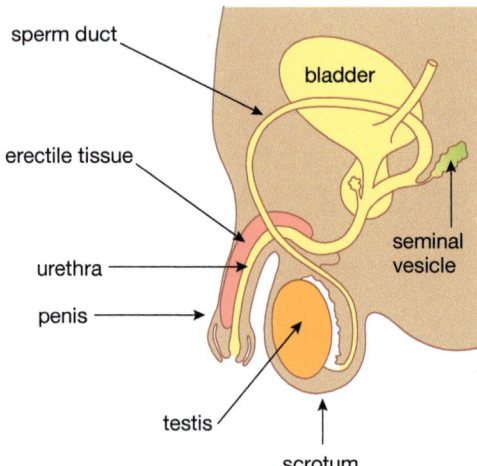

▲ Figure 11.8 The human male reproductive system.

During **sexual intercourse** the man's penis becomes erect (bigger and harder) and is moved repeatedly in and out of the woman's vagina. Sperm pass along the sperm duct and are released into the woman's vagina in a special fluid from the seminal vesicles. This fluid, together with the sperm, is called **semen**. Release of semen is called **ejaculation**. Semen provides nutrients for the sperm, and allows them to swim towards the oviducts to meet the egg.

Each month, an egg is released into an **oviduct** from one of the ovaries. This is called **ovulation**. If an egg is present in the oviduct, it may be fertilised by sperm introduced during intercourse. The zygote formed begins to develop into an embryo, which sinks into ('implants' in) the lining of the uterus. Here, the embryo develops an organ called the **placenta**, which allows it to exchange substances with the mother's blood. The placenta also maintains the embryo's position in the uterus. As the embryo develops, it becomes more and more complex. When it becomes recognisably human, we no longer call it an embryo but a fetus (Figure 11.9).

> **DID YOU KNOW?**
> The oviduct is also known as the fallopian tube.

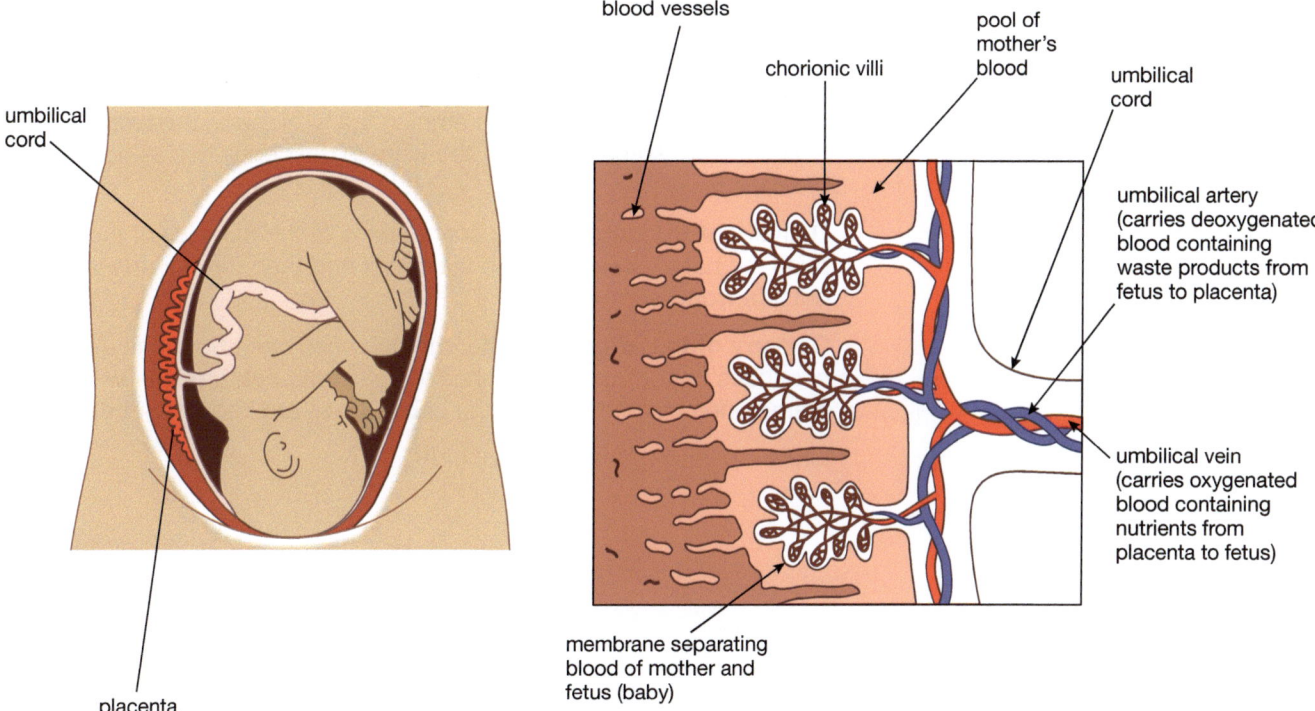

▲ Figure 11.9 The position of the placenta and fetus just before birth, and the structure of the placenta.

Note that the mother and the embryo have separate blood circulatory systems. In the placenta, the mother's blood (maternal blood) and the blood of the fetus (fetal blood) are close together but do not mix. Exchange of substances takes place by diffusion across the membranes separating the maternal blood from the fetal blood.

The maternal blood flows from the mother's circulatory system through the blood spaces surrounding the villi of the placenta, and then re-enters the mother's blood vessels. Blood from the fetus flows through two arteries in the **umbilical cord**, through the capillary networks in the villi and back to the fetus through the umbilical vein.

The growing embryo or fetus needs oxygen and nutrients such as glucose for respiration, and other nutrients, such as amino acids, for growth. It cannot provide for itself, so the fetal blood in the villi is low in these substances. The mother's blood contains high concentrations of oxygen and nutrients, so they diffuse down a concentration gradient from maternal to fetal blood. Conversely, the fetal blood contains high concentrations of carbon dioxide and waste products such as urea. These diffuse in the opposite direction, from fetal to maternal blood.

The placenta exchanges many other substances. These include antibodies from the mother, which will help to protect the baby from infections after it is born. The placenta also secretes female hormones, in particular **progesterone**, which help to maintain the pregnancy until the baby is ready to be born.

During pregnancy, a membrane called the **amnion** encloses the developing embryo. The amnion secretes a fluid called **amniotic fluid**, which protects the developing embryo against sudden movements and bumps. At the end of nine months of development, there is no room left for the fetus to grow and it sends a hormonal 'signal' to the mother to begin the birth process. Figure 11.9 shows the position of a human fetus just before birth.

There are three stages to the birth of a child.

- **Dilation of the cervix:** The **cervix** is the 'neck' of the uterus. It gets wider (dilates) to allow the baby to pass through. The muscles of the uterus contract quite strongly and the amnion tears, allowing the amniotic fluid to escape. (In some countries, the woman describes this as her 'waters breaking'.)
- **Delivery of the baby:** Strong contractions of the muscles of the uterus push the baby's head through the cervix and then through the vagina to the outside world.
- **Delivery of the afterbirth:** After the baby has been born, the uterus continues to contract and pushes the placenta out, together with the membranes that surrounded the baby. These are known as the afterbirth.

Figure 11.10 shows the stages of birth.

> **DID YOU KNOW?**
> Just before birth, the fetus takes up so much room that many of the mother's organs are moved out of position. The heart is pushed upwards and rotates so that the base points towards the left breast.

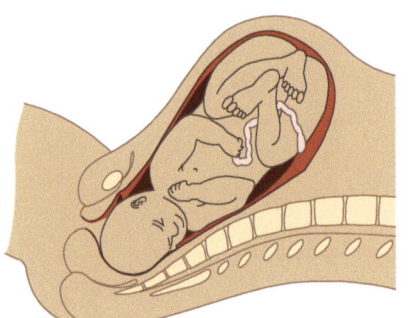
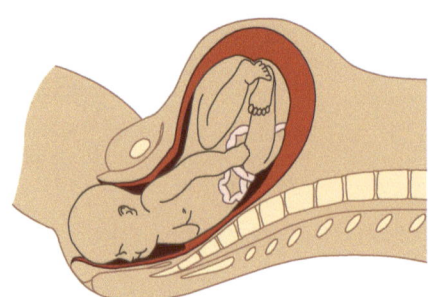
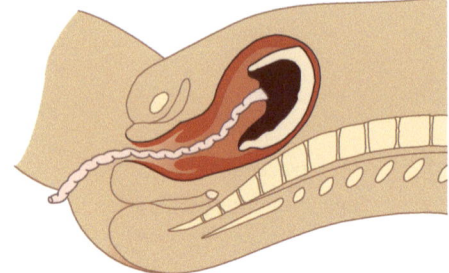

1 The baby's head pushes against the cervix; the mucus plug dislodges and the waters break.

2 The uterus contracts to push the baby out through the vagina.

3 The placenta becomes detached from the wall of the uterus and is expelled through the vagina as the afterbirth.

▲ Figure 11.10 The stages of birth.

BREASTFEEDING

Within hours of its birth, a baby will be fed by its mother on milk from her **mammary glands**. Babies continue to feed on milk for several months, until they start to eat semi-solid foods. During pregnancy, the mother's breasts increase in size as the mammary glands grow new tissue. They start to produce milk under the influence of a hormone called **prolactin** from the mother's pituitary gland. Another hormone from the pituitary, called **oxytocin**, stimulates the release of milk from the mammary glands.

While she is breastfeeding, the balance of the mother's diet has to change to accommodate the needs of milk production. Her energy intake from food needs to increase by about 25%, together with increases in protein, calcium and vitamins in her diet.

> **EXTENSION WORK**
> When the baby sucks at its mother's nipple, a nervous reflex feeds back to the pituitary, stimulating it to release more oxytocin. This in turn causes the release of more milk. This is an example of a *positive feedback* system. You have seen several examples of negative feedback – positive feedback is much less common in the body.

There are several advantages to breastfeeding. Breast milk is the perfect food for healthy growth of the baby. It contains antibodies, which help to protect the baby against infectious diseases. Breastfeeding also helps to form an emotional bond between the mother and her offspring. While it is certainly true that breastfeeding is best for both mother and baby, some mothers will not be able to breastfeed, for medical or other reasons. The alternative is to feed the baby a special artificial milk preparation, called 'formula' milk.

HORMONES CONTROLLING REPRODUCTION

When a baby is born, it is recognisable as a boy or girl by its sex organs. The presence of male or female sex organs is known as the primary sex characteristics. However, like most animals, humans are unable to reproduce when they are young. During their teens, changes happen to boys and girls that lead to sexual maturity; this means that they are able to have babies. These changes are controlled by hormones, and the time when they happen is called puberty. Puberty involves two developments. The first is that the gametes (eggs and sperm) start to mature and be released. The second is that the bodies of both sexes adapt to allow reproduction to take place. These events are started by hormones released by the pituitary gland (see Table 9.2 on page 163). These hormones are called **follicle stimulating hormone** (**FSH**) and **luteinising hormone** (**LH**).

In boys, FSH stimulates sperm production, while LH instructs the testes to secrete the male sex hormone, testosterone. Testosterone controls the development of the male **secondary sexual characteristics**. These include growth of the penis and testes, growth of facial and body hair, muscle development and 'breaking' of the voice, when the voice gets lower in tone (Table 11.2).

In girls, the pituitary hormones control the release of a female sex hormone called oestrogen, from the ovaries. Oestrogen produces the female secondary sexual characteristics, such as breast development and the beginning of menstruation ('periods').

> **DID YOU KNOW?**
> FSH and LH are known as gonadotrophic hormones. 'Gonads' is a biological term for the sex organs. 'Gonadotrophic' means that these hormones stimulate the development of the sex organs.

> **DID YOU KNOW?**
> Sperm production is most efficient at a temperature of about 34 °C, just below the core body temperature (37 °C). This is why the testes are outside the body in the scrotum, where the temperature is a little lower.

▼ Table 11.2 Changes at puberty.

In boys	In girls
sperm production starts	the menstrual cycle begins, and eggs are released by the ovaries every month
growth and development of male sexual organs	growth and development of female sexual organs
growth of armpit and pubic hair, and chest and facial hair (beard)	growth of armpit and pubic hair
increase in body mass; growth of muscles, e.g. chest	increase in body mass; development of 'rounded' shape to hips
voice breaks	voice deepens without sudden 'breaking'
sexual 'drive' develops	sexual 'drive' develops
	breasts develop

The age when puberty takes place can vary a good deal, but it is usually between about 11 and 14 years in girls and 13 and 16 years in boys. It takes several years for puberty to be completed. Some of the most complex changes take place in girls, with the start of menstruation.

HORMONES AND THE MENSTRUAL CYCLE

'Menstrual' means 'monthly', and in most women the **menstrual cycle** takes about a month, although it can take as little as two weeks or as long as six weeks (Figures 11.11 and 11.12). In the middle of the cycle is an event called ovulation, which is the release of a mature egg cell (ovum).

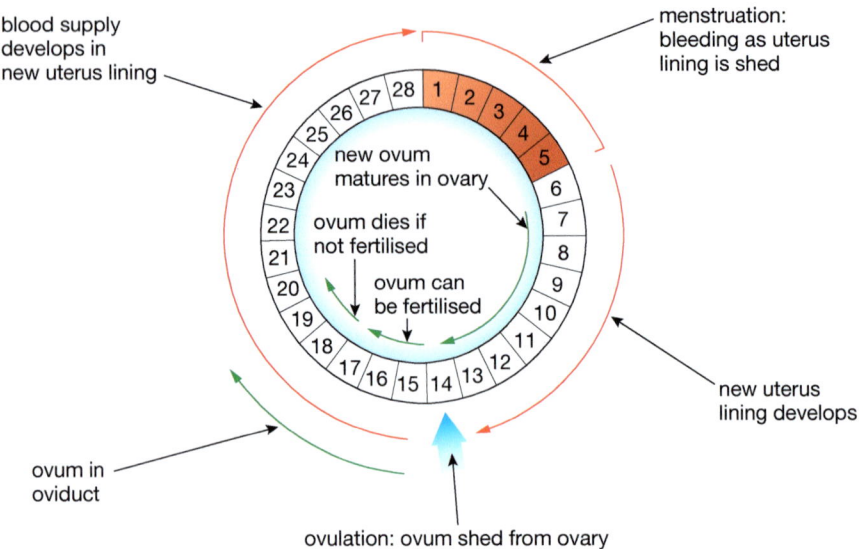

▲ Figure 11.11 The menstrual cycle.

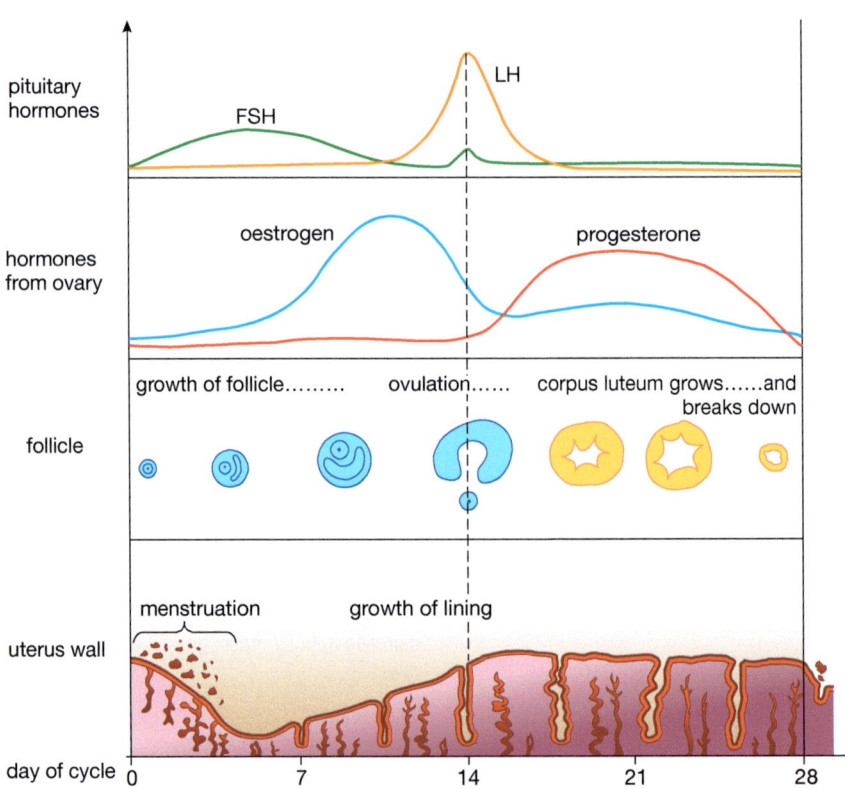

▲ Figure 11.12 Changes taking place during the menstrual cycle.

One function of the cycle is to control the development of the lining of the uterus (womb), so that if the egg is fertilised, the lining will be ready to receive it. If the egg is not fertilised, the lining of the uterus is lost from the woman's body as the flow of menstrual blood and cells of the lining, called a **period**.

A cycle is a continuous process, so it does not really have a beginning, but the first day of menstruation is usually called day 1.

Inside a woman's ovaries are hundreds of thousands of cells that could develop into mature eggs. Every month, one of these grows inside a ball of cells called a **follicle** (Figure 11.13). The pituitary hormone that switches on the growth of the follicle is called 'follicle stimulating hormone'. At the middle of the cycle (about day 14) the follicle moves towards the edge of the ovary and the egg is released as the follicle bursts open. This is the moment of ovulation.

While this is going on, the lining of the uterus has been repaired after menstruation and has thickened. This change is caused by the hormone oestrogen, which is secreted by the ovaries in response to FSH. Oestrogen also has another job. It slows down production of FSH, while stimulating secretion of LH. It is a peak of LH that causes ovulation.

After the egg has been released, it travels down the oviduct to the uterus. It is here in the oviduct that fertilisation may happen, if sexual intercourse has taken place. What is left of the follicle now forms a structure in the ovary called the **corpus luteum**. The corpus luteum makes another hormone called progesterone. Progesterone completes the development of the uterus lining by thickening and maintaining it, ready for the fertilised egg to sink into it and develop into an embryo. Progesterone also inhibits (prevents) the release of FSH and LH by the pituitary gland, stopping ovulation.

If the egg is not fertilised, the corpus luteum breaks down and stops making progesterone. The lining of the uterus then breaks down and passes out through the woman's vagina during menstruation. If, however, the egg is fertilised, the corpus luteum continues making progesterone, the lining is not released, and menstruation does not happen. The first sign that a woman is pregnant is when her monthly periods stop. Later on in pregnancy, the placenta secretes progesterone, which takes over the role of the corpus luteum.

> **DID YOU KNOW?**
> A small percentage of women can sense the exact moment that ovulation happens, as the egg bursts out of an ovary.

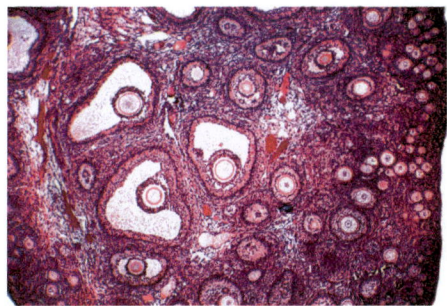

▲ Figure 11.13 Eggs developing inside the follicles of a human ovary.

> **DID YOU KNOW?**
> 'Corpus luteum' is Latin for 'yellow body'. A corpus luteum appears as a large yellow swelling in an ovary after the egg has been released. The growth of the corpus luteum is under the control of luteinising hormone (LH) from the pituitary.

CONTRACEPTION

The word 'conception' means 'becoming pregnant'. **Contraception** means avoiding conception – that is, using ways to avoid becoming pregnant. Not having sexual intercourse is the most certain method of contraception, but many couples do want to have sex without the woman becoming pregnant. One way of doing this is to use the so-called 'natural' method. This uses knowledge of the menstrual cycle to avoid intercourse at times when sperm could possibly fertilise an egg. Fertilisation is most likely to take place in the middle of the menstrual cycle, either side of the day of ovulation. Sperm and eggs live for only a few days, so sperm released into the vagina four days before ovulation will have died by the time ovulation occurs. Two or three days after ovulation, an unfertilised egg will have been lost through the vagina. By avoiding intercourse during this period, and limiting it to the rest of the menstrual cycle (the 'safe period'), a woman can usually avoid becoming pregnant (Figure 11.14).

The disadvantage of the 'safe period' method is that it is not very reliable. For this method to be successful, the woman's menstrual cycle must be regular, and she must know when ovulation occurs in her cycle. There is usually a slight increase in a woman's body temperature around the time of ovulation, so she can take her temperature daily to try to judge when she is ovulating. This is not completely accurate, though, and the 'safe period' method results in many unplanned pregnancies. Some couples, however, rely on the 'safe period' method, usually because they have moral or religious reasons for not using other means of contraception.

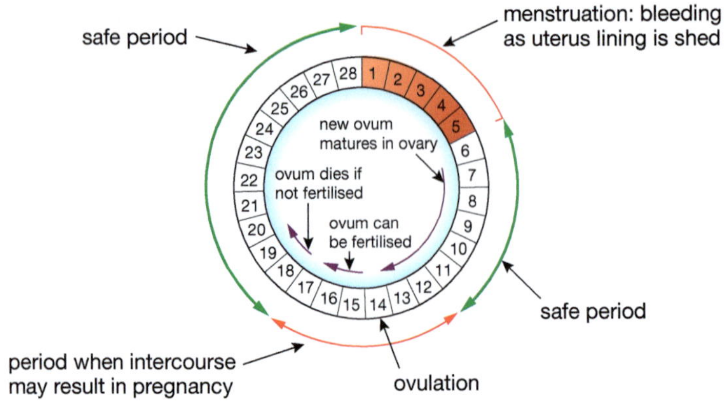

Figure 11.14 The 'safe period'.

Another 'natural' method is where the man withdraws his penis from the woman's vagina before ejaculation. As with the 'safe period' method, this has the advantage that it does not involve using artificial contraception. But again, the 'withdrawal' method is unreliable – some sperm may be released before ejaculation, or the man may not withdraw in time.

There are four other categories of contraception:

- barrier methods
- intrauterine devices (IUDs)
- hormonal methods
- sterilisation.

BARRIER METHODS

Barrier methods use some kind of barrier to prevent sperm reaching the egg. The **condom** is a sheath of very fine rubber that fits over the penis and collects the semen before it enters the vagina (Figure 11.15). Another version, called the **femidom**, is similar, but inserted into the vagina before intercourse. Condoms and femidoms are generally easy to obtain and use, and have the added advantage that they are the only methods of contraception which give protection against sexually transmitted diseases.

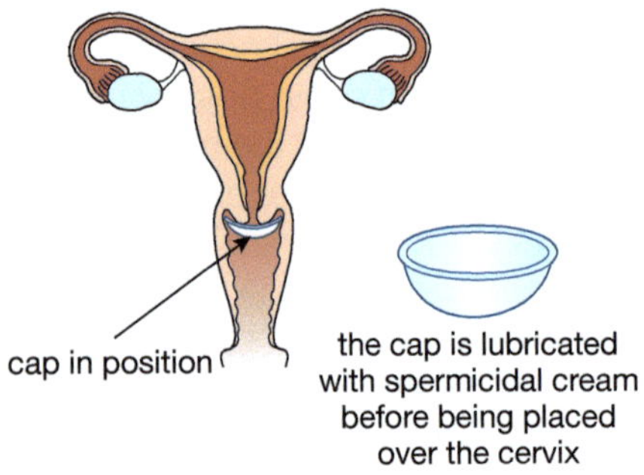

Figure 11.15 Condoms are easily obtained from clinics or pharmacists.

Figure 11.16 The contraceptive cap (diaphragm).

Another barrier method is the **diaphragm** or **cap** (Figure 11.16). This is a dome-shaped piece of rubber that a woman inserts into her vagina before intercourse. The cap covers the cervix, preventing sperm from entering the uterus. A **spermicidal cream** containing a chemical that kills sperm is used with the cap as extra protection against pregnancy (spermicidal creams also increase the reliability of condoms). The cap is left in position for 6 hours after intercourse. Caps differ in size, and the woman needs a medical examination by a doctor or nurse to select the correct size of cap to use.

INTRAUTERINE DEVICES

Intrauterine devices (IUD or coil) are small pieces of plastic or copper of various shapes. An IUD is inserted through the cervix into the uterus (Figure 11.17). The copper IUD works by preventing a fertilised egg from implanting in the lining of the uterus. Some IUDs contain the hormone progesterone, which thickens the mucus in the cervix, stopping sperm from getting through. One disadvantage of the IUD is that it has to be fitted by a doctor.

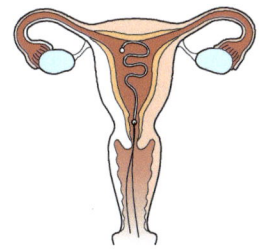
IUD in place in uterus

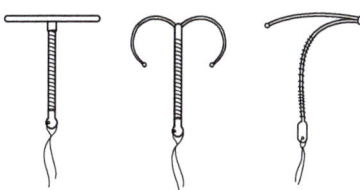
some different styles of IUD

▲ Figure 11.17 Intrauterine devices.

HORMONAL METHODS

Hormonal methods usually mean taking the oral **contraceptive pill**. The *combined pill* contains a mixture of oestrogen and progesterone. These hormones prevent the production of FSH and LH from the pituitary gland. This means that the follicles inside the ovary do not develop, and ovulation does not take place. Without an egg being released, a woman cannot become pregnant. She takes the pill for 21 days of the menstrual cycle (Figure 11.18) and then stops taking it for the last 7 days. During this time, she will have a period. Hormonal contraceptives can also be given by injection or inserted under the skin as a contraceptive 'implant'.

> **DID YOU KNOW?**
> The combined pill has been linked to a number of health problems in women, the main one being an increased risk of a blood clot (thrombosis) forming in blood vessels, which can be fatal. The increased risk is very slight for most women, but is higher if they are smokers. The mini-pill is not quite as reliable as the combined pill that contains oestrogen, but it does not increase the chances of blood clots developing.

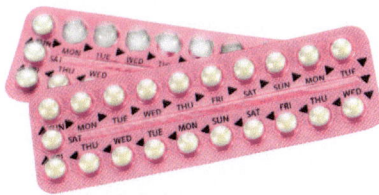

▲ Figure 11.18 Synthetic hormones can be used to prevent pregnancy. The pills are marked with the days of the week, so the woman can check whether she has remembered to take a pill each day.

The *mini-pill* contains only progesterone. It works by causing a thickening of mucus in the cervix, which acts as a barrier to sperm.

STERILISATION

Sterilisation is a surgical operation that can be carried out on men to prevent sperm passing to the penis, or on women to prevent eggs passing to the uterus. This makes them unable to have children (sterile). Sterilisation is a method that is normally only used by couples who have produced the number of children they want and do not want to use other methods of contraception to prevent further pregnancies. The operation is usually irreversible, so the decision to opt for sterilisation must be considered carefully.

Male sterilisation is called **vasectomy**. Under local anaesthetic, the sperm ducts are cut and tied so that no sperm can get through. After the operation, the man can still ejaculate but the semen will contain no sperm. In women, the oviducts are cut by a similar operation, called **tubal ligation**.

Table 11.3 compares the main methods of contraception in order of their rates of failure, and lists some of the advantages and disadvantages of each method.

▼ Table 11.3 The main methods of contraception.

Method	Failure rate (%)*		Some advantages and disadvantages
	Careful use	Typical use	
none – intercourse without contraception	85	85	
withdrawal	4	27	High failure rate.
'safe period'	9	25	High failure rate. Woman needs to have a regular cycle, and to keep records of her cycle.
diaphragm with spermicidal cream	6	16	Medical examination needed to select correct size. Not simple to use – must be inserted before intercourse and left in place for 6 hours afterwards.
condom	2	15	Generally easy to obtain and use. Gives protection against sexually transmitted diseases. May slip off during intercourse.
mini-pill	0.5	8	Low failure rate if used carefully. Must be taken every day, at the same time each day, to be effective.
combined pill	0.3	8	Low failure rate if used carefully. Must be taken every day. Links to some health problems.
intrauterine device	0.6	0.8	Must be fitted by a doctor. Can cause heavier periods.
tubal ligation	0.5	0.5	Very low failure rate. Operation usually not reversible.
vasectomy	0.1	0.1	Very low failure rate. Operation usually not reversible.

*The values for 'failure rate' are the number of pregnancies that result per 100 women per year. There are two sets of numbers. The 'careful use' column shows the failure rate when the contraceptives are used carefully, exactly as they should be. These values are from medical trials under controlled conditions. However, people are human, and make mistakes. The rates in the 'typical use' column are from surveys of couples that use the contraceptive methods normally, sometimes without taking the same degree of care as in the medical trials. This results in a higher rate of failure.

> **DID YOU KNOW?**
> Even with a high sperm count, up to 40% of sperm are not able to function properly. Some lack a head. Others have the appearance of a broken neck, with the head at 90° to the tail. Some even have two tails. Sperm like this cannot swim or fertilise an egg.

FERTILITY TREATMENT

Some couples have difficulty in achieving a pregnancy, even after years of trying. This medical problem is called **infertility**. There are many different reasons for infertility. For example, the man may not produce enough sperm, the woman's eggs may not develop properly, or there may be a problem with achieving fertilisation.

Up to 40% of infertility problems are due to a low **sperm count**, so this is the first thing that doctors will check. The man's sperm count must be above 20 million sperm cells per cm^3 of semen, and the majority of the cells should show normal development.

If the sperm count is normal, the doctors will perform tests on the woman to see if she is ovulating normally. If she is not, there are various treatments, depending on the nature of the problem. There may be a physical problem, such as a blocked oviduct or a cyst on the ovary. In these cases, surgery may be the answer.

Sometimes a woman does not produce enough follicle stimulating hormone (FSH) to start egg development in the ovary. This can be treated by giving her injections of FSH or other hormones, to stimulate ovulation. Care has to be taken, because too much FSH can produce multiple ovulations (many eggs are released at once). If all these eggs are fertilised, several embryos will develop in the uterus, leading to 'multiple births' (Figure 11.19).

Although the children shown in Figure 11.19 were born healthy, multiple births may be premature (early). Even if the fetuses survive the full term of pregnancy (9 months), the mass of each baby is likely to be very low, and they may have health problems.

Sometimes **artificial insemination (AI)** is used. Healthy sperm are placed in a woman's uterus at the time of ovulation, to increase the chances of fertilisation. The woman's partner can donate the sperm, or sperm from another man can be used.

When all other methods of overcoming infertility problems have been tried, the last alternative is **in vitro fertilisation (IVF)**. 'In vitro' is Latin for 'in glass', meaning fertilisation is carried out in a Petri dish or a similar container, outside the woman's body. Sperm is added to the egg, or a sperm can even be injected into an egg to fertilise it. The embryo's growth is monitored for a few days to check that all is well, then it is implanted into the woman's uterus to continue to grow normally. This is now a common medical procedure that has produced many healthy 'test-tube babies'.

▲ Figure 11.19 Hormones can be used to help women become pregnant. Infertility treatment often results in the birth of twins or triplets.

GROWTH AND DEVELOPMENT

A human develops from a fertilised egg into an embryo, then into a fetus. After birth, the baby grows into a child, then an adolescent, when it becomes sexually mature. Finally, he or she becomes an adult. This is the normal pattern of human development, the sequence through which the body changes during life. The instructions for development are carried in our genes. Most of these instructions are the same in everyone – well over 99% of our genes are the same in all other human beings. These genes control the basic growth pattern to produce a human body.

The remainder – a small fraction of the total number of genes – control the development of features that make each of us unique. As you have seen, this genetic variation comes from new assortments of genes being produced as a result of meiosis and fertilisation.

Genes act on growth and development in many different ways, for example, controlling the sex of the baby. Sometimes genes produce their effects indirectly, by controlling the body's ability to make hormones, which in turn coordinate growth and development. For example, growth hormone, made by the pituitary gland, increases the rate of synthesis of proteins in muscles, and the growth of bone tissue. Thyroxine from the thyroid gland controls the metabolic rate (see Chapter 9).

Testosterone from the testes stimulates the development of the male secondary sex characteristics, and oestrogen from the ovaries does the equivalent in girls. There are over a dozen different hormones that affect growth in some way.

THE PATTERN OF HUMAN GROWTH

If you plot a graph of the height of a boy or girl against time, the graph is not a straight line. Instead it shows growth 'spurts' – periods of rapid growth and development (Figure 11.20). The changes are easier to see if you plot the gain in height against the age of the child (Figure 11.21).

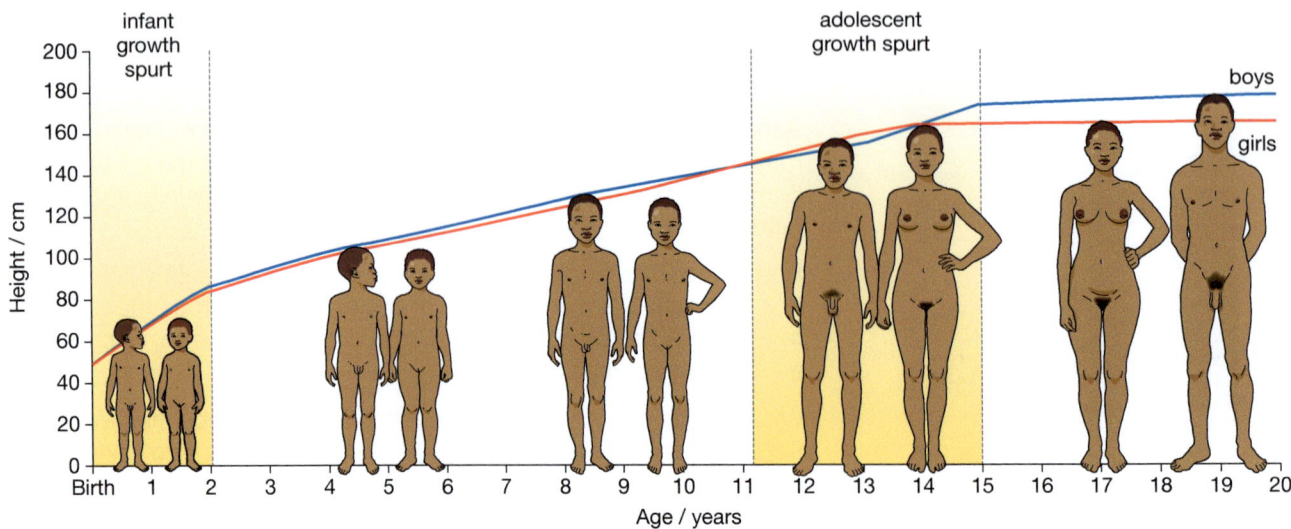

▲ Figure 11.20 Growth in boys and girls, showing changing rates of growth and growth spurts.

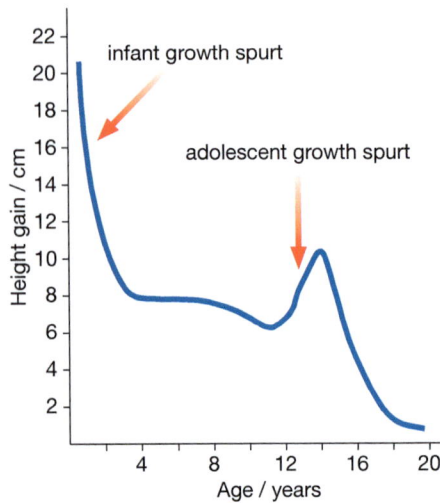

▲ Figure 11.21 Height gain per year for boys from birth to age 18 (the curve for girls is similar but with an adolescent spurt about 18 months earlier).

A healthy child is always growing – the curve in Figure 11.21 is always greater than zero. But the rate of growth changes dramatically between birth and adolescence. Growth is fastest soon after birth. This is called the infant growth spurt. The rate then decreases rapidly until the child is about 4 years old. Between about 4 and 10 years, the rate of growth still decreases, but more slowly. At puberty, there is an increase in the rate of growth – the adolescent growth spurt. This happens rather earlier in girls than in boys. Growth in both sexes has more or less finished by the late teens.

There is also a change in body proportions during growth. For example, a baby's head is very large in proportion to its body (Figure 11.22).

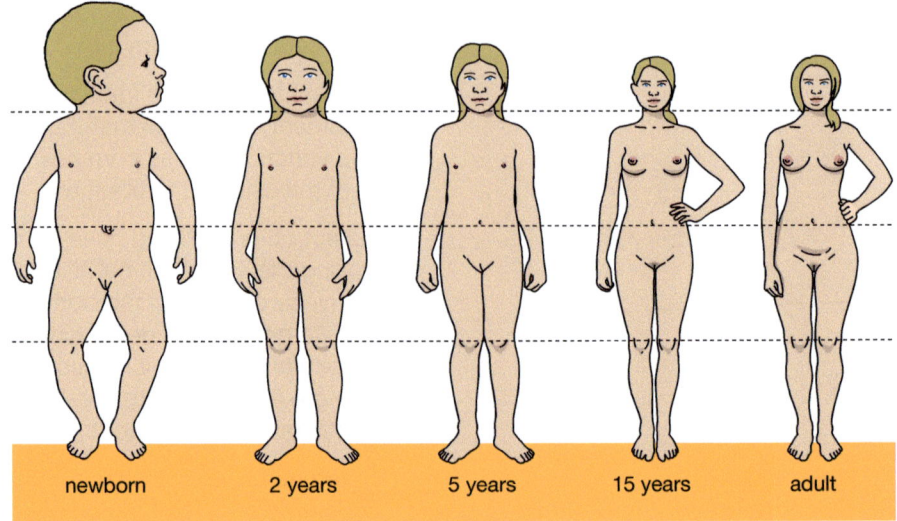

▲ Figure 11.22 Changes in proportions of the body with age.

Different organs grow at different rates (Figure 11.23). Some organs like the kidney and liver keep pace with the rate of growth of the body as a whole. Others, such as the reproductive organs, grow much more slowly at first, and only catch up with the rest of the body after puberty, when the sex hormones have their effect. The brain and skull grow very rapidly early on, reaching 90% of their full size by the age of 6. At birth a baby's brain makes up 13% of its body weight, but by adulthood this has fallen to only 2%. The reason for the rapid early growth of the brain and head is mainly that the brain does not increase its number of cells much. By the age of two, you have as many cells in your brain as you will have as an adult, and a full-sized brain needs a full-sized head!

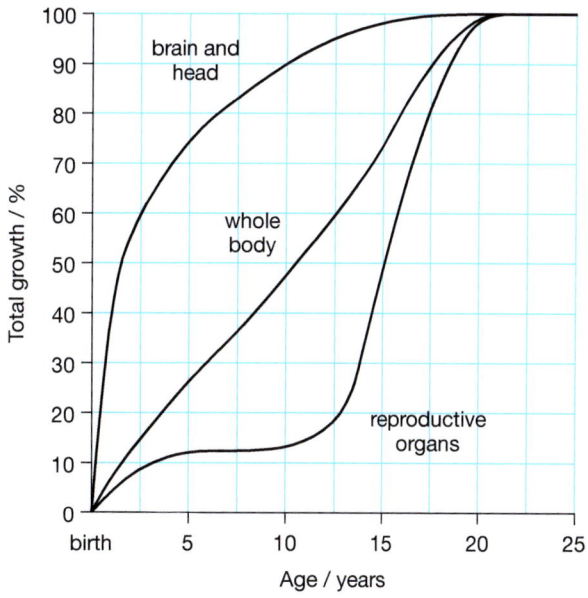

▲ Figure 11.23 Growth curves for different parts of the body compared with overall body growth.

CHAPTER QUESTIONS

SKILLS ANALYSIS

1 Cells can divide by mitosis or by meiosis. Human cells contain 46 chromosomes. The graphs show the changes in the number of chromosomes per cell as two different human cells undergo cell division.

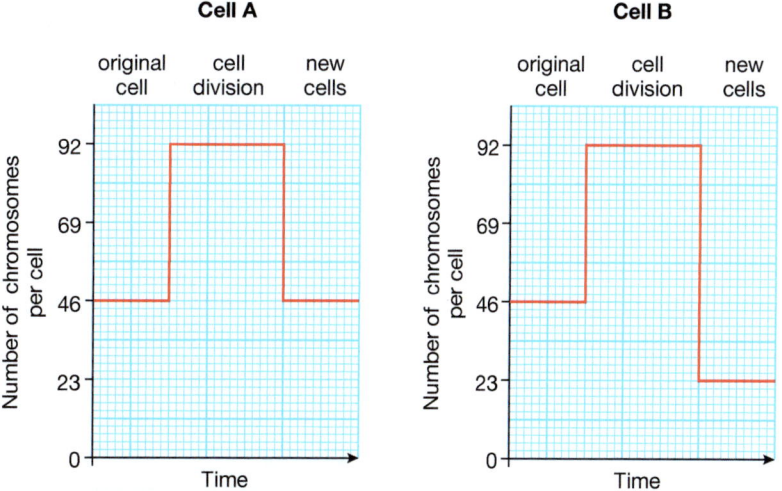

a Which of the two cells, A or B, is dividing by meiosis? Explain how you arrived at your answer.

SKILLS REASONING

b Explain the importance of meiosis, mitosis and fertilisation in maintaining the human chromosome number constant at 46 chromosomes per cell, generation after generation.

SKILLS CRITICAL THINKING

c State three differences between mitosis and meiosis.

SKILLS ANALYSIS

2 The diagram shows a baby about to be born.

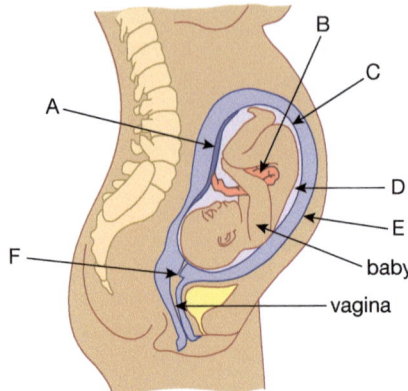

a Name parts A to E on the diagram.
b What is the function of A during pregnancy?
c What must happen to D and E just before birth?
d What must E and F do during birth?

3 a The diagram shows the female reproductive system.

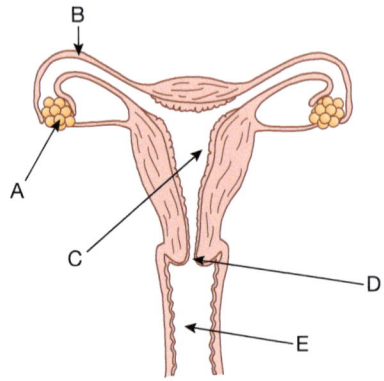

Which letter represents:
i the site of production of oestrogen and progesterone
ii the structure where fertilisation usually occurs
iii the structure that must dilate when birth commences
iv the structure that releases eggs?

SKILLS CRITICAL THINKING

b The graph shows the changes in the thickness of the lining of a woman's uterus over 100 days.

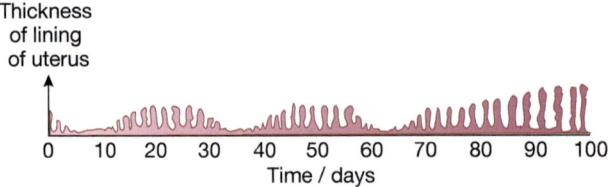

SKILLS ANALYSIS

i Name the hormone that causes the thickening of the uterine lining.
ii Use the graph to determine the duration of this woman's menstrual cycle. Explain how you arrived at your answer.
iii From the graph, deduce the approximate day on which fertilisation leading to pregnancy took place. Explain how you arrived at your answer.

SKILLS CRITICAL THINKING

iv Why must the uterus lining remain thickened throughout pregnancy?

SKILLS INTRAPERSONAL

4 The number of sperm cells per cm³ of semen (the fluid containing sperm) is called the 'sperm count'. Some scientists believe that human sperm counts have decreased over the last 50 years. They think that this is caused by a number of factors, one of which is drinking water polluted with oestrogen and other chemicals. Carry out a research project to find the evidence to support this theory. Summarise your findings in about 2–3 pages.

SKILLS INTERPRETATION

5 Construct a table of the hormones involved in the menstrual cycle. Use these headings:

Name of hormone	Place where the hormone is made	Function(s) of the hormone

SKILLS CRITICAL THINKING

6 a Briefly explain how each of the following methods of contraception works.
   i the diaphragm (cap)
   ii the intrauterine device (IUD)
   iii the contraceptive pill (combined pill)

b What are the advantages and disadvantages of using the following methods of contraception?
   i the condom
   ii the 'safe period'.

SKILLS ANALYSIS

7 The graph shows the pattern of growth of different parts of the body between birth and 25 years of age.

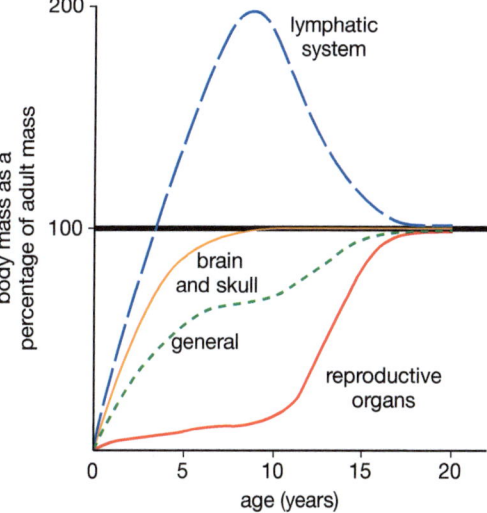

a i Describe how the pattern of growth of the brain and head differs from the general growth of the body between birth and 25 years of age.

ii Suggest reasons for these differences.

b Describe the pattern of growth of the reproductive organs between 10 and 25 years of age.

c Explain the cause of the growth pattern you described in your answer to part (b).

12 HEREDITY

The science of heredity is called genetics. It studies the sources of genetic variation and how the characteristics of an organism are inherited, that is passed from generation to generation. Genetic crosses follow simple mathematical rules and can be explained using diagrams. In this chapter, you will learn how to interpret genetic crosses and find out about the inheritance of some conditions caused by faulty genes.

LEARNING OBJECTIVES

- Know that genes exist in alternative forms called alleles, which give rise to differences in inherited characteristics
- Know the meaning of the terms dominant, recessive, homozygous, heterozygous, phenotype, genotype and codominance
- Understand patterns of monohybrid inheritance using a genetic diagram and know how to predict the probabilities of outcomes from the diagram
- Understand how to interpret family pedigrees
- Understand the role of multiple alleles in the inheritance of ABO blood groups
- Know that the sex of a person is controlled by a pair of chromosomes, XX in a female and XY in a male
- Explain how the sex of offspring is determined at fertilisation, using a genetic diagram
- Describe the causes and effects of inherited conditions such as haemophilia and red-green colour blindness (sex-linked inheritance), polydactyly (a dominant allele) and cystic fibrosis (a recessive allele)
- Describe the research into the use of viruses in gene therapy to treat cystic fibrosis
- Understand that variation within a species can be genetic, environmental or a combination of both

GENES AND ALLELES

Genes are sections of DNA that control the production of proteins in a cell (see Chapter 1). Each protein contributes towards a particular body feature. Sometimes the feature is visible, such as eye colour or skin pigmentation. Sometimes the feature is not visible, such as the type of haemoglobin in red blood cells or the type of blood group antigen on the red blood cells.

Some genes have more than one form. We can see this with the genes controlling certain facial features (Figure 12.1).

The gene for earlobe attachment has two forms, which produce the characteristics 'attached earlobe' and 'free earlobe'. These alternative forms of the gene are called alleles. Homologous chromosomes carry genes for the same features in the same sequence, but the alleles of the genes may not be the same (Figure 12.2). The DNA in the two chromosomes is not quite identical (exactly the same).

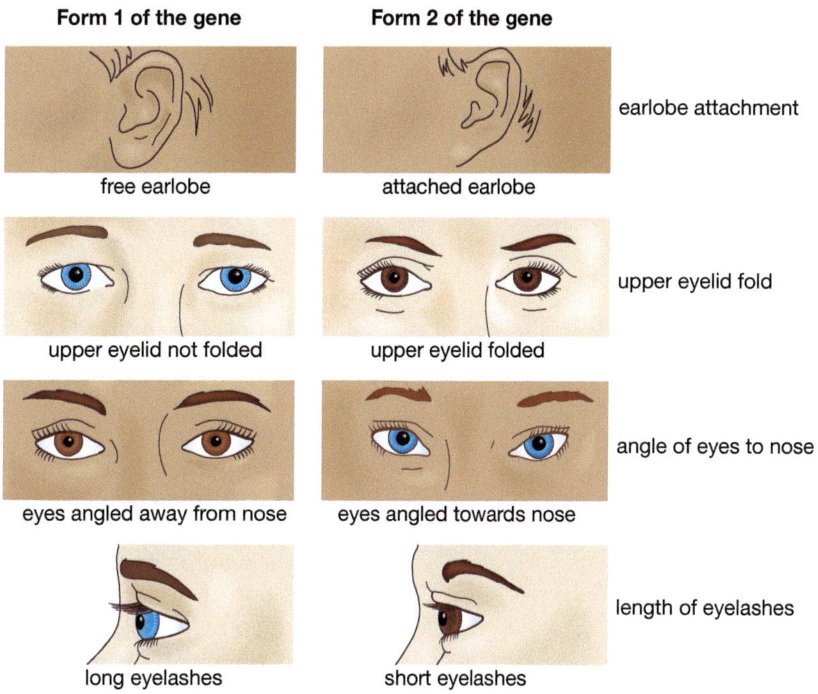

Figure 12.1 Four facial features, each controlled by two alternative forms of a gene.

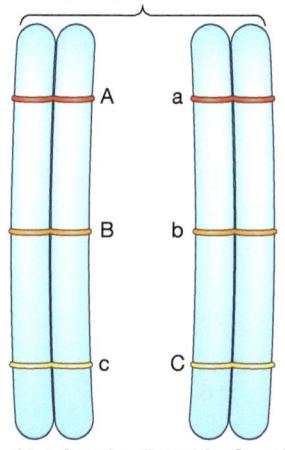

Figure 12.2 **A** and **a**, **B** and **b**, **C** and **c** are different alleles of three genes. Each pair of alleles controls the same feature but the individual alleles code for alternative ways in which the feature can be expressed. (Note that each chromosome is made of two identical chromatids.)

Suppose that, for the gene controlling earlobe attachment, a person has one allele for attached earlobes and one for free earlobes. What happens? Is one earlobe free and the other attached? Are they both partly attached? Neither. In this case, both earlobes are free. The 'free' allele is **dominant**. This means that it will show its effect, whether or not the allele for 'attached' is present. The allele for 'attached' is called **recessive**. The recessive allele will only show up (be expressed) in the appearance of the person if there is no dominant allele present. The scientific way to say that a gene 'shows up' in the appearance of a person is to say that it is 'expressed'.

DID YOU KNOW?
Some of the examples that follow are about genes in animals, but exactly the same principles apply to humans (or plants). The laws of genetics are the same in most organisms.

GENETIC CROSSES

We usually show the dominant allele of a gene with a capital letter (e.g. A) and the recessive allele with the corresponding small letter (a). Consider the coats of guinea pigs (Figure 12.3). Several features of their coats are controlled by single genes with dominant and recessive alleles. For example, short hair is dominant to long hair, straight hair is dominant to curls and the presence of rosettes is dominant to smooth hair.

Let us look at the genetics of the last characteristic. The allele for rosettes (R) is dominant over the allele for smooth hair (r). There are three possible combinations of alleles (Table 12.1).

Table 12.1 The genotypes controlling presence or absence of rosettes in guinea pigs.

Genotype	Description of genotype	Appearance of guinea pig (phenotype)
RR	homozygous dominant	rosettes
rr	homozygous recessive	smooth
Rr	heterozygous	rosettes

Figure 12.3 The coats of guinea pigs can be many different colours, with long or short hair and the presence or absence of circular whorls called rosettes.

> **KEY TERM**
>
> The noun from the word homozygous is 'homozygote'. You can say that an animal is homozygous or 'a homozygote'. Similarly, an animal can be heterozygous or 'a heterozygote'.

There are some new terms here that need to be explained:

- The **genotype** is the genetic make-up of an organism (RR, rr or Rr).
- The **phenotype** is the appearance of an organism (rosettes or smooth hair).
- **Homozygous** means that the two alleles of the gene are the same (RR or rr).
- **Heterozygous** means that the two alleles of the gene are different (Rr).

Remember that the dominant allele always shows itself (is expressed) if it is present. This means that the heterozygote and dominant homozygote have the same phenotype – both RR and Rr guinea pigs have rosettes. The recessive allele is only expressed in the recessive homozygote (rr).

CONSTRUCTING GENETIC DIAGRAMS

We can use the symbols for alleles to explain how they are passed on to the offspring of a cross between two animals. If we are only dealing with one gene, this is called a **monohybrid cross**. These diagrams depend on the fact that meiosis separates each pair of homologous chromosomes, so a gamete only receives one copy of each allele. The alleles are brought back together again at fertilisation. We will look at diagrams of some crosses. Firstly, consider a cross between two guinea pigs, one that is homozygous dominant and one that is homozygous recessive (Figure 12.4).

> **KEY TERM**
>
> The first generation produced by crossing the parents is called the **F1 generation**. This stands for 'first filial' generation, from the Latin for 'son of'. The next generation, produced by crossing the animals from the F_1 generation, is called the **F2 generation**.

Phenotype of parents	rosettes	smooth	Explanation
Genotypes of parents	RR	rr	Each parent has two copies of only one allele
Gametes (eggs and sperm)	R	r	Meiosis produces gametes with one allele
Genotype of first (F_1) generation	Rr		F_1 are all heterozygous
Phenotype of F_1 generation	all have rosettes		The R allele is expressed – all offspring have rosettes

▲ Figure 12.4 Cross between a guinea pig that is homozygous dominant for rosettes and one that is homozygous recessive for smooth hair (RR × rr). Note that the sex of the parents does not matter – the male guinea pig could be either RR or rr and vice versa. The cross would produce the same results either way.

Now look at what happens when two of the heterozygous animals from the F_1 generation are crossed (Figure 12.5).

Phenotype of parents	rosettes	rosettes	Explanation
Genotypes of parents (F_1)	Rr	Rr	Each parent has a copy of both alleles
Gametes (eggs and sperm)	R or r	R or r	Half of the gametes from each parent carry the R allele and half the r allele
Genotype of second (F_2) generation	RR Rr Rr rr		The ratio of the genotypes in the F_2 is the result of random fertilisation of eggs by sperm
	1RR : 2 Rr : 1 rr		
Phenotypes of F_2 generation	3 rosettes : 1 smooth		The *probability* is that the offspring will be in this ratio

▲ Figure 12.5 A cross between two heterozygous guinea pigs (Rr × Rr).

If you construct a genetic diagram like this, using arrows, it is easy to make mistakes. A better way is to use a diagram called a **Punnett square** (Figure 12.6).

Gametes	R	r
R	RR rosettes	Rr rosettes
r	Rr rosettes	rr rosettes

▲ Figure 12.6 The cross between two heterozygous guinea pigs drawn as a Punnett square.

KEY POINT

Ratios and probabilities can be written as a percentage or as a decimal fraction of 1. For example, a 3 : 1 ratio can be written as 75% : 25% or 0.75 : 0.25. A 50% probability can be written as 0.5.

The Punnett square produces the same expected ratio in the F_2 – 3 with rosettes : 1 smooth.

There is an important point here. These are only the *expected* ratios, based on the laws of probability. What this means is that if the cross produced a *large number* of offspring, the probability is that their genotypes would be in the ratio 1RR : 2Rr : 1rr or their phenotypes would be in the ratio 3 with rosettes : 1 smooth.

If you carried out the same cross lots of times with many pairs of heterozygous guinea pigs, you could produce 400 baby guinea pigs. With a large number like this, the most likely outcome would be that you would get 300 animals with rosettes and 100 with smooth coats. However, the laws of probability mean it is quite likely that the numbers would be 307 : 93, or 295 : 105 – close to the expected ratio, but not exactly the same.

With a small number of offspring, the ratios are likely to be different from the expected values. Just by chance, a pair of heterozygous guinea pigs could produce four babies with rosettes, or even four smooth ones. Fertilisation of an egg by a sperm is a random event, so it is impossible to be certain about the outcome.

PEDIGREES

Writing out a genetic cross is a useful way of showing how genes are passed through one or two generations, starting from the parents. However, if we want to show a family history of a genetic condition, we need more than this. We can use a diagram called a **pedigree**. A pedigree is a 'family tree' showing the inheritance of a gene.

Polydactyly is an inherited condition in which a person develops extra digits (fingers or toes) on the hands or feet. It is determined by a dominant allele. The recessive allele causes the normal number of digits to develop.

If we use the symbol D for the polydactyly allele and d for the normal-number allele, the possible genotypes and phenotypes are:

- DD – person has polydactyly (has two dominant polydactyly alleles)
- Dd – person has polydactyly (has a dominant polydactyly allele and a recessive normal allele)
- dd – person has the normal number of digits (has two recessive, normal-number alleles).

You might wonder why we do not use P and p to represent the alleles. This is because the letters P and p look very similar and could easily be confused.

The pedigree for polydactyly is shown in Figure 12.7.

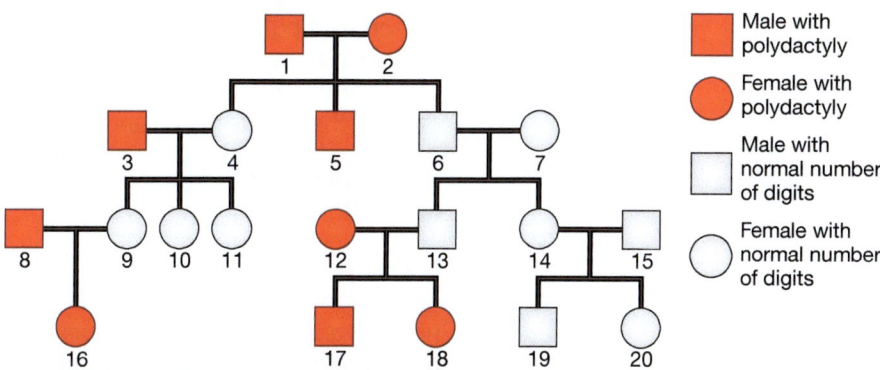

▲ Figure 12.7 A pedigree showing the inheritance of polydactyly in a family.

A pedigree contains a lot of information. In this example:

- four generations are shown (individuals are arranged in four horizontal lines)
- individuals 4, 5 and 6 are the children of individuals 1 and 2 (a family line connects each one directly to 1 and 2)
- individual 4 is the first-born child of 1 and 2 (the first-born child is shown to the left, then the second born to the right of this, then the third born and so on)
- individuals 3 and 7 are not children of 1 and 2 (no family line connects them directly to 1 and 2)
- individuals 3 and 4 are father and mother of the same children – as are 1 and 2, 6 and 7, 8 and 9, 12 and 13, 14 and 15 (a horizontal line joins them).

It is usually possible to work out which allele is dominant from a pedigree. You look for a situation where two parents show the same feature and at least one child shows the contrasting feature. In Figure 12.7, individuals 1 and 2 both have polydactyly, but children 4 and 6 do not. There is only one way to explain this:

- The normal alleles in 4 and 6 can only have come from their parents (1 and 2), so 1 and 2 must both carry normal alleles.
- 1 and 2 show polydactyly, so they must have polydactyly alleles as well.
- If they have both polydactyly alleles *and* normal alleles but show polydactyly, the polydactyly allele must be the dominant allele.

Now that we know which allele is dominant, we can work out most of the genotypes in the pedigree. All the people with the normal number of digits *must* have the genotype dd (if they had even one D allele, they would show polydactyly). All the people with polydactyly must have at least one polydactyly allele (they must be either DD or Dd).

From here, we can begin to work out the genotypes of the people with polydactyly. To do this, we need to remember that people with the normal number of digits must inherit one 'normal number' allele from each parent, and also that people with the normal number of digits will pass on one 'normal-number' allele to each of their children.

From this we can say that any person with polydactyly who has children with the normal number of digits must be heterozygous (the child must have inherited one of their two 'normal-number' alleles from this parent). Also, any person with polydactyly who has one parent with the normal number of digits must also be heterozygous (the 'normal-number' parent can only have passed on a 'normal-number' allele). Individuals 1, 2, 3, 16, 17 and 18 fall into one or other of these categories and must be heterozygous.

We can now add this genetic information to the pedigree. This is shown in Figure 12.8.

We are still uncertain about individuals 5, 8 and 12. They could be homozygous or heterozygous. For example, individuals 1 and 2 are both heterozygous. Figure 12.9 shows the possible outcomes from a genetic cross between them. Individual 5 could be any of the outcomes indicated by the shading. It is impossible to distinguish between DD and Dd.

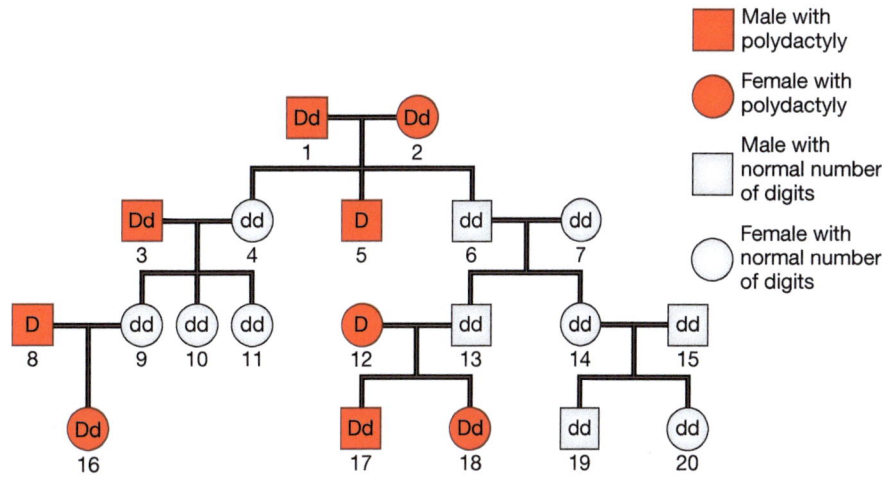

▲ Figure 12.8 A pedigree showing the inheritance of polydactyly in a family, with details of genotypes added.

Genotypes of parents Dd Dd

Gametes (D) and (d) (D) and (d)

Genotypes of children

	female gametes (D)	female gametes (d)
male gametes (D)	DD	Dd
male gametes (d)	Dd	dd

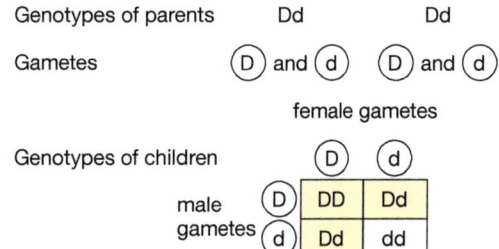

▲ Figure 12.9 Possible outcomes from a genetic cross between two parents, both heterozygous for polydactyly.

CODOMINANCE

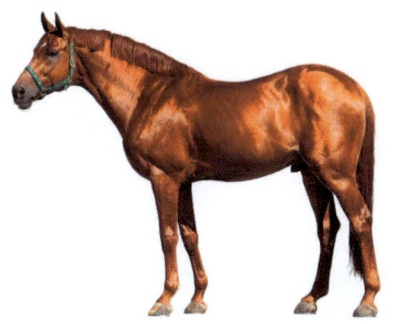

▲ Figure 12.10 A red roan horse produced from a cross between a chestnut horse and a white horse.

So far, all the examples of genetic crosses we have seen have involved *complete* dominance, where one dominant allele completely hides the effect of a second, or recessive allele. However, there are many genes with alleles that *both* contribute to the phenotype. If two alleles are expressed in the same phenotype, this is called **codominance**. The alleles are **codominant**.

If a chestnut (red-brown) horse is crossed with a white horse, all the foals resulting from the cross will be an intermediate colour, called red roan (Figure 12.10). The appearance of a third phenotype shows that there is codominance. We can represent the alleles for coat colour with symbols:

- R = allele for chestnut (red-brown) hair
- W = allele for white hair.

Figure 12.11 shows the cross between the parent animals. Note that the alleles for chestnut and white hair are given *different* letters, since one is not dominant over the other.

Genotypes of parents RR WW

Gametes all (R) all (W)

Genotypes of offspring all RW

	(R)	(R)
(W)	RW	RW
(W)	RW	RW

▲ Figure 12.11 Crossing a chestnut horse (genotype RR) with a white horse (genotype WW) produces a third phenotype, red roan (genotype RW).

When red roan horses are crossed together, all three phenotypes reappear, in the expected ratio:

1 chestnut: 2 red roan : 1 white (Figure 12.12).

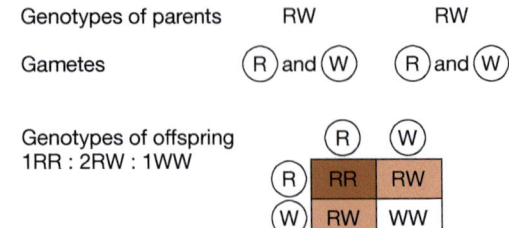

▲ Figure 12.12 Crossing red roan horses.

In fact, most genes do not show complete dominance. Genes can show a range of dominance, from complete dominance as in rosette/smooth hair in guinea pigs through to equal dominance as in the horse coat colour, where the new phenotype is halfway between the other two.

ABO BLOOD GROUPS

The inheritance of human ABO blood groups also shows codominance. However, the pattern of inheritance in blood groups is more complex than for coat colour in horses, as three different alleles are involved. When there are more than two alleles of one gene, this is known as the inheritance of **multiple alleles**.

The blood group of a person is the result of the presence or absence of two antigens, the **A** antigen and the **B** antigen, on the surface of the red blood cells. There are three alleles involved in the inheritance of these antigens.

- I^A – determines the production of the **A** antigen
- I^B – determines the production of the **B** antigen
- I^o – determines that neither antigen is produced.

A person inherits two alleles of the gene. The alleles I^A and I^B are codominant, but I^o is recessive to both. The possible genotypes and phenotypes are shown in Table 12.2.

▼ Table 12.2 Genotypes in blood groups.

Genotype	Antigen produced	Blood group
$I^A I^A$, $I^A I^o$	A	A
$I^B I^B$, $I^B I^o$	B	B
$I^A I^B$	A and B	AB
$I^o I^o$	neither	O

Parents who are heterozygous for blood group A and blood group B could produce four children, each with a different blood group (Figure 12.13).

You can interpret pedigrees of blood groups in the same way as other pedigrees. For example, in Figure 12.14, what are the blood groups of individuals 5 and 8?

Phenotypes of parents	Blood group A	Blood group B
Genotypes of parents	$I^A I^o$	× $I^B I^o$
Gametes	I^A and I^o	I^B and I^o

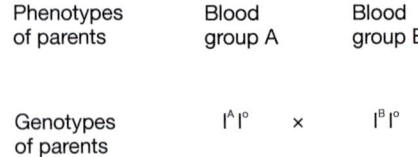

Phenotypes of children   1AB : 1A : 1B : 1O

▲ Figure 12.13 Possible offspring from two parents heterozygous for blood groups A and B.

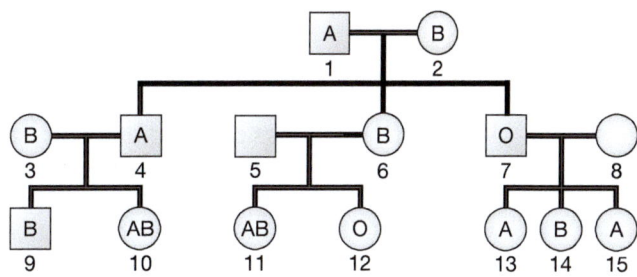

▲ Figure 12.14 Pedigree of human blood groups.

Individual 12 has two I⁰ alleles (since she is blood group O), so she must inherit one of them from individual 5. Individual 11 must inherit her I^A allele from individual 5, as her other parent is blood group B. Individual 5 therefore has the genotype I^AI^o and so must be blood group A.

Individuals 7 and 8 produce children with blood group A and blood group B. Individual 7 is blood group O and so both the I^A and I^B alleles must come from individual 8. Individual 8 is blood group AB.

SEX DETERMINATION

Our sex – whether we are male or female – is not under the control of a single gene. It is determined by the X and Y chromosomes – the **sex chromosomes**. All human cells contain 44 non-sex chromosomes. All cells of females (except the egg cells) also contain two X chromosomes. All cells of males (except the sperm) also contain one X and one Y chromosome. Our sex is effectively determined by the presence or absence of the Y chromosome. The full set of chromosomes of a man is shown in Figure 12.15.

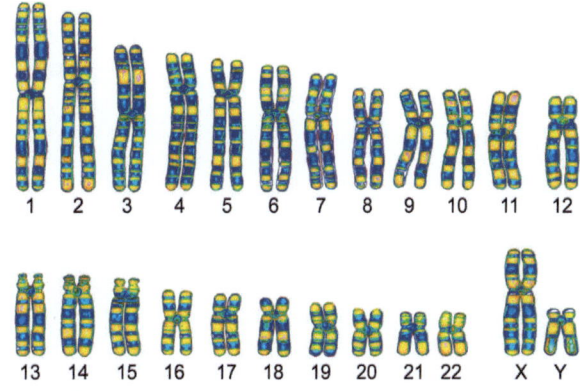

▲ Figure 12.15 The chromosomes of a man. Note that the full (diploid) set contains two of each of the chromosomes 1 to 22, plus the two sex chromosomes. The chromosomes of a woman are the same except that she has two X chromosomes. A picture of all the chromosomes in a cell is called a karyotype.

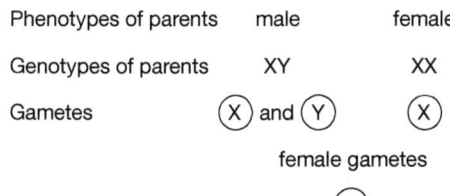

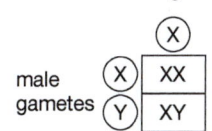

▲ Figure 12.16 Determination of sex in humans.

The inheritance of sex follows the pattern shown in Figure 12.16. In any one family, however, this ratio may well not be seen. Remember that predicted genetic ratios usually only happen when large numbers of offspring are involved. Overall, the ratio of male to female births is 1 : 1.

SEX-LINKED GENES

The sex chromosomes do not only determine sex. They also carry genes for other characteristics. These are called **sex-linked genes**. The Y chromosome is smaller than the X chromosome, so it contains fewer genes. This means that for some genes, a male will only have one allele of a pair present (on his X chromosome).

HAEMOPHILIA

An example of this is the gene that causes a blood disorder called **haemophilia**. When a healthy person's skin is cut, a clot forms. This prevents loss of blood and entry of bacteria (see Chapter 6). Clotting is a complex process, involving many chemicals in the blood. The commonest type of haemophilia is caused by a gene mutation that affects the production of one of these chemicals – a protein in the blood plasma. Without this protein, the blood of a person with haemophilia does not clot. They may need blood transfusions after minor injuries, and injections of the missing clotting factor.

The allele for haemophilia is recessive, and is given the symbol h. The allele for normal blood clotting is dominant (H). Since the gene is found only on the X chromosome, a man needs to inherit only one allele of the gene to have the disease. The genotype for this is shown as X^hY – notice there is no allele for this gene on the Y chromosome. A woman, with two X chromosomes, would need to inherit two copies of the faulty allele, which is shown as X^hX^h. If a woman has only one copy of the haemophilia allele (X^HX^h), she will not have the disease, because of the presence of the dominant H allele on one of her X chromosomes. However, she can pass the haemophilia allele to her children, so she is called a **carrier**.

> **DID YOU KNOW?**
> There are several types of haemophilia. They all have a genetic cause and are due to a lack of blood-clotting factors. Two forms are due to a sex-linked gene on the X chromosome.

Boys normally inherit the recessive allele from a carrier mother. This means it is possible for two healthy parents to have a son with haemophilia (Figure 12.17).

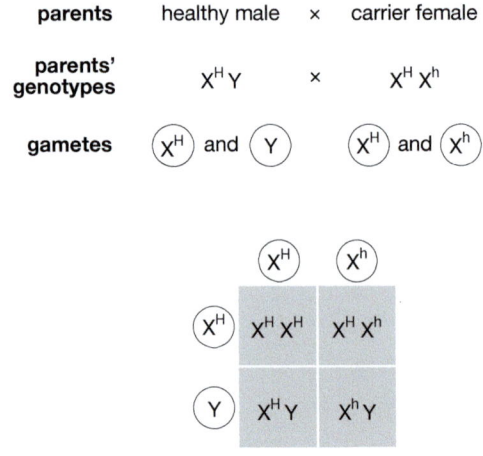

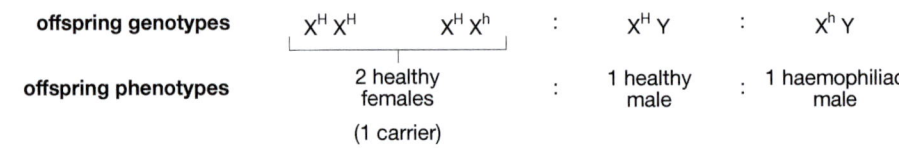

▲ Figure 12.17 Inheritance of haemophilia from a healthy father and an unaffected (carrier) mother.

For a girl to have haemophilia, she would have to be the daughter of a haemophiliac father and a carrier mother (Figure 12.18). This is possible, but much less likely. This means that haemophilia is much more common in boys than in girls. The most common form of haemophilia affects about one in 5 000–10 000 males.

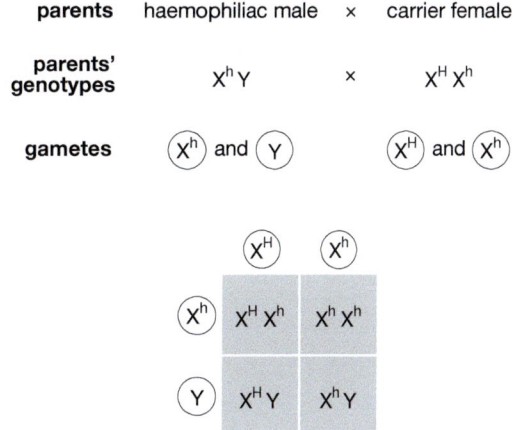

▲ Figure 12.18 Inheritance of haemophilia from a haemophiliac father and an unaffected (carrier) mother.

RED-GREEN COLOUR BLINDNESS

Another example of sex-linkage is the gene that causes **red-green colour blindness**. It is inherited in the same way as haemophilia, and is more common in boys than in girls. About 6% of boys have the condition, compared with 0.4% of girls. The normal (dominant) allele of the gene causes a protein to be produced that forms the pigment in the cones of the eye that detects green light (see Chapter 8). The mutant (recessive) allele does not cause this pigment to be formed. People without the normal allele cannot tell the difference between similar shades of green and red (Figure 12.19).

▲ Figure 12.19 A test for red-green colour blindness. People with normal colour vision can see a figure 8. People with red-green colour blindness only see a figure 3. People who are colour blind cannot distinguish very well between similar shades of red and green (or the orange and yellow wavelengths in between these colours).

An example of a cross involving colour blindness is shown in Figure 12.20.

Let: G = normal allele resulting in production of the pigment that detects green light
g = allele resulting in no production of the pigment that detects green light

parents	normal male	×	carrier female
parents' genotypes	$X^G Y$	×	$X^G X^g$
gametes	X^G and Y	×	X^G and X^g

	eggs	
sperm	X^G	X^g
X^G	$X^G X^G$	$X^G X^g$
Y	$X^G Y$	$X^g Y$

offspring genotypes	$X^G X^G$	$X^G X^g$	$X^G Y$	$X^g Y$
offspring phenotypes	normal female	carrier female	normal male	colour-blind male

▲ Figure 12.20 The inheritance of red-green colour blindness – a cross between a man with normal colour vision and a woman who is a carrier of the mutant allele. There is a 50% probability of a son inheriting the mutant allele from his mother and being colour blind. In this cross, there is a 50% probability that a daughter will be a carrier of the mutant allele.

CYSTIC FIBROSIS

Mucus is a fluid that is produced by cells in glands throughout the body, for example in the airways leading to the lungs, in the lining of the digestive system and in the reproductive organs. It acts as a lubricant. **Cystic fibrosis (CF)** is a disease caused by a mutated gene that affects the production of mucus. The gene has a high mutation rate, and if the mutation occurs in the cells producing eggs or sperm, the faulty allele may be passed to a person's children.

The dominant allele of this gene allows the production of normal mucus. The mutated recessive allele, however, results in the production of a thick, sticky (viscous) mucus. This does not flow easily like normal mucus and causes many problems. In the lungs, it is not easily removed by the cilia lining the airways. The build-up of mucus causes difficulties with breathing and gas exchange. Bacteria become trapped in the mucus and cause chest infections. CF causes a wide range of problems in other parts of the body.

> **DID YOU KNOW?**
> Over recent years, research has shown that there are over 1500 mutations that can affect the CF gene. The faulty gene affects a protein in the cell membrane of the mucus-producing cells. This protein is responsible for transporting ions such as chloride across the membrane (see Chapter 1, page 26). As yet, how this causes the symptoms of CF is not well understood.

> **DID YOU KNOW?**
> Some other problems that result from CF are:
> – The viscous mucus blocks the intestine, which causes poor absorption of nutrients and a low rate of growth.
> – The mucus blocks the pancreatic duct, so that enzymes from the pancreas cannot reach the small intestine. This affects the digestion of carbohydrates, lipids and proteins.
> – The mucus blocks the tubes leading from the testes, so most males with cystic fibrosis are infertile.

Because of the symptoms, people with cystic fibrosis often die young. However, modern treatments with drugs and other medications, as well as advances in physiotherapy, have greatly increased the average lifespan of people with CF. Gene therapy (see below) may offer a cure in the future.

For a person to be affected by CF, they must inherit a recessive allele from each parent, so both the mother and father must carry at least one recessive allele. Most commonly, each parent will be heterozygous for the CF condition (Figure 12.21). There is a 1 in 4 (25%) chance of a child from these parents developing cystic fibrosis.

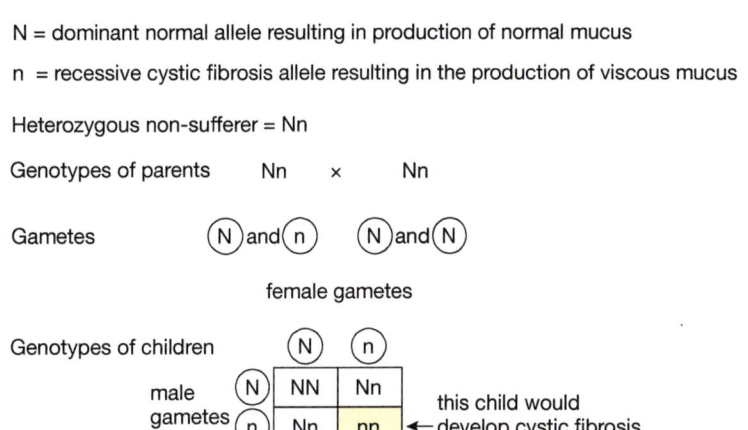

▲ Figure 12.21 The inheritance of cystic fibrosis.

USING GENE THERAPY TO TREAT CYSTIC FIBROSIS

Gene therapy is a technique by which a patient can be treated for a genetic disorder by having a functioning gene transferred into their DNA. It is possible to use a harmless virus as a 'carrier' or vector for transferring the gene. Research has been carried out in which scientists aimed to place a copy of the gene that produces normal mucus into affected cells. This has been tried in laboratory animals and in trials on human volunteers. The results so far have been disappointing, mainly because few cells have been found to take up the virus and express the gene – the lungs' defence system against infection stopped the virus from entering the affected cells.

In 2015 another method was tried, using a different vector. This involved using minute droplets of lipid enclosing the gene, which were sprayed into the patient's lungs. This method showed a small but significant improvement in lung function.

Research in the use of gene therapy to treat cystic fibrosis continues, in the hope that one day it may provide a cure for this terrible disease.

GENES AND ENVIRONMENT BOTH PRODUCE VARIATION

Identical twins are formed from the same zygote. When the zygote divides by mitosis, the first two cells formed do not stay together, but separate. Each cell then continues to divide and produces an embryo. Because they each come from the same zygote, they are genetically identical. Identical twins usually look very alike, and often develop similar talents. However, they never look exactly the same. This is especially true if, for some reason, they live apart when they are growing up (for example, if members of a pair of twins are adopted by different sets of parents). The differences are due to their *environment* affecting their appearance. In other words, both genes and environment have an effect on human variation.

A good example of this is adult human body mass (weight) or height. A person's growth is affected by many genes. There are genes that influence protein synthesis in muscles, bone development, production of hormones, etc. These will all have an effect on the growth of the body, but growth will take place only if the person has access to a healthy balanced diet. Their weight will largely depend on environmental factors – in this case, availability of food.

Skin colour is inherited (Figure 12.22) but it is also affected by the environment, in that exposure to the sun increases the amount of melanin in the skin, causing it to darken.

One of the most controversial issues in human variation is how much of human intelligence is genetic or environmental. Does an intelligent child get her genes for intelligence from her parents, or is much of it a result of the environment in which she grew up? This is sometimes called the nature/nurture argument. It has never been answered satisfactorily, and probably never will be. It is certainly true, however, that a child's intellectual development is affected by access to books and a good education, as well as a healthy diet and good medical care.

▲ Figure 12.22 These children have inherited different combinations of genes for skin colour from their parents.

LOOKING AHEAD – DIHYBRID INHERITANCE

For your International GCSE, you only have to deal with the inheritance of single genes (monohybrid inheritance). If you study biology beyond this level, you will probably have to construct diagrams to show the inheritance of two genes. This is called dihybrid inheritance.

For example, guinea pigs can have short hair or long hair, and dark eyes or pink eyes. The allele for short hair (H) is dominant to the allele for long hair (h), and the allele for dark eyes (E) is dominant to the allele for pink eyes (e).

If a guinea pig that is homozygous dominant for both hair length and eye colour (HHEE) is crossed with one that is homozygous recessive for both features (hhee), all the offspring in the F_1 generation will be heterozygous for both characteristics. They will all have short hair and dark eyes (Figure 12.23).

Genotypes of parents:	HHEE	×	hhee
Genotypes of gametes:	all (HE)		all (he)
Genotype of F_1		all HhEe	
Phenotype of F_1		all short hair, dark eyes	

▲ Figure 12.23 Cross between a guinea pig that is homozygous dominant for hair length and eye colour, with one that is homozygous recessive for both genes. Note that the gene is not sex linked, so either parent could be the homozygous dominant one.

KEY POINT

Note that in a dihybrid cross, each gamete contains *two* alleles – one for each gene.

Assuming the genes for hair length and eye colour are on different chromosomes, then gametes with four possible combinations of alleles can be produced from the F_1 genotype (HhEe). They are (HE), (He), (hE) and (he). The combinations are a result of meiosis (see Chapter 11).

If two F$_1$ guinea pigs are crossed, the possible outcomes in the F$_2$ are as shown in Figure 12.24.

	female gametes			
male gametes	HE	He	hE	he
HE	HHEE short hair, dark eyes	HHEe short hair, dark eyes	HhEE short hair, dark eyes	HhEe short hair, dark eyes
He	HHEe short hair, dark eyes	HHee short hair, pink eyes	HhEe short hair, dark eyes	Hhee short hair, pink eyes
hE	HhEE short hair, dark eyes	HhEe short hair, dark eyes	hhEE long hair, dark eyes	hhEe long hair, dark eyes
he	HhEe short hair, dark eyes	Hhee short hair, pink eyes	hhEe long hair, dark eyes	hhee long hair, pink eyes

▲ Figure 12.24 Dihybrid cross between two double heterozygous guinea pigs.

The expected ratio of phenotypes in the F$_2$ is:

9 short hair, dark eyes : 3 short hair, pink eyes : 3 long hair, dark eyes : 1 long hair, pink eyes

9 : 3 : 3 : 1 is the usual ratio obtained from a dihybrid cross between two double heterozygotes.

CHAPTER QUESTIONS

SKILLS REASONING

1 In guinea pigs, the allele for short hair is dominant to that for long hair.
   a Two short-haired guinea pigs were bred and their offspring included some long-haired guinea pigs. Explain these results.
   b How could you find out if a short-haired guinea pig was homozygous or heterozygous for hair length?

SKILLS ANALYSIS

2 The diagram shows the inheritance of phenylthiocarbamide (PTC) tasting in a family. Although PTC has a very bitter taste, some people cannot taste it.

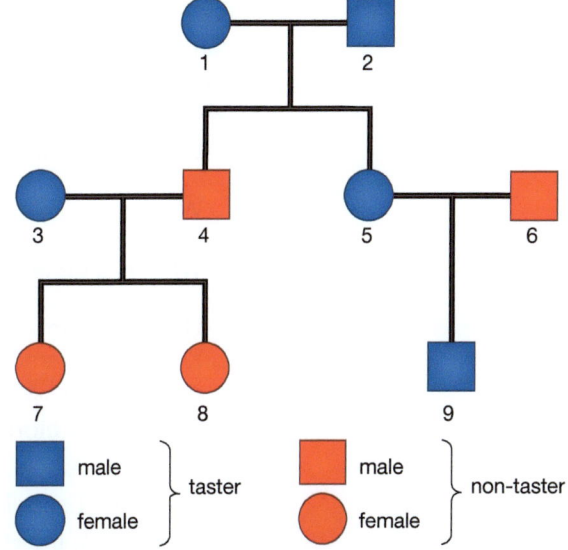

a What evidence in the diagram suggests that the allele for PTC tasting is dominant?

b Using T to represent the tasting allele and t to represent the non-tasting allele, give the genotypes of individuals 3 and 7. Explain how you arrived at your answers.

c Why can we not be sure of the genotype of individual 5?

d If individuals 3 and 4 had another child, what is the chance that the child would be able to taste PTC? Construct a genetic diagram to show how you arrived at your answer.

3 The inheritance of ABO blood groups is controlled by three alleles of the same gene. The alleles are given the symbols I^A, I^B and I^o. Alleles I^A and I^B are codominant. Both alleles I^A and I^B are dominant to allele I^o.

a Explain what is meant by the terms:
  i allele
  ii codominant.

b Complete the table to show all the possible genotypes for each blood group. One has been done for you.

Blood group	Genotype
A	$I^A I^A$
B	
AB	
O	

c A woman with blood group A married a man with blood group B. They had one child with blood group A, and another child with blood group O. Construct a genetic diagram to show the results of this cross. Indicate the phenotypes of the children on your diagram.

4 A couple you know have four children, all girls. They insist that there is a high probability their next child will be a boy. Do you agree with them? What is the probability that their next child will be a boy? Draw a diagram to explain this.

5 Haemophilia is a sex-linked genetic disorder. It was present in several European royal families during the 19th and 20th centuries, and seems to have started as a new mutation with Britain's Queen Victoria. Although her husband Prince Albert did not have haemophilia, Queen Victoria was a carrier and passed it to her son, Prince Leopold, who was a haemophiliac. Victoria and Albert's oldest daughter, Victoria, the Princess Royal, escaped having the faulty gene.

a What are the symptoms of haemophilia?

b What is a sex-linked gene?

c What is a gene mutation?

d Using the symbol X^H for the normal allele on the X chromosome, X^h for the haemophilia allele on the X chromosome and Y for the Y chromosome, state the genotypes of:
  i Queen Victoria
  ii Prince Albert
  iii Prince Leopold
  iv Victoria, the Princess Royal.

6 Cystic fibrosis is an inherited condition. The diagram shows the incidence of cystic fibrosis in a family over four generations.

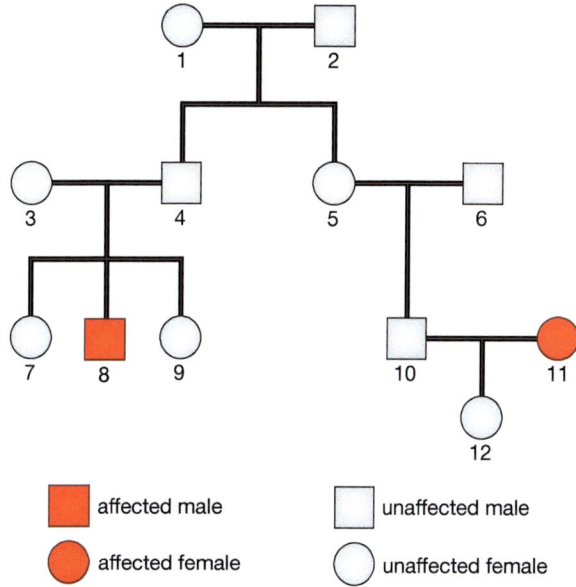

a What evidence in the pedigree suggests that cystic fibrosis is determined by a recessive allele?
b What are the genotypes of individuals 3, 4 and 11? Explain your answers.
c Draw genetic diagrams to work out the probability that the next child born to individuals 10 and 11 will:
 i be male
 ii have cystic fibrosis.

13 MICROORGANISMS

You have seen that some diseases are caused by an unhealthy lifestyle, such as smoking, drinking too much alcohol or having a poor diet. Other diseases are genetic, caused by faulty genes. These are examples of non-infectious diseases. Some diseases are infectious – they are caused by pathogenic microorganisms. In this chapter, you will learn about these microorganisms and how they cause disease. You will also look at some of the roles of non-pathogenic bacteria.

LEARNING OBJECTIVES

- Know that diseases are caused by pathogenic microorganisms
- Describe the structure, nutrition and reproduction of bacteria, including the interpretation of bacterial growth curves
- Describe the structure and reproduction of viruses
- Understand the general course of a disease – infection, incubation and symptoms
- Describe the methods of transmission, treatment and prevention of spread of Ebola and HIV (human immunodeficiency virus, the virus that causes AIDS)
- Explain the importance of oral rehydration therapy
- Know the methods of transmission, treatment and prevention of spread of cholera and gonorrhoea
- Describe the methods of transmission, treatment and prevention of spread of athlete's foot
- Explain the role of the mosquito in the transmission of malaria, and the housefly in the transmission of typhoid
- Describe the treatment and prevention of spread of malaria and typhoid
- Understand the antibody–antigen reaction
- Understand the differences between natural and artificial immunity, and active and passive immunity
- Explain how vaccines work to prevent the spread of disease
- Know the sources and roles of antibiotics
- Explain how resistant pathogens such as MRSA arise and why they are a cause for concern
- Describe how to investigate the effect of antibiotics and antibacterial agents on the growth of bacterial cultures
- Understand the role of non-pathogenic bacteria and fungi as decomposers, useful to humans in the decomposition of organic matter
- Know the processes of sewage treatment in a modern sewage works and a pit latrine, including the role of aerobic and anaerobic microorganisms in the breakdown of sewage

WHAT IS A DISEASE?

It is not easy to define the word 'disease'. Disease does not just mean the absence of health – if you are less fit than you could be, or cannot sleep because you are thinking about an approaching exam, you might be considered unhealthy, but this does not mean you have a disease. One definition of disease is 'a condition with a specific cause in which part or all of the body functions abnormally and less efficiently'. A disease may be caused by unhealthy activities, such as smoking or drinking alcohol; it can be genetic, such as the mutations responsible for haemophilia or cystic fibrosis; or, most commonly, the cause is a **microorganism**. Microorganisms are responsible for causing *infectious* diseases – those that can be transmitted from one person to another.

DID YOU KNOW?
Microorganisms are very small organisms that can only be seen through a microscope.

DID YOU KNOW?
Protozoa are single-celled (unicellular) organisms with cells that are similar in structure and function to the cells of animals.

Microorganisms that cause diseases are known as pathogens or pathogenic organisms. There are several different types, including **bacteria**, **viruses**, some **fungi** and **protozoa**. Table 13.1 and Figure 13.1 give details of some pathogenic organisms.

▼ Table 13.1 The main types of pathogenic microorganism and some diseases they cause.

Type of microorganism	How the microorganism causes disease	Examples of diseases caused
bacteria	Bacteria release poisons called toxins as they multiply. The toxins affect cells in the region of the infection and sometimes in other parts of the body as well.	typhoid, tuberculosis (TB), gonorrhoea, cholera, pneumonia
viruses	Viruses enter a living cell and disrupt the metabolic systems of that cell. The genetic material of the virus takes over the cell and instructs it to produce more viruses.	influenza ('flu), poliomyelitis (polio), human immunodeficiency virus (HIV), measles, rubella, common cold, Ebola
fungi	When fungi grow in or on the body, their fine threads (hyphae) secrete digestive enzymes onto the tissues, breaking them down. Growth of hyphae also physically damages the tissues. Some fungi secrete toxins. Others cause an allergic reaction.	athlete's foot, thrush, 'ringworm' (a skin disease)
protozoa	There is no set pattern as to how protozoa cause disease.	malaria, trypanosomiasis (sleeping sickness)

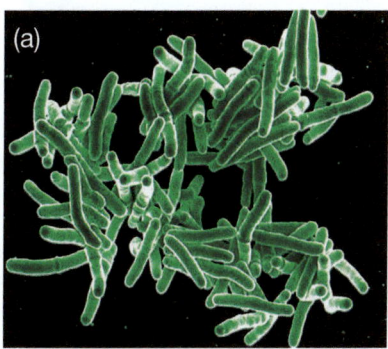
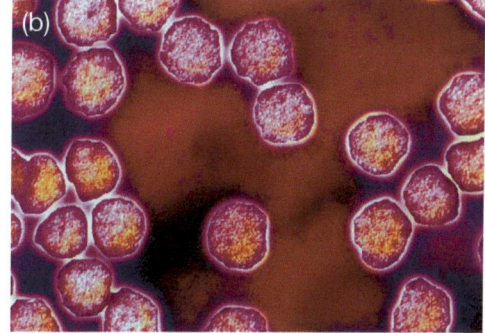
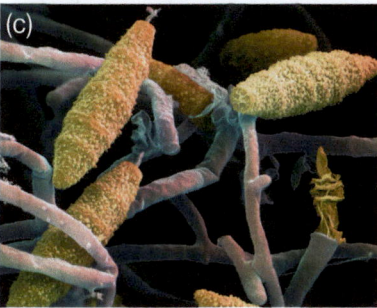
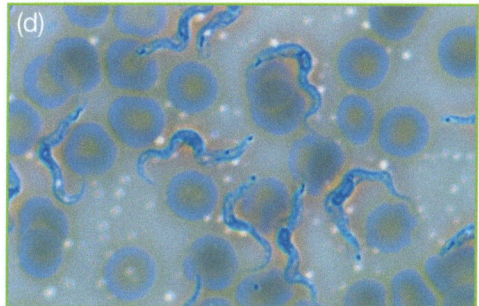

▲ Figure 13.1 Some pathogenic microorganisms: (a) the bacterium that causes tuberculosis; (b) the rubella virus; (c) the mould fungus responsible for ringworm; (d) blood containing the protozoan *Trypanosoma gambiense* – the cause of sleeping sickness.

We will now look in more detail at the structure and function of two of these groups – bacteria and viruses.

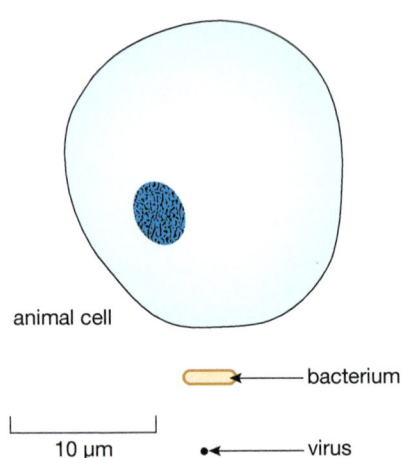

▲ Figure 13.2 A bacterium is much smaller than an animal cell. The relative size of a virus is also shown.

BACTERIA

Bacteria are small single-celled organisms. Their cells are much smaller than those of animals or plants and they have a much simpler structure. To give you some idea of their size, a typical animal cell might be 10 to 50 µm in diameter (1 µm, or one micrometre, is a millionth of a metre). Compared with this, a typical bacterium is only 1 to 5 µm in length and its volume is thousands of times smaller than that of the animal cell (Figure 13.2).

There are three basic shapes of bacteria – spheres, rods and spirals – but they all have a similar internal structure (Figure 13.3).

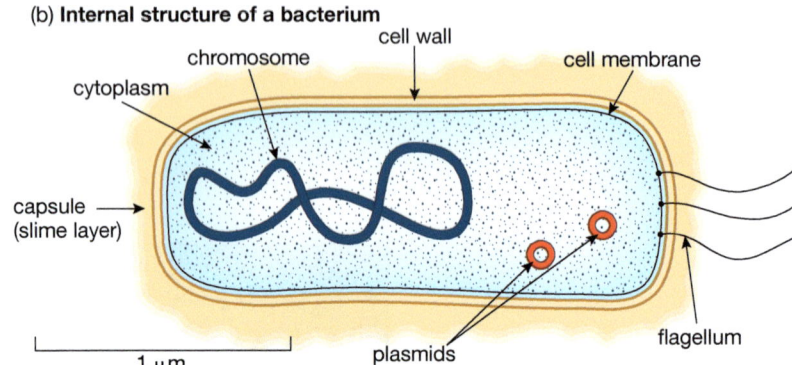

▲ Figure 13.3 (a) Different shapes of bacterial cells. (b) The general structure of a bacterium.

All bacteria are surrounded by a **cell wall**, which protects the bacterium and keeps the shape of the cell. Some species have another layer outside this wall, called a **capsule** or slime layer. Both give the bacterium extra protection. Underneath the cell wall is the cell membrane. The middle of the cell is made of cytoplasm. One major difference between a bacterial cell and the more complex cells of animals and plants is that the bacterium has no nucleus. Instead, its genetic material (DNA) is in a single chromosome loose in the cytoplasm, in the shape of a circular loop.

Some bacteria can swim, and are propelled through water by corkscrew-like movements of structures called **flagella** (singular **flagellum**). However, most bacteria do not have flagella and cannot move by themselves. Other structures present in the cytoplasm include plasmids. These are small, circular rings of DNA, carrying some of the bacterium's genes. About three-quarters of all known species of bacteria contain plasmids.

Most species of bacteria are harmless to humans. Some free-living species contain chlorophyll in their cytoplasm and can make their own food by photosynthesis, like plants. Most bacteria, along with fungi, are important **decomposers** (see later in this chapter). They feed on dead organisms and waste products in the soil and recycle them. They are also important in sewage treatment. Some bacteria are even used by humans to make foods such as cheese and yoghurt. Relatively few species are pathogens (Figure 13.4).

> **REMINDER**
>
> We have an important use for plasmids. They are used as a tool in genetic engineering, to transfer genes into bacteria. Genetically modified (GM) bacteria are used to make useful products, such as human insulin for treating diabetes (see Chapter 9).

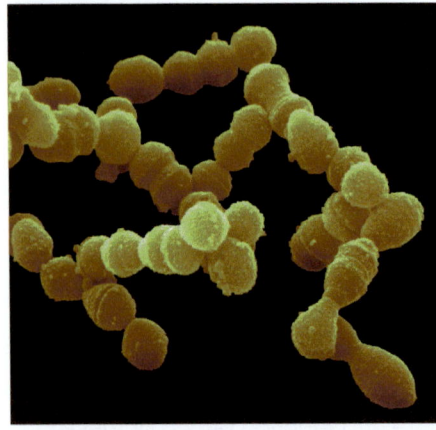

▲ Figure 13.4 Rounded cells (cocci) of the bacterium *Streptococcus pneumoniae*, one cause of the lung disease pneumonia.

BACTERIAL GROWTH CURVES

Bacteria reproduce asexually by a process called **binary fission**, which means 'splitting into two'. The bacterial cell grows until it has doubled in length. In the cytoplasm, the single chromosome is copied, and the two copies separate to opposite ends of the cell. The cell membrane folds inwards, forming a double layer across the middle of the cell. Two cell walls are then made, between the two membranes. Finally the two daughter cells separate and grow until they are ready to divide again (Figure 13.5).

When scientists want to carry out research on bacteria, they can grow or culture them in Petri dishes containing **agar**. Alternatively, they may grow them in a liquid called a 'broth'.

> **DID YOU KNOW?**
> Students sometimes think that binary fission is the same as mitosis. It is not – binary fission is a much simpler process. The cell does not go through the complex stages that happen during mitosis.

> **DID YOU KNOW?**
> Agar is a jelly made from seaweed. Nutrients are added to the agar to supply food for the bacteria. Agar is useful for growing bacteria because they cannot feed on the agar itself.

Bacteria can reproduce very quickly. When they are supplied with nutrients and favourable conditions of temperature and pH, some species can divide every 15–20 minutes. This means that thousands of bacteria can be produced from one cell within a few hours. After a few days, a culture may contain many millions of bacteria. We can count the number of cells in a broth and plot their numbers as a growth curve (Figure 13.6).

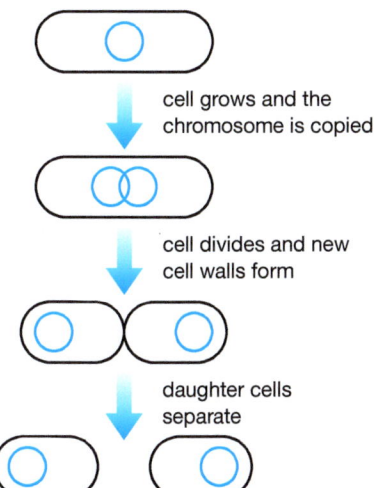

▲ Figure 13.5 Bacteria reproduce by binary fission.

> **DID YOU KNOW?**
> The growth phase is also called the *exponential phase*. This is because the cell numbers double in unit time intervals (e.g. every 20 minutes). In maths, this is called an exponential increase.

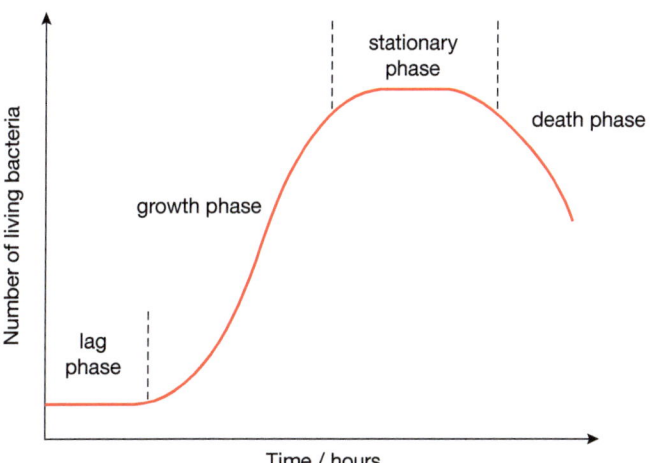

▲ Figure 13.6 A typical bacterial growth curve showing lag, growth, stationary, and death phases.

The curve starts with a lag phase, where the bacteria adjust to their new environment. They may need to switch on genes needed to make enzymes to digest the nutrients in the broth. Growth then accelerates into the growth phase, where the cells divide in equal time intervals. Towards the end of the growth phase, growth slows down and cells start to die. The bacteria enter the stationary phase, when the rate of formation of new cells equals the rate of cell death. This is followed by the death phase, when the number of cells that die overtakes the number of new cells.

What causes the stationary and death phases? There are several possible **limiting factors**. As they grow and reproduce, the bacteria may use up the food or oxygen in the broth, or they may produce poisonous waste products. Lack of nutrients or build-up of waste will slow down the rate of reproduction and cause the death of more cells. A limiting factor is something that slows down (limits) a process, e.g. a nutrient that is in short supply.

Despite the simple structure of the bacterial cell, it is still a living cell that performs the normal processes of life, such as respiration, feeding, excretion, growth and reproduction. This is very different from the next group, the viruses, which are much simpler microorganisms.

VIRUSES

All viruses are **parasites**, and can only reproduce inside living cells. There are many different types of virus. Some live in the cells of animals or plants, and there are even viruses which infect bacteria. Viruses are much smaller than bacterial cells – most are between 0.01 and 0.1 µm in diameter (see Figure 13.2).

> **DID YOU KNOW?**
> A parasite is an organism that lives in or on another organism, called its host, and causes it harm.

Viruses are not made of cells. A virus particle is very simple. It has no nucleus or cytoplasm, and is composed of a core of genetic material surrounded by a protein coat (Figure 13.7). The genetic material can be either DNA or RNA (see Chapter 1). In either case, the genetic material makes up just a few genes – all that is needed for the virus to reproduce inside its host cell.

Sometimes a virus has a membrane called an envelope around it, but the virus does not make this. Instead it is 'stolen' from the surface membrane of the host cell.

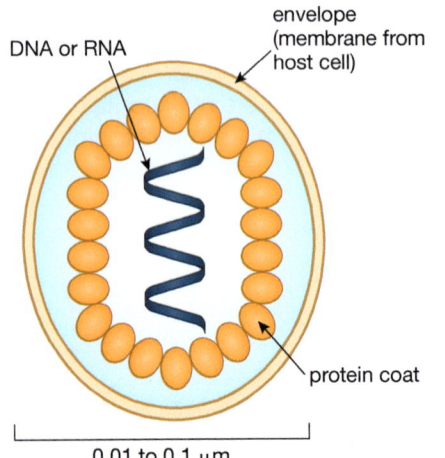

▲ Figure 13.7 The structure of a typical virus, such as the type causing influenza ('flu).

Viruses do not feed, respire, excrete, move, grow or respond to their surroundings. They do not display any of the normal characteristics of living things except reproduction, and they can only do this parasitically. This is why most biologists do not consider viruses to be living organisms. You can think of them as being on the border between an organism and a non-living chemical.

A virus reproduces by entering the host cell and taking over the host's genetic machinery to make more virus particles. After many virus particles have been made, the host cell dies and the particles are released to infect more cells. Many human diseases are caused in this way, such as influenza ('flu'). Other examples include colds, measles, mumps, polio and rubella ('German measles'). Of course, the reproduction process does not continue forever. Usually, the body's immune system destroys the virus and the person recovers. Sometimes, however, a virus cannot be destroyed by the immune system quickly enough, and it may cause permanent damage or death. With other infections, the virus may attack cells of the immune system itself. This is the case with **HIV** (the **human immunodeficiency virus**), which causes the condition called **AIDS** (**acquired immune deficiency syndrome**).

THE COURSE OF A DISEASE

The general course of a disease starts with **infection** by the pathogenic organism, followed by an **incubation period**, until a person shows the **symptoms** of the disease.

Infection means the transfer of a pathogen to a person. Table 13.2 summarises the main ways in which people can become infected by various diseases.

Table 13.2 Methods of infection by various diseases.

Method of infection	How transmission happens	Examples of diseases
droplet infection	Many of the microorganisms transmitted in this way cause respiratory diseases (diseases that affect the airways of the lungs). The organisms are carried in tiny droplets through the air when an infected person coughs or sneezes. They are inhaled by other people.	common cold, influenza, tuberculosis, pneumonia
drinking contaminated water	The microorganisms transmitted in this way often infect regions of the gut. When a person drinks unclean water containing the organisms, they colonise a suitable area of the gut and reproduce. They are passed out with faeces and find their way back into the water.	cholera, typhoid, polio
eating contaminated food	Most food poisoning is bacterial but some viruses are transmitted this way. The organisms initially infect a region of the gut.	typhoid, polio, salmonellosis, listeriosis, botulism
direct contact	Many skin infections, such as athlete's foot, are spread by direct contact with an infected person or contact with a surface carrying the organism. Some diseases (e.g. Ebola) are transmitted by direct contact with bodily fluids (e.g. blood, faeces and vomit).	athlete's foot, ringworm, Ebola
sexual intercourse	Organisms infecting the sex organs can be passed from one sexual partner to another during intercourse. Some (such as the fungus thrush, which causes candidiasis) are transmitted by direct body contact. Others (such as the AIDS virus) are transmitted in semen or vaginal secretions. Some (such as syphilis) can be transmitted in saliva.	chlamydia, syphilis, AIDS, gonorrhoea
blood-to-blood contact	Many sexually transmitted diseases can also be transmitted in this way. Drug users sharing an infected needle can transmit AIDS.	AIDS, hepatitis B
animal vectors	Many diseases are transmitted by insect bites. Mosquitoes spread malaria and tsetse flies spread sleeping sickness. In both cases, the pathogen is transmitted when the insect bites humans in order to feed on blood. Flies can carry microorganisms from faeces onto food.	malaria, sleeping sickness, typhoid, salmonellosis

After infection there is an incubation period. This is the time between when a person is first infected with the pathogen and when they first show signs and symptoms of the disease. During the incubation period, the infected person may not feel sick, but could be infectious to others. Incubation periods vary greatly between diseases, from hours to months. Some diseases, such as leprosy, can have an incubation period lasting years. The incubation period is followed by a time when the patient shows symptoms, when they feel unwell and go to the doctor.

> **DID YOU KNOW?**
> To a doctor, 'signs' and 'symptoms' of a disease have slightly different meanings.
> – A *sign* is visible to other people. It can be seen, heard or measured. For example, a doctor might listen to a patient's chest with a stethoscope to hear signs of a chest infection, or measure their blood pressure to check for a heart problem.
> – A *symptom* is not usually visible to other people. It is what the patient is experiencing as a result of the disease, such as pain, chills, dizziness or nausea. The symptoms are the first thing that a patient notices that make them go to the doctor.
>
> Non-medical people like ourselves normally call both 'symptoms'.

Clearly there are very many diseases you could study. We are going to look at a small number in detail. Two are caused by viruses (HIV and Ebola), three by bacteria (cholera, typhoid and gonorrhoea), one by a fungus (athlete's foot) and one by a protozoan (malaria). You will learn about the method of transmission, treatment and prevention of spread of each of these diseases.

> **DID YOU KNOW?**
>
> Some diseases are only found in certain parts of the world. For example, malaria is common in tropical and subtropical countries. If a disease is always present in the population of a particular geographical area, it is called *endemic*. This must not be confused with an *epidemic*, which is a widespread outbreak of an infectious disease, with many people becoming infected at the same time, spreading over a wide area. If the disease spreads across the world, we call it a *pandemic*. There have been 'flu pandemics in 1918 (Spanish 'flu), 1957 (Asian 'flu), 1968 (Hong Kong 'flu) and 2009 (swine 'flu). Over 50 million people died from the Spanish 'flu pandemic in 1918, more than were killed in the whole of the First World War (1914–1918).

HIV/AIDS

AIDS (acquired immune deficiency syndrome) is one of the world's most significant killers. AIDS is not actually a disease, but a 'syndrome', which is a set of symptoms caused by a medical condition. AIDS is caused by a virus called HIV, which stands for human immunodeficiency virus. Figure 13.8 shows the structure of the virus.

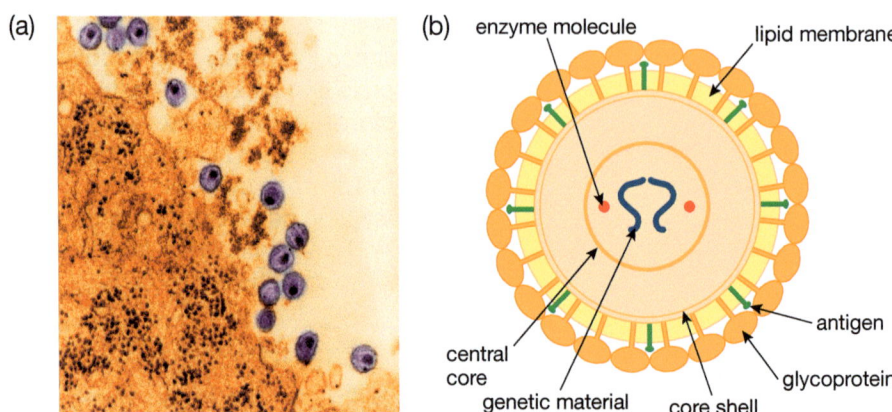

▲ Figure 13.8 (a) The human immunodeficiency virus (purple particles) on the surface of a human cell. (b) The structure of the human immunodeficiency virus.

HIV initially infects a type of white blood cell, a lymphocyte called a T-helper cell. The role of these cells is to help other lymphocytes to become active and start fighting infections. A typical infection with HIV follows this course:

1. The genetic material of HIV is RNA. The virus makes DNA from the RNA, and this DNA becomes incorporated into the DNA of the T-helper cell.
2. At some point in the infection, the HIV DNA is activated. It then instructs the lymphocyte to make HIV proteins and more RNA.
3. The HIV proteins and RNA are assembled into new virus particles.
4. Some of the HIV proteins end up as antigens on the surface of the cell.

> **DID YOU KNOW?**
> Part of this immune response to infection by HIV involves the production of antibodies to destroy the virus. At this stage, the person is said to be HIV positive, because their blood gives a positive result when tested for HIV antibodies.

5. These HIV proteins are recognised by the person's immune system as being 'foreign' to the body.
6. The lymphocyte is destroyed by the immune system.
7. The assembled virus particles escape to infect other lymphocytes.
8. The cycle repeats itself for as long as the body can replace the lymphocytes that have been destroyed.
9. Eventually the body will not be able to replace the lymphocytes as quickly as they are being destroyed.
10. The number of free viruses in the blood increases rapidly and HIV may infect other areas of the body, including the brain.
11. The immune system is severely damaged, and other disease-causing microorganisms infect the body.
12. Death is usually the result of 'opportunistic' infection by TB and pneumonia, because the immune system can no longer destroy the organisms that cause these diseases. Death can also be caused by rare forms of cancer, such as Kaposi's sarcoma (Figure 13.9).

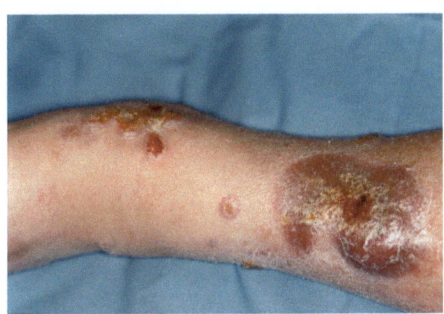

▲ Figure 13.9 Skin lesions on the leg of a person suffering from Kaposi's sarcoma. This cancer is caused by an opportunistic virus and is much more common in AIDS patients than in the rest of the population.

> **DID YOU KNOW?**
> The period during which the body replaces the lymphocytes as fast as they are destroyed is called the latency period. It can last for up to 20 years. The person shows no symptoms of AIDS during this period, but will be highly infectious to others.

Figure 13.10 shows how the levels of virus particles, HIV antibodies and T-helper cells in the blood change during the course of an HIV infection.

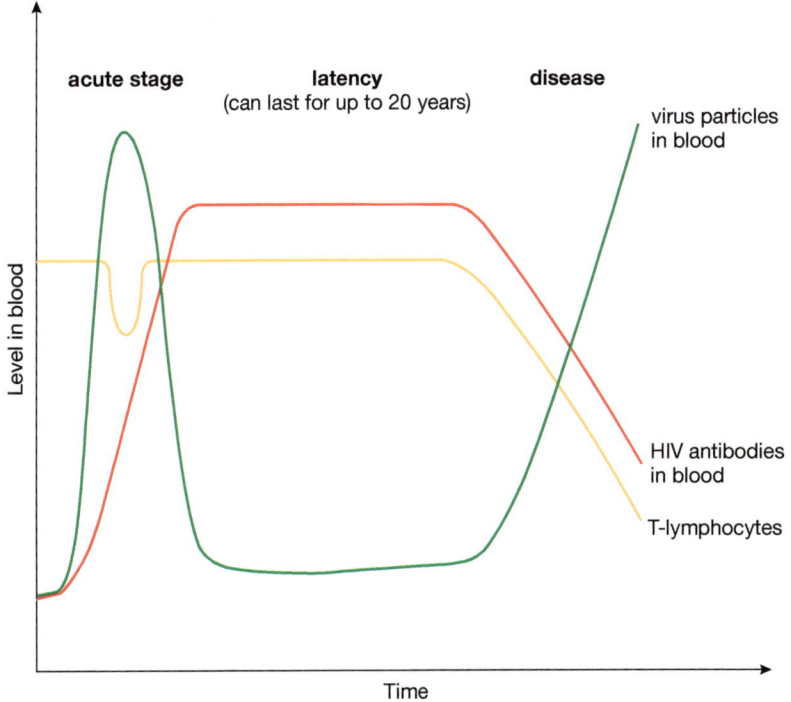

▲ Figure 13.10 Changes taking place in the blood of a person infected with HIV that lead to the development of AIDS.

> **DID YOU KNOW?**
> The real significance of AIDS is that the cells being destroyed are the cells needed to help the other lymphocytes destroy the infected cells.

The main ways that HIV is transmitted are:

- by unprotected sexual intercourse (i.e. without using a condom)
- by blood-to-blood contact, for example, when drug users share an infected needle, or if infected blood is given to a patient during a transfusion.

HIV cannot be transmitted by kissing, sharing from a cup or glass, giving blood, skin-to-skin contact, or sitting on the same toilet seat as an infected person.

As yet, there is no effective vaccine against HIV, although a number of antiviral drugs can delay the onset of AIDS.

The transmission of HIV can be controlled by a number of measures. These include:

- use of condoms – although transmission can still occur, its occurrence is greatly reduced
- drug users using new, sterile needles, and not sharing needles
- limiting the number of sexual partners
- testing the blood used for transfusions.

EBOLA

Ebola is a severe, often fatal disease, caused by a virus. The disease first appeared in Africa in 1976, and there have been several outbreaks since, the largest in 2014 in the African countries of Guinea, Liberia and Sierra Leone.

The incubation period for Ebola is 2 to 21 days, and humans are not infectious until they develop the symptoms. The first symptoms are fever, muscle pain, headache and a sore throat. These are followed by vomiting, diarrhoea, a rash, and damaged liver and kidney function. In some patients, there is internal and external bleeding (e.g. blood oozing from the gums and blood in the faeces). Laboratory tests show low white blood cell and platelet counts.

It is believed that the natural hosts of the virus are African fruit bats. Ebola was probably introduced to the human population through contact with the blood or tissues of the bats or other infected animals, such as chimpanzees, monkeys, antelope and porcupines. Ebola then spreads from person to person. This happens by direct contact with the body fluids (e.g. blood, saliva, semen or faeces) of infected people. This contact may occur through broken skin, or through *mucus membranes* such as the mouth and intestine. Ebola can also be transmitted from infected materials such as bedding. The disease is highly infectious, and even well-protected health workers who practise strict control precautions have become infected.

At the moment, there is no proven drug treatment for Ebola, and no vaccine to prevent people becoming infected. However, drug therapies and vaccines are rapidly being developed, and trials of two vaccines began in Africa in 2015. Treatment consists of caring for patients and treating the symptoms, which increases the chances of survival. One method used is oral rehydration therapy (see below); patients can also be given fluids intravenously (the fluids are transferred directly into a vein).

Prevention and control rely on a package of measures: dealing with individual patients, checking their contacts, having laboratory testing facilities, and organising safe burials of people who have died from the disease. The key to success is engaging the help of the people where the outbreaks happen and making them aware of the risk factors. The main prevention measures are:

- reducing the risk of transmission of the virus from animals – e.g. avoiding contact with fruit bats and monkeys, wearing gloves and thoroughly cooking meat to destroy the virus
- reducing the risk of transmission from infected patients – e.g. wearing gloves and protective clothing, regular hand washing and other hygiene measures
- reducing the risk of transmission by sexual intercourse – survivors of the disease are advised to practise 'safe sex' using condoms for 12 months after the disease, or until their semen twice tests negative for Ebola virus
- containing the outbreak – including safe burial of the dead, identifying and monitoring patient contacts, and isolating sick people to prevent spread of the disease.

ORAL REHYDRATION THERAPY

The correct balance of water and ions in the blood and tissue fluid is essential for the healthy functioning of the body. A simple gut infection by harmful microorganisms can result in diarrhoea and vomiting. If these symptoms continue, the patient's tissues may become dehydrated and the normal balance of salts will be upset. In severe cases, this leads to brain and kidney damage, and eventually death.

It is a sad fact that, despite all the advances of modern medicine, one of the biggest killers in the world today is diarrhoea. This particularly affects babies and young children, who are not able to adjust to rapid dehydration. In many less developed countries around the world, diarrhoea still kills millions of people every year.

There is a simple answer that saves lives – **oral rehydration therapy**. This uses a pack containing a solution of salts and glucose, which is mixed with sterile water (Figure 13.11). If a child is suffering from diarrhoea, they are fed the solution (Figure 13.12). This helps to rehydrate their tissues, and can prevent further damage or death, even if the diarrhoea has not stopped.

Oral rehydration therapy was first used in the 1970s. As a result, by 2009 the number of infant deaths around the world from diarrhoea had been cut from 5 million to 1.5 million. Medical workers aim to reduce this figure further still, by providing easier access to this cheap, simply-prepared treatment.

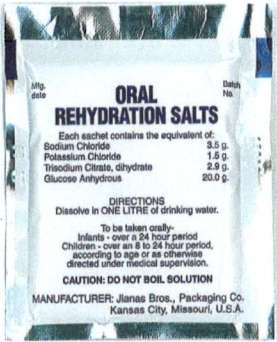

▲ Figure 13.11 An oral rehydration pack.

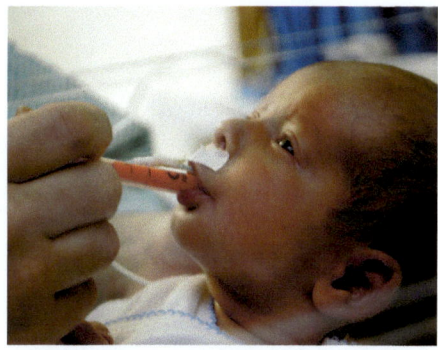

▲ Figure 13.12 An infant being given oral treatment for dehydration.

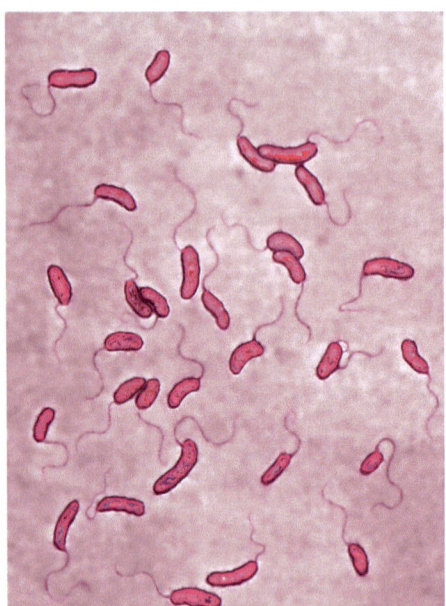

▲ Figure 13.13 A light microscope image of *Vibrio cholerae*, the bacterium that causes cholera. Note that each cell has a flagellum for swimming.

CHOLERA

Cholera is caused by the bacterium *Vibrio cholerae* (Figure 13.13).

The disease is mainly transmitted from person to person through contaminated drinking water. It is also passed on if infected people handle food without washing their hands. In some countries, it is transmitted by eating undercooked seafood caught in waters polluted with sewage (toilet waste). Cholera is endemic in less developed countries where untreated sewage is discharged into water supplies (Figure 13.14). In more economically developed countries, where there is efficient sewage treatment (see later in this chapter) and drinking water is purified and treated with chlorine to kill microorganisms, cholera is almost unknown. Three-quarters of infected people are carriers. They show no symptoms but can pass the infection to others.

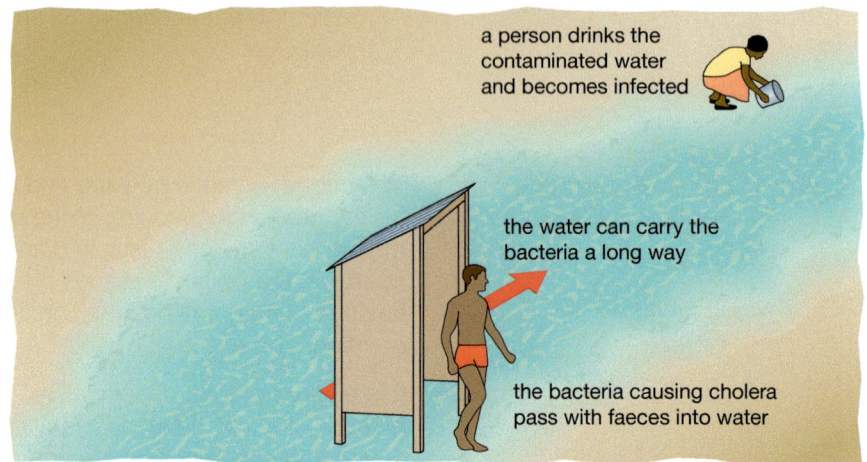

▲ Figure 13.14 How cholera is spread.

Symptoms of cholera can appear a few hours to a few days after infection. The cholera bacteria enter the gut. If they survive the acid conditions in the stomach, they reach the small intestine, where they reproduce and secrete a toxin that upsets the functioning of the epithelium lining the intestine. The toxin causes loss of water and salts from the blood into the intestine, resulting in severe diarrhoea, so that the body rapidly becomes dehydrated. The patient feels weak, with a rapid heart rate, low blood pressure and muscle cramps. The loss of fluid and shock can result in death within hours.

Cholera is very easily treated, and most people who receive the correct treatment recover quickly. A person who shows the symptoms must be given oral rehydration therapy immediately. This restores the osmotic balance in their blood and tissue fluid. It is a sad fact that people (particularly vulnerable people, such as children and the elderly) still die from the disease, because this is easily preventable.

Cholera is prevented from spreading by providing clean drinking water, building sanitation facilities (lavatories), and constructing sewage treatment systems. Other preventative measures include good hygiene, such as washing hands and cooking food thoroughly.

TYPHOID

Typhoid (also known as typhoid fever) is caused by a bacterium called *Salmonella typhi*. As with cholera, it is transmitted by drinking water contaminated with human faeces, or by flies transferring the bacterium from faeces to food (Figure 13.15). Before the twentieth century, typhoid

▲ Figure 13.15 Houseflies do not bite humans, but will feed on human food if it is left uncovered. They are attracted to animal or human faeces, and transmit many bacteria and viruses on their body or in their saliva. The microorganisms are passed to humans when they eat the food. Many serious diseases, such as typhoid, cholera, diphtheria, meningitis and polio, are transmitted in this way.

> **DID YOU KNOW?**
> Megacities are cities with populations of over 10 million people, such as Mexico City, Mumbai, Karachi or Buenos Aires. Many megacities are surrounded by shanty towns, where poor people live in overcrowded and unsanitary conditions, and disease is very common. One-sixth of the world's population now lives in these conditions.

was endemic throughout the world and typhoid epidemics were common. As a result of modern sanitation and sewage systems, it is now rare in the developed world. In developing countries, however, where there is poor sanitation, it still causes illness and death to thousands of people. The World Health Organization estimates that there are over 20 million cases of typhoid every year, and over 200 000 deaths from the disease.

Typhoid outbreaks often occur in the slum areas of 'megacities' in Africa, South America or the Far East, where sanitation is poor or non-existent (Figure 13.16). Typhoid epidemics also happen when a country experiences a disaster such as war, famine, earthquake or floods. These result in people living in temporary camps, where the crowded conditions and lack of sanitation allow the bacterium to spread.

▲ Figure 13.16 Typhoid outbreaks are common in many parts of the developing world, especially in the slum areas of 'megacities' such as Mumbai.

The incubation period of the disease is about 2 weeks. It starts with 'flu-like symptoms: a high fever, headaches, a cough and generally feeling unwell. The disease develops over the following weeks, when the patient suffers stomach cramps, constipation or diarrhoea, vomiting and delirium (mental confusion). Diarrhoea leads to severe dehydration. About 10–20% of sufferers develop life-threatening symptoms, when the bacteria attack the lining of the intestine, causing bleeding and even holes in the gut wall. Toxins released by the bacteria can cause inflammation of the heart and multiple organ failure.

Vaccines against typhoid are available, and antibiotics such as penicillin are effective against the bacteria. Of course, these are often not available to the poor people most likely to suffer from the disease. Oral rehydration therapy is very effective in treating the effects of the dehydration caused by diarrhoea.

Good sanitation and hygiene are essential in preventing the disease. It is only spread in places where human faeces or urine come into contact with food or drinking water. Unfortunately, this is likely to continue when so many people of the world live in overcrowded, unhygienic conditions.

GONORRHOEA

Gonorrhoea is a sexually transmitted disease (STD). It is common throughout the world. Gonorrhoea is caused by the bacterium *Neisseria gonorrhoeae*, which is passed from person to person during sexual intercourse. Symptoms usually appear a few days after infection, although about half of all infected people, mainly women, show no symptoms at all. Most men show symptoms, including discharge of pus from the penis and pain when urinating. There are also general symptoms, such as fever and headaches. Some women have a discharge from the vagina, or bleeding between menstrual periods. If the disease is untreated it becomes much more serious, with the infection spreading to the uterus, oviducts and ovaries. This can lead to a woman becoming infertile.

It is possible to cure gonorrhoea using antibiotics. Often, a person with gonorrhoea will be infected with other bacterial STDs, so combinations of antibiotics are usually prescribed. The main method of prevention is to avoid sexual intercourse with people who might have the infection, or to use a condom. Condoms are 99% effective in preventing transmission of the disease. As with so many bacterial diseases, the bacterium has started to evolve resistance to some antibiotics (see the section on MRSA below). The World Health Organization is worried that antibiotic resistance will soon make it very difficult to treat the disease.

ATHLETE'S FOOT

A few human diseases are caused by fungi. Those fungi that are pathogenic mostly cause infections of the skin or nails, or the moist mucus membranes of the mouth or vagina. One common fungal disease is **athlete's foot**. It is caused by several species of fungi (Figure 13.17). Athlete's foot usually grows on the warm, moist skin between the toes. Fungal spores may be transferred to the skin from the air or floor – a common place where the infection is picked up is in sports changing rooms. The fungus feeds on the outer layers of the skin. A bad infection causes sore, raw patches that may become infected by other organisms (Figure 13.18).

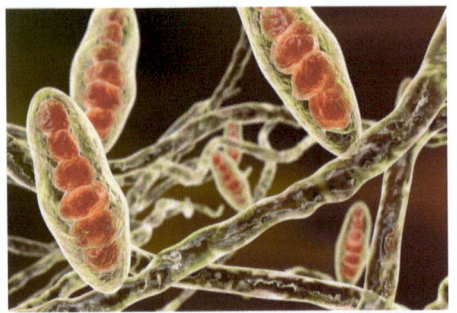

▲ Figure 13.17 A mould fungus that is one cause of athlete's foot. The fungus consists of thin threads called hyphae, which feed on the skin tissue. The infection is spread by the spores of the fungus (red).

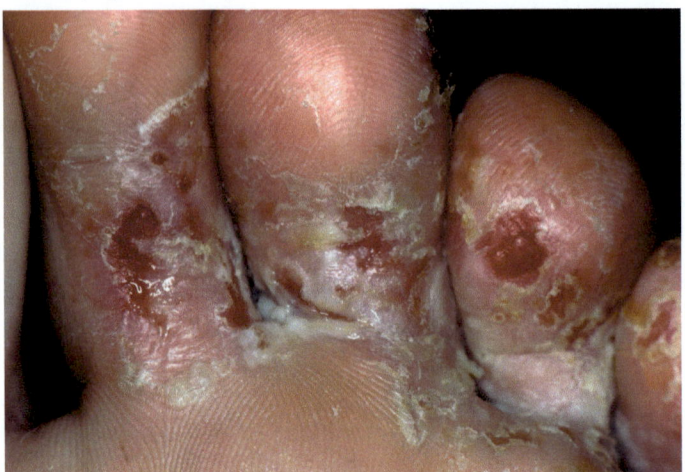

▲ Figure 13.18 A severe case of athlete's foot between the toes.

Many cases of athlete's foot do not need treatment: 30–40% of athlete's foot infections are cured by the body's immune system. The rest are easily dealt with using antifungal drugs, which can be applied to the skin or taken by mouth. There are several hygiene measures people can take to prevent athlete's foot, for example:

- avoid changing rooms or wear protective foot covers when in a changing room
- wear cotton socks and loose-fitting shoes to stop the feet becoming moist
- keep toenails cut short.

After an infection, it is virtually impossible to kill the fungal spores in shoes and socks so, to avoid re-infection, these should not be worn again.

> **DID YOU KNOW?**
> Athlete's foot is much less common in countries where people do not wear shoes.

MALARIA

Malaria is caused by five different species of a protozoan pathogen called *Plasmodium* (Figure 13.19), which is passed to humans via an insect host – the mosquito.

The female *Anopheles* mosquito transmits *Plasmodium* in her saliva when she sucks blood from humans (Figure 13.20).

Figure 13.21 shows the life cycle of the *Anopheles* mosquito and Figure 13.22 shows the life cycle of the malarial parasite.

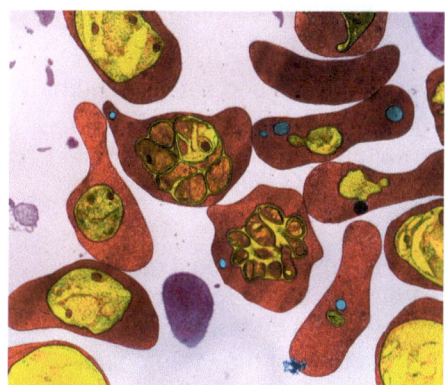

▲ Figure 13.19 Red blood cells infected with the malaria parasite *Plasmodium*.

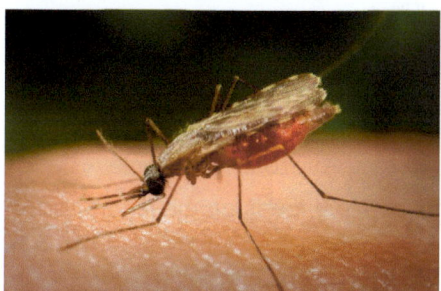

▲ Figure 13.20 A female *Anopheles* mosquito sucking human blood.

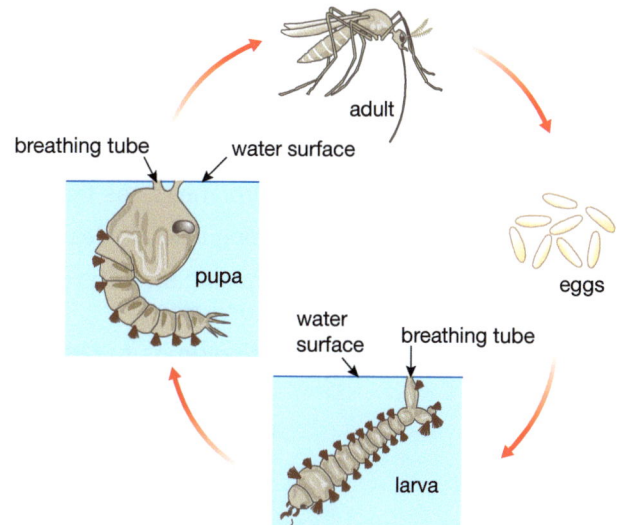

▲ Figure 13.21 Life cycle of the *Anopheles* mosquito.

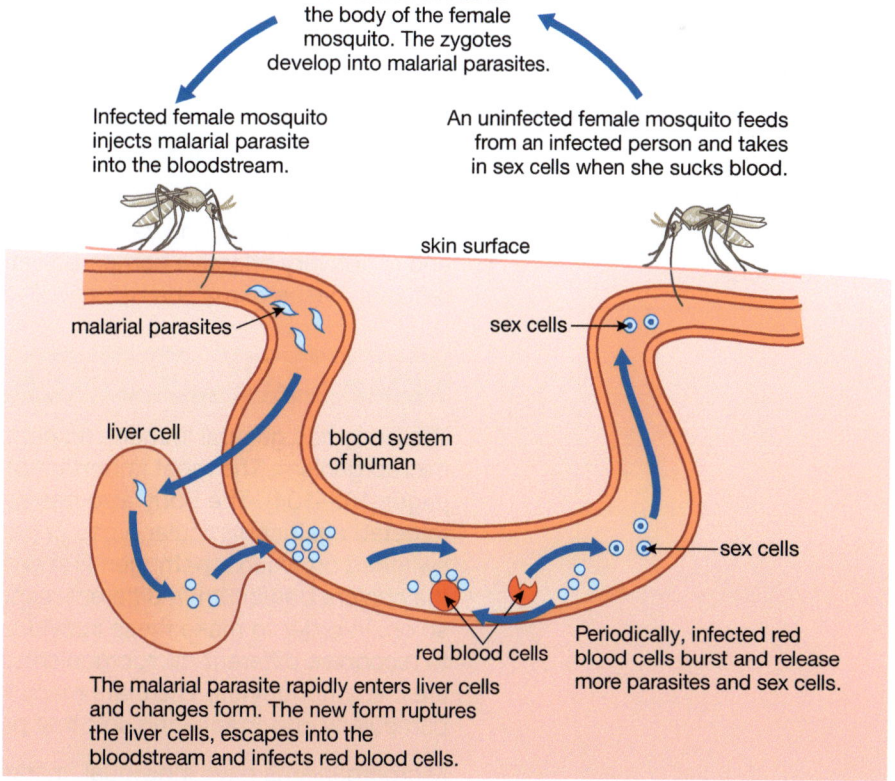

▲ Figure 13.22 Life cycle of the malarial parasite.

Malaria is widespread in tropical and sub-tropical parts of the world, and is estimated to cause the death of several million people every year. It also makes hundreds of millions of people so ill that they cannot work – contributing to poverty in the developing countries where it is endemic.

The malaria parasite spends anywhere between 2 weeks and several months in the person's liver before the next stage in the life cycle infects red blood cells. The symptoms of malaria then appear – alternating cold sweats and fever, vomiting, joint pains and anaemia. Severe malaria causes coma and death, especially in young children.

If the life cycle of either the malarial parasite or the mosquito host can be broken at any point, then the transmission of malaria can be controlled. Controlling the number of mosquitoes means there will be fewer insects to transmit the protozoan pathogen. Controlling numbers of the *Plasmodium* parasite means there will be fewer opportunities for mosquitoes to take in the parasite and transmit it to other humans.

Control measures that have been tried include:

- using insecticides to kill the adult mosquitoes
- draining swamps and pools that form the natural habitat of the larvae of the mosquitoes
- using drugs to target the various stages of the protozoan's life cycle
- introducing a fish called *Tilapia* into ponds. *Tilapia* feeds on the larvae of mosquitoes
- using insect repellents, wearing long-sleeved shirts and sleeping under mosquito nets to prevent bites from the adult mosquitoes.

A number of drugs are used to treat malaria. The oldest treatment is *quinine*, a natural substance extracted from the bark of the cinchona tree. Modern drugs are used to kill the pathogen or prevent it reproducing in the body. Other drugs reduce the symptoms of the disease. Travellers to countries where malaria is endemic are advised to take anti-malarial drugs for up to several weeks before they travel. They should take preventative measures such as using mosquito nets while they are in the country, and should continue to take anti-malarial drugs for a time after they return from the trip.

Research is under way to develop an effective vaccine against *Plasmodium*. As of 2016 there is only one vaccine approved for use. It requires four injections and is not very effective compared with vaccines against other diseases, reducing cases of malaria by only 26–50%.

> **DID YOU KNOW?**
> Malaria is a difficult disease for our immune system to attack, because the protozoan parasite spends much of its time 'hidden' inside red blood cells, and changes form several times inside the body.

DEFENCE AGAINST DISEASE – IMMUNITY

The body has general immune responses that it uses against any microorganism. The most important of these is phagocytosis (see Chapter 6, pages 103–104). The body also has *specific* immune responses, which are targeted against particular types of pathogen. Specific immune responses are switched on when a pathogen first enters the body. There is one response to the cholera bacterium, a different response to the common cold virus, and so on. In order to make these individual responses, our body has to be able to recognise different microorganisms. The cells in our blood that can do this are the lymphocytes. There are many kinds of lymphocyte, but one type is particularly important – the sort that produces antibodies.

Microorganisms have individual 'marker chemicals' or antigens on their surface. Lymphocytes have receptor proteins on their surface that can bind

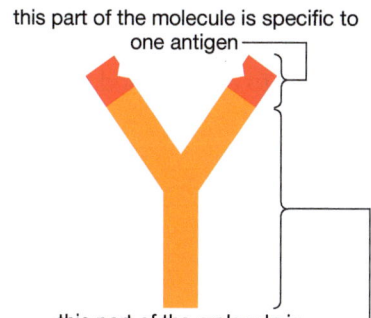

▲ Figure 13.23 The structure of an antibody molecule.

> **EXTENSION WORK**
>
> The lymphocytes that make antibodies are called B-lymphocytes. Other types of lymphocyte do not make antibodies. T-lymphocytes attach directly to virus-infected cells and cancer cells and destroy them in various ways. Some make a hole in the cell so that the contents leak out and the cell dies. Others activate a process called 'programmed cell death', which is part of the genetic code of every cell. T-helper cells produce chemicals that help T-lymphocytes to carry out their job. (The T-helper cells are the ones attacked by the HIV virus.)

with the antigens on the surface of the microorganisms. Each lymphocyte has slightly different receptor proteins from other lymphocytes and so can bind with different antigens and recognise a different microorganism.

When a lymphocyte binds with an antigen, it becomes *activated* and starts to divide rapidly. This will eventually produce millions of the same type of lymphocyte, all able to recognise the same type of microorganism.

When certain lymphocytes are activated, they start to produce antibodies, which are released into the blood. Antibodies are protein molecules shaped like a letter 'Y', which bind to the antigens on the surface of a microorganism. Each antibody is specific to a particular microorganism (Figure 13.23).

There are two general purposes of antibodies:

- **Neutralising the target threat:** Some antibodies bind to a target bacterium and stop it infecting more cells. Some antibodies cause bacteria to burst open. Others bind to bacterial toxins in the blood, which neutralises them.

- **Recruiting other cells and chemicals to combat the threat:** Some antibodies act as markers to attract phagocytes to a bacterium. Others cause bacteria to clump together. In this form, they are inactive and can easily be killed by phagocytes.

Antibodies can also identify mutated cancer cells or cells that have been attacked by viruses, marking them for destruction.

Figure 13.24 shows how antigens on two different species of bacteria cause lymphocytes to produce two different antibodies. The figure also shows one way in which antibodies can deal with the bacteria – by causing them to clump together.

Some of the activated lymphocytes do not get involved in killing microorganisms at this stage. Instead, they develop into **memory cells**. Memory cells make us immune to a disease in the future. These cells remain in the blood for many years, in some cases a lifetime. If the same microorganism re-infects us, the memory cells start to reproduce and make antibodies. This

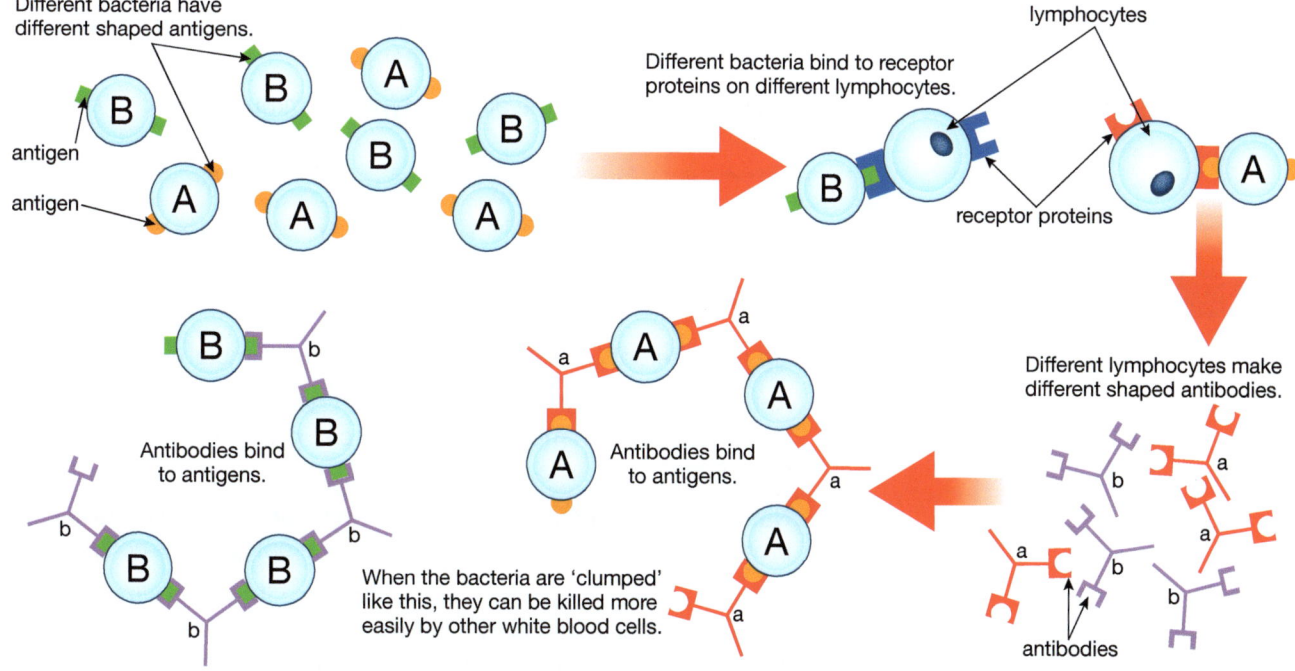

▲ Figure 13.24 How lymphocytes produce specific antibodies and destroy bacteria by clumping them.

REMINDER

Note that it is the memory cells that remain in the blood for long periods of time, not the antibodies.

secondary immune response is much faster and more effective than the first (primary) response. The antibodies quickly rise to high levels in the blood, and the microorganisms are killed before they have time to multiply and cause disease. This is shown as a graph in Figure 13.25.

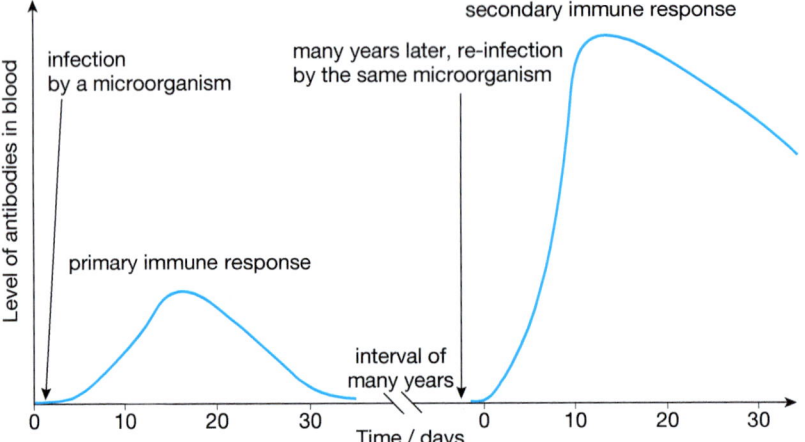

▲ Figure 13.25 Graph showing the levels of antibody production in the primary and secondary immune responses. Many years may pass after the primary response before we are re-infected and need to make a secondary response.

TYPES OF IMMUNITY

The responses described above are known as *natural active* immunity. 'Active' means that the antibodies are made by our own bodies, and 'natural' means in response to an infection. Immunity can be either active or passive, and both types of immunity can be gained naturally or artificially.

- **Active immunity** happens when we produce our own antibodies to an antigen. This is **natural** if it happens in response to an infection, and **artificial** if it is the result of a **vaccination** (see below).

- **Passive immunity** is when the body does not produce its own antibodies, but receives them from somewhere else. This is natural when the antibodies are received by a fetus across the placenta during pregnancy, and natural when received by a baby in its mother's milk. It is artificial as a result of injection of antibodies from another person.

It is easiest to compare the different types of immunity using a table (Table 13.3).

▼ Table 13.3 Types of immunity.

	Active (body makes antibodies)	Passive (body receives antibodies)
Natural	natural active immunity – results from an infection	natural passive immunity – antibodies pass from mother to fetus across the placenta, or from mother to baby in milk
Artificial	artificial active immunity – results from vaccination	artificial passive immunity – antibodies are given by injection

There are different advantages to each type of immunity:

- Active immunity lasts much longer than passive immunity, because active immunity results in memory cells that can produce a secondary immune response.

- Natural passive immunity is important in a baby, because it gives the child protection when its immune system is developing and is not yet fully working.

- Artificial passive immunity is useful because it can give immediate protection against a disease. For example, during measles outbreaks, injections of antibodies are sometimes given to protect people who are not immune but have been exposed to infection.

VACCINATION

Vaccination is active artificial immunity. We can be made immune to a disease without actually becoming infected by that disease. In a vaccination, a person is given an 'agent' that carries the same antigens as a particular disease-causing microorganism. The vaccination is usually given by injection (Figure 13.26), although some vaccines can be given orally (by mouth). Lymphocytes recognise the antigens and multiply exactly as if that microorganism had entered the bloodstream. They form memory cells and make us immune to the disease. The vaccine that is given may be one of the following:

- a harmless strain of the actual microorganism, e.g. the vaccines for polio, tuberculosis and measles
- dead microorganisms, e.g. the vaccines for typhoid and whooping cough
- modified bacterial toxins, e.g. the toxins of the tetanus and diphtheria bacteria
- the antigens on their own, e.g. the influenza (flu) vaccine
- harmless bacteria, genetically modified to carry the antigens of a different pathogen, e.g. the hepatitis B vaccine.

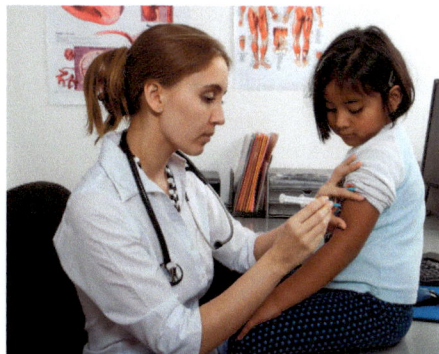

▲ Figure 13.26 A vaccine is injected to stimulate an immune response to a particular disease-causing organism.

A vaccination may give protection for many years, even a lifetime. Some vaccinations are not so effective and extra 'booster' injections have to be given to maintain a person's immunity.

Some diseases are endemic in different parts of the world. Before they travel to foreign countries, people need to take advice from their doctor about any vaccinations that may be necessary.

Vaccination programmes have sometimes been effective in completely wiping out a disease. The best-known example is smallpox. Smallpox is a highly infectious disease caused by a virus. It once affected ten million people in thirty countries around the world (Figure 13.27).

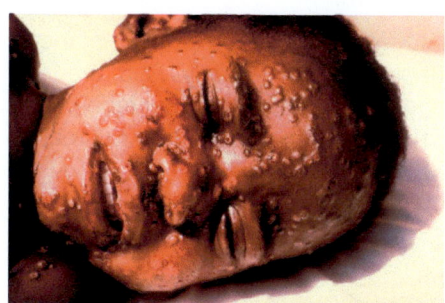

▲ Figure 13.27 A boy with smallpox. His face is covered with skin lesions called pustules, which are a characteristic sign of the disease.

In the 1960s, the World Health Organization started a worldwide programme with the aim of completely eradicating smallpox. The programme included vaccinations, isolation of patients with the disease and observation to detect new cases. The programme was successful. The last case of smallpox was identified in 1977, and in 1980 the world was officially declared free of the disease.

ANTIBIOTICS

Antibiotics are chemicals that kill microorganisms, or reduce their growth. They are mainly used in medicine to treat bacterial infections, although a few antibiotics are effective against fungal pathogens. Antibiotics do not work on viruses, so they are no use in treating any disease caused by a virus.

Natural antibiotics are produced by bacteria and fungi. They are used by the bacteria and fungi to kill other microorganisms. This gives the organism making the antibiotic an advantage when it is competing with other species for nutrients.

The first antibiotic was discovered in 1929. It is made by the mould fungus *Penicillium*, and is called **penicillin**. Penicillin kills bacteria. It was first isolated and used to treat bacterial infections in the 1940s. Since then, other natural antibiotics have been discovered, such as streptomycin and chloramphenicol (both made by bacteria). Many more artificial antibiotics have been made. Some are semi-synthetic, chemically-altered versions of natural antibiotics; others are fully synthetic chemicals that have been 'designed' in a laboratory.

Antibiotics that kill bacteria are called *bactericidal*. Antibiotics that just stop them reproducing are called *bacteriostatic*. There are also *broad spectrum* antibiotics, which act against a range of bacteria, while others are more limited in their target. Table 13.4 shows three antibiotics and the ways in which they act.

▼ Table 13.4 The sources, modes of action and effects of three types of antibiotic.

Example	Mode of action	Effect	Type of antibiotic
penicillin (from the mould fungus *Penicillium*)	Interferes with manufacture of bacterial cell wall	Weakens cell wall – water enters by osmosis and bursts the cell (osmotic lysis)	bactericidal
nalidixic acid (a synthetic antibiotic)	Interferes with DNA replication	Bacteria are not killed but cell division is impossible, so they cannot multiply	bacteriostatic
tetracycline (from a bacterium called *Streptomyces*)	Interferes with protein synthesis	Stops production of enzymes	bactericidal

ANTIBIOTIC-RESISTANT BACTERIA

> **DID YOU KNOW?**
> Some patients expect to be given antibiotics to treat illnesses such as cold and 'flu. These infections are caused by viruses, so antibiotics are useless against them.

The use of antibiotics has increased dramatically, particularly over the last 20 years. We now almost expect to be given an antibiotic for even the most minor of illnesses. This can be dangerous, as it leads to the development of bacterial resistance to antibiotics, so that the antibiotics are no longer effective in preventing bacterial infection. Figure 13.28 shows antibiotics being tested for their effectiveness in killing a species of bacterium.

Resistance starts when a random gene mutation gives a bacterium resistance to a particular antibiotic. In a place where the antibiotic is widely used, the new resistant bacterium has a big advantage over non-resistant bacteria of the same type. The resistant strain of bacterium will survive and multiply in greater numbers than the non-resistant type.

As you have seen, bacteria reproduce very quickly. A bacterium that divides every 20 minutes produces 72 generations in a single day. This soon results in a population of millions of resistant bacteria.

A particularly worrying example of a resistant bacterium is **MRSA**. MRSA stands for **methicillin-resistant Staphylococcus aureus**. It has been called a 'super bug' because it is resistant to many antibiotics (including methicillin, a type of penicillin that is no longer used). It is a major problem in hospitals, where it is responsible for many infections that are difficult to treat.

▲ Figure 13.28 This photo shows a colony of bacteria growing on a Petri dish of nutrient agar. The circular white discs contain different antibiotics. The clear areas around the discs show where the bacteria have been killed by the antibiotic diffusing into the agar. If the antibiotic does not work against the bacterium (e.g. due to resistance), there will be no clear area around the disc.

Resistant bacteria will not be killed by the antibiotic, so the antibiotic is no longer effective in controlling the disease.

Bacterial resistance to antibiotics was first noticed in hospitals in the 1950s, and has grown to be a major problem today. There are now many resistant strains of different bacteria, which make it very difficult to control infections.

Doctors are now more reluctant to prescribe antibiotics. They know that using them less will mean the bacteria with resistance have less of an advantage and will not become as widespread.

> **DID YOU KNOW?**
> Some people talk about bacteria becoming *immune* to antibiotics. This is a misunderstanding. Immunity happens in people – we become immune to microorganisms that infect us, as a result of the immune response. Bacteria become *resistant* to antibiotics.

ACTIVITY 1

▼ PRACTICAL: INVESTIGATING THE EFFECT OF ANTIBIOTICS ON THE GROWTH OF BACTERIA

You can test the effect of antibiotics on the growth of bacteria using paper discs containing the antibiotic, as shown in Figure 13.28.

A Petri dish is prepared, containing a nutrient agar culture of a harmless bacterium such as *Staphylococcus albus*.

The points of a pair of metal forceps are sterilised by briefly placing them in the flame of a Bunsen burner. The sterilised forceps are used to place a paper disc containing penicillin on the surface of the agar. It is important that the lid of the Petri dish is opened as briefly as possible, to prevent other microorganisms entering the dish. Further discs are added, containing other antibiotics (such as streptomycin) or different concentrations of antibiotic. A Control disc containing no antibiotic is also applied to the agar. About four discs per Petri dish is a good number to show the results of the investigation. The position of each type of disc is noted.

The lid of the Petri dish is closed and fixed to the base with sticky tape. The dish is labelled with its contents and the date, and placed in an incubator (a low temperature oven) set at 30 °C.

After 48 hours, the Petri dishes are removed from the incubator and examined (without opening them!). The results should look like the dish in Figure 13.28. The diameters of any clear areas around the discs are noted. The effectiveness of an antibiotic is proportional to the diameter of the clear area.

Discs can be used to test the effects of other chemicals on the growth of bacteria, for example, disinfectants, mouthwashes or antiseptics. Blank paper discs can be made by using a piece of filter paper and a hole punch.

Safety Note: Avoid skin contact with the nutrient agar culture and the antibiotic discs; handle these with forceps. Do not seal the Petri dish around its circumference. Instead, use strips of sticky tape to hold the lid to the base in three or four places.

DECOMPOSERS

Microorganisms are not all pathogens. In nature, the waste products of animals, as well as the bodies of dead plants and animals, must be broken down and recycled. This is carried out by microorganisms called decomposers. Decomposers feed on the dead remains, breaking down complex organic materials into simpler substances, which they release back into the environment.

Decomposers are mainly fungi and bacteria in the soil. We can look at their key role in recycling by taking the example of carbon, one of the main elements present in living organisms. Humans and other animals respire, releasing carbon into the air as carbon dioxide gas. Plants make their food by the process of **photosynthesis**, which uses up the carbon dioxide again. Without decomposers, however, carbon would remain 'trapped' in the bodies of animals and plants when they die. As a result of the feeding and respiration of decomposers, carbon dioxide is produced, which can be used again by plants. There is a *cycle* of carbon in the environment (Figure 13.29).

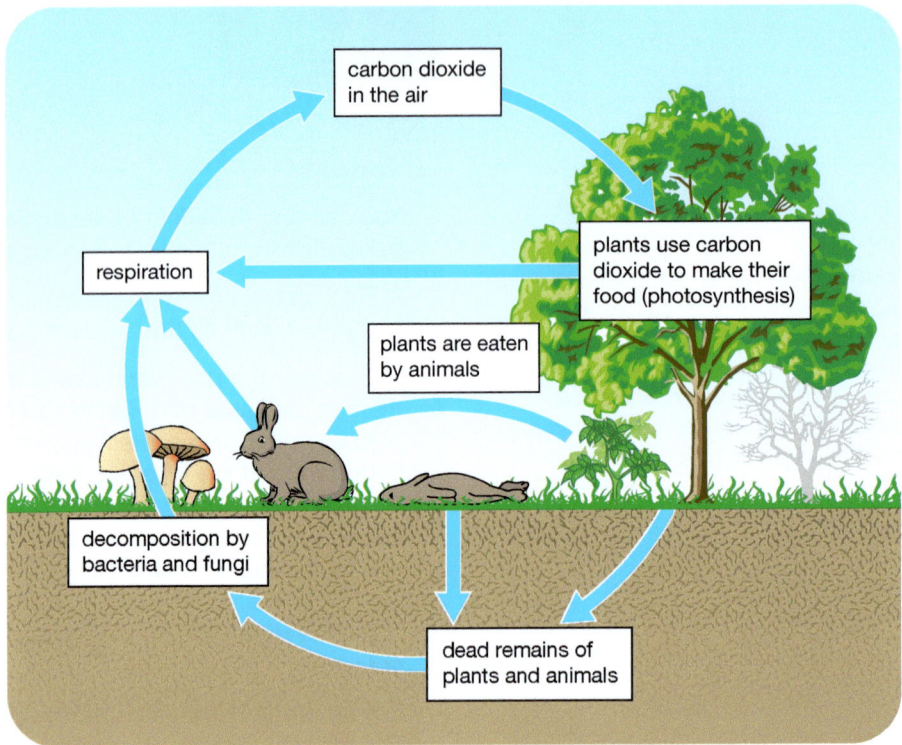

▲ Figure 13.29 The cycle of carbon depends on decomposers.

Fungi and bacteria are also involved in recycling other elements in the ecosystem, such as nitrogen. For example, proteins contain nitrogen. The proteins in the bodies of dead animals and plants are broken down into ammonia and then converted into nitrates. Nitrates are essential mineral ions that plants need to make proteins again.

These natural recycling processes involving decomposers are used in the treatment of sewage.

SEWAGE TREATMENT

Sewage is wet waste from houses, factories and farms. Developed countries have very large sewage treatment works, and industrial and agricultural sewage is usually dealt with separately from household sewage. Household sewage consists of waste water from kitchens and bathrooms, and contains human urine and faeces, as well as chemicals such as soaps and detergents. It is carried away in pipes called sewers, and is treated before it enters rivers or the sea.

If untreated sewage is discharged into waterways, it causes two main problems:

- Sewage contains pathogenic bacteria and viruses, which are a health hazard. It is the cause of transmission of many diseases, such as cholera and typhoid.

- Aerobic bacteria from the sewage use up dissolved oxygen in the water as they break down the organic matter. Lowered oxygen levels result in the death of animals such as fish.

The aim of sewage treatment is to remove solid and suspended organic matter and pathogenic microorganisms, so that a cleaner liquid waste can be released into waterways.

Two methods of sewage treatment are the **percolating filter method** and the **activated sludge method**. The two methods are compared in Figures 13.30 and 13.31.

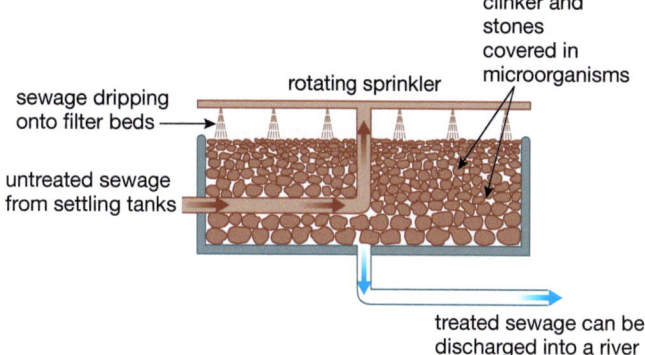

▲ Figure 13.30 The percolating filter method.

- Sewage arrives for treatment, is screened to remove large objects, and stands in large settling tanks to allow other solid material to settle out.
- It is then sprayed through a pipe rotating over the filter bed, and passes through the filter. Aerobic bacteria, fungi and protozoa in the filter oxidise any organic matter.
- Treated sewage is then released into a waterway.

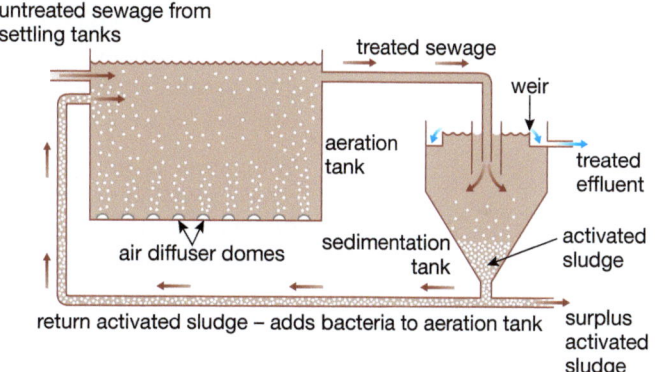

▲ Figure 13.31 The activated sludge method.

- Sewage arrives for treatment, is screened to remove large objects, and stands in large settling tanks to allow other solid material to settle out.
- It is then passed into an aeration tank, where it is 'activated' by oxygen being pumped in. Aerobic bacteria in the tank oxidise the organic material.
- From here it passes to a sedimentation tank, where the activated sludge settles out.
- Some is returned to the aeration tank, and the purified effluent is released.

Both methods of sewage treatment rely on a complex system of aerobic bacteria, fungi, protozoa and other organisms to digest the sewage. Fungi and bacteria start the breakdown process, converting proteins into ammonia and then into nitrates. In turn, protozoa and larger **invertebrate** animals, such as worms, feed on the bacteria and fungi as well as on the organic matter. The end result is an effluent (outflow) that contains much less organic matter, and far fewer pathogenic microorganisms. This effluent can be safely released into a river.

The waste sludge that accumulates in the settling tanks must be treated further to reduce the organic matter and pathogens it contains, before it can be disposed of. There are various ways of achieving this. The commonest method is *anaerobic* digestion by microorganisms in a fermentation tank. This produces *biogas* as a waste product. Biogas is a mixture of methane and carbon dioxide. The methane is produced by anaerobic bacteria called *methanogens*, which means 'methane makers'. Biogas can be used as a fuel in electricity generators or for heating. After anaerobic digestion, the dry, solid material left over can be used for fertiliser or disposed of in landfill sites.

PIT LATRINES

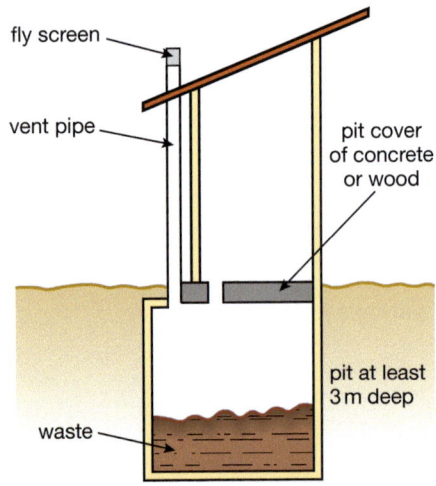

▲ Figure 13.32 A ventilated improved pit latrine.

In areas of the world where there are no sewers or sewage treatment, the *composting toilet* is a simple way of dealing with human waste. There are a variety of types, but they all use the same principle. The toilet consists of a hole in the ground over a sunken pit. In the pit, the faeces and urine are broken down by microorganisms. The simplest type of composting toilet is the **pit latrine**, where the user squats over a shallow trench 1–2 metres deep. An improved version is called the ventilated improved pit latrine, or VIP (Figure 13.32). This has a covered pit at least 3 metres deep, and a vent pipe leading from the pit, to take away odours. The pipe is covered with a fly screen. The vent pipe acts like a chimney, producing a suction effect that draws clean air through the latrine. Flies are attracted to the faeces to lay their eggs. The fly screen helps to prevent the flies from entering the latrine, and any flies that do get into the pit are drawn upwards through the chimney, and prevented from leaving the pit by the fly screen. Since they cannot escape, the flies eventually die and fall back into the pit. This means that they are prevented from spreading diseases. Earth, sand or sawdust may be added to the pit after use, to prevent smells and further discourage flies.

With some larger pit latrines, the large waste pit means that natural decomposition proceeds quickly enough for the pit not to need emptying. Smaller pits have to be emptied; alternatively, they can be filled in with earth periodically and replaced by a new pit.

CHAPTER QUESTIONS

SKILLS ANALYSIS

1 The diagram below shows the structure of a typical bacterium.

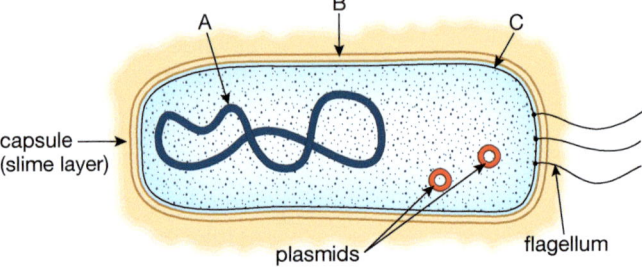

   a Identify structures A, B and C.

SKILLS CRITICAL THINKING

   b What is the function of the flagellum?

   c 'All bacteria are pathogens'. Briefly discuss whether this statement is true.

SKILLS INTERPRETATION

2 a Draw a diagram to show the structure of a typical virus particle.

SKILLS CRITICAL THINKING

   b Is a virus a living organism? Explain your answer.

   c Explain the statement, 'Viruses are all parasites'.

SKILLS CRITICAL THINKING

3 a List four ways in which infectious diseases can be transmitted. In each case, give an example of a disease transmitted in that way.
   b Give two examples of insects that act as vectors for disease. In each case, explain how the insect is responsible for the spread of disease.
   c For each of the insect vectors you named in (b), describe two measures that can be taken to control the spread of the disease. Explain how each method works.

4 AIDS (acquired immune deficiency syndrome) results in many human deaths. It is caused by the human immunodeficiency virus (HIV).
   a Explain how viruses use living cells to reproduce.
   b State two ways in which HIV can enter the body.
   c What does being 'HIV positive' mean?

SKILLS ANALYSIS

   d The graph shows the changes in the numbers of T-helper cells during an AIDS infection.

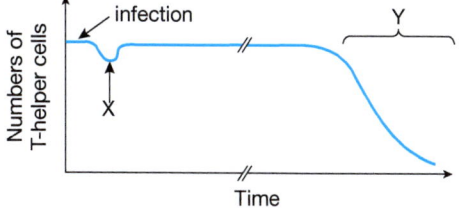

Suggest explanations for the changes in numbers of T-helper cells at:
   i   the period marked X
   ii the period marked Y.

SKILLS REASONING

   e Explain why people with AIDS are more at risk of becoming ill with infections such as pneumonia.

SKILLS CRITICAL THINKING

5 a Describe the cause and method of transmission of cholera.
   b Explain the use of oral rehydration therapy to treat cholera.

6 Gonorrhoea is a sexually transmitted disease caused by a bacterium. The graph shows the number of cases of gonorrhoea treated in UK clinics between 1925 and 1990.

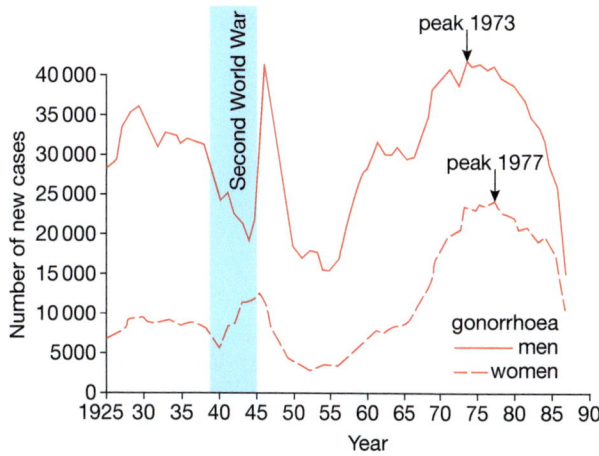

   a What is meant by a sexually transmitted disease?

SKILLS ANALYSIS

   b Describe the general trends in the number of cases of gonorrhoea from 1925 to 1990.

SKILLS REASONING

c Suggest reasons for the increases in the number of cases of gonorrhoea during the 1960s.

d Suggest how the increase in AIDS cases in the 1980s may be linked to the decrease in the number of cases of gonorrhoea.

e Gonorrhoea can usually be cured by taking penicillin.

  i Describe how penicillin kills the bacteria that cause gonorrhoea.

SKILLS CRITICAL THINKING

SKILLS REASONING

  ii Suggest why penicillin is not always effective in the treatment of gonorrhoea.

SKILLS CRITICAL THINKING

7 The diagram shows the levels of antibodies produced by a person during a first infection and later, when re-infected by the same microorganism.

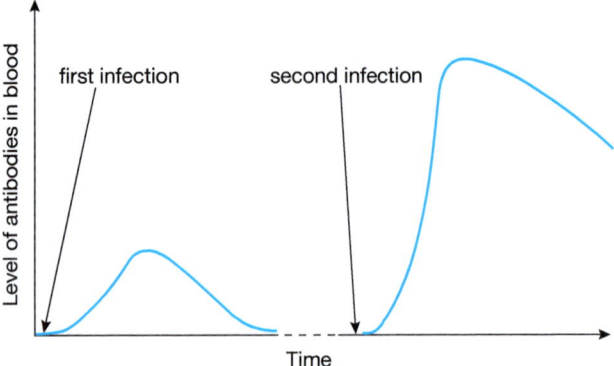

a State the names of the first and second responses by the body to infection.

b Describe the differences between the two responses in terms of their size, speed and duration.

c Explain how the first response is produced.

8 a What is an antibiotic?

b Describe two ways that antibiotics work.

9 The flowchart below shows stages in the percolated filter method of sewage treatment.

untreated sewage
↓
screening
↓
settling tank
↓
filter bed
↓
treated effluent discharged into river

a What is the function of the settling tank?

b Sludge from the settling tank can be treated by anaerobic fermentation. Name two useful products of this process.

SKILLS REASONING

c Why is it important that conditions are aerobic in the filter bed?

d Explain why it is important that untreated sewage does not enter a river or the sea.

UNIT QUESTIONS

SKILLS REASONING

1 A species of mammal has 32 chromosomes in its muscle cells. Which row in the table below shows the number of chromosomes in the mammal's skin cells and sperm cells?

	skin cells	sperm cells
A	32	32
B	16	16
C	16	32
D	32	16

(Total 1 mark)

SKILLS CRITICAL THINKING

2 Which of the following is *not* a function of the ovaries?

A The secretion of progesterone

B The production of eggs

C The secretion of oestrogen

D The site of fertilisation

(Total 1 mark)

3 Which of the following organs produce the hormone progesterone?

1 placenta
2 ovary
3 uterus
4 pituitary gland

A 1 and 2

B 2 and 3

C 3 and 4

D 2 and 4

(Total 1 mark)

SKILLS PROBLEM SOLVING

4 A woman's first day of menstruation was on 1 June. Assuming she has a 28-day menstrual cycle, when is she most likely to ovulate?

A 7 June

B 10 June

C 14 June

D 21 June

(Total 1 mark)

SKILLS CRITICAL THINKING

5 Which of the following is true of dominant alleles?

A A dominant allele is expressed if present with a recessive allele

B They determine the most favourable of a pair of alternative features

C They are inherited in preference to recessive alleles

D They are only expressed if present as a pair

(Total 1 mark)

SKILLS PROBLEM SOLVING

6 In guinea pigs, the allele for coloured eyes is dominant to the allele for pink eyes. A guinea pig heterozygous for eye colour is crossed with a guinea pig with pink eyes. What would be the expected ratio of genotypes in the offspring?

A 2 : 1

B 1 : 1

C 1 : 0

D 3 : 1

(Total 1 mark)

7 The allele for yellow coat colour in mice (Y) is dominant to the allele for non-yellow coat colour (y).

Mice with the genotype yy have non-yellow coats.

Mice with the genotype Yy have yellow coats.

Mice with the genotype YY die as embryos.

Two heterozygous mice were crossed. What is the probability that a surviving mouse in the F_1 generation will be yellow?

A 0.00

B 0.25

C 0.50

D 0.67

(Total 1 mark)

8 Alleles B and b are codominant. Two heterozygous individuals were crossed. What would be the expected ratio of phenotypes in the F_1 generation?

A 1 : 1

B 3 : 1

C 1 : 2 : 1

D 1 : 1 : 1 : 1

(Total 1 mark)

SKILLS CRITICAL THINKING

9 What type of microorganism causes typhoid?

A bacterium

B fungus

C protozoan

D virus

(Total 1 mark)

10 Which of the following diseases is *not* caused by a virus?

A influenza

B measles

C malaria

D Ebola

(Total 1 mark)

UNIT 4 — UNIT QUESTIONS

SKILLS CRITICAL THINKING

11 What is the name of the drugs that destroy bacteria?

A antibodies

B antibiotics

C antigens

D vaccines

(Total 1 mark)

12 Which of the following is a correct definition of vaccination?

A Injecting a dead form of the disease to stimulate an immune response

B Injecting an inactive form of the disease to stimulate an immune response

C Injecting a harmful toxin of the pathogen to stimulate an immune response

D Injecting an inactive form of the pathogen to stimulate an immune response

(Total 1 mark)

SKILLS ANALYSIS

13 The graph shows some of the changes taking place during the menstrual cycle.

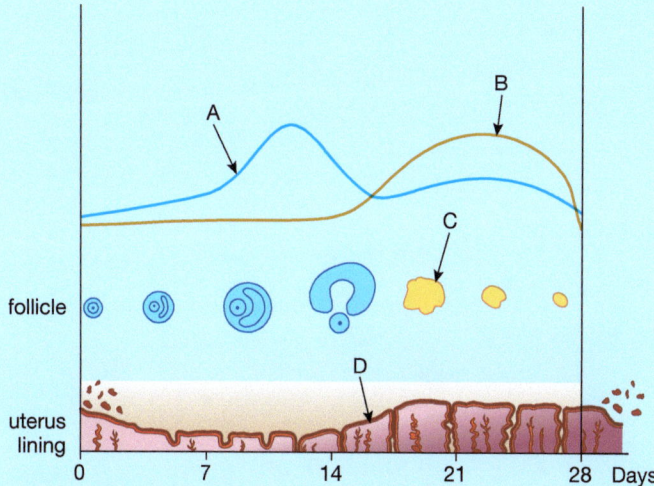

a Identify the two hormones produced by the ovary, which are shown by the lines A and B on the graph. (2)

SKILLS CRITICAL THINKING

b Name the structure C. (1)

c What is the purpose of the thickening of the uterus lining at D? (1)

d When is sexual intercourse most likely to result in pregnancy, at day 6, 10, 13, 20 or 23? (1)

e Why is it important that the level of progesterone remains high in the blood of a woman during pregnancy? (2)

f How does her body maintain a high level of progesterone:

　i just after she becomes pregnant? (1)

　ii later on in pregnancy? (1)

(Total 9 marks)

14 One of the genes for colour vision is found on the X chromosome, but is not present on the Y chromosome. The dominant allele of this gene produces normal colour vision, while the recessive allele causes red-green colour blindness.

a Using the symbol X^A for the dominant allele on the X chromosome, X^a for the recessive allele on the X chromosome, and Y for the Y chromosome, write down the genotypes of:
  i   a colour-blind man
  ii a colour-blind woman
  iii a woman who is a carrier for the colour blindness gene. (3)

b A couple who both have normal colour vision have a child with red-green colour blindness. Draw a diagram to show how this happened. What sex is the child? (4)

(Total 7 marks)

15 Garlic is said to protect the body from harmful bacteria. You are provided with cultures of bacteria on agar plates. Design an investigation to test the hypothesis 'Garlic juice kills bacteria.'

(Total 6 marks)

APPENDICES

APPENDIX A: A GUIDE TO EXAM QUESTIONS ON EXPERIMENTAL SKILLS

Copies of official specifications for all Edexcel qualifications may be found on the Edexcel website, www.edexcel.com. Past papers, marks schemes and examiners' reports are also available on the website.

WHY IS THIS APPENDIX IMPORTANT?

This appendix is designed to help you gain the 20% of marks allotted to the questions on experimental skills. These skills are tested in both Paper 1 and Paper 2, 'mixed up' with theory questions. The practicals that you should know about are included in the biology content of the specification and fully described in this book. In the exams, you will be asked questions based on these practicals, although the apparatus shown may be slightly different. You may also be asked questions about unfamiliar biological investigations. Don't be worried by these – you are not expected to have carried them out. You just have to be able to apply your knowledge and understanding to interpret experimental design or to analyse data.

Questions in the exams will generally cover the following areas:

- Understanding safety precautions
- Recognising apparatus and understanding how to use it
- Recalling tests for certain substances
- Manipulating data
- Plotting graphs
- Understanding experimental design
- Planning experiments.

1. SAFETY PRECAUTIONS

As part of a question, you may be asked to comment on appropriate safety precautions. The most obvious precaution is to *wear eye protection*. Eye protection must be worn whenever chemicals or Bunsen burners are used. This will apply to many experiments or investigations, with some exceptions, such as measuring heart rate. Some other examples of safety precautions, and the reasons for them, are shown in the table below.

Precaution	Reason
Wash hands after handling biological material, such as enzymes	To avoid contamination
Keep flammable liquids such as ethanol away from naked flames	To avoid the liquid catching fire
Take care with fragile glassware such as pipettes, microscope cover slips, etc.	To avoid cutting yourself
Do not touch electrical apparatus (e.g. a microscope with built-in lamp) with wet hands	To avoid getting an electric shock
Use a water bath to heat a test tube of water, rather than heating it directly in a Bunsen flame	To avoid the heated liquid jumping out of the tube

2. RECOGNISING APPARATUS AND DEMONSTRATING UNDERSTANDING OF HOW TO USE IT

One of the simplest types of question in the exams will require you to recognise common pieces of laboratory apparatus, such as a Bunsen burner, thermometer, measuring cylinder and stopwatch. You will also need to recognise particular 'biological' apparatus such as a microscope, bench lamp and so on, and know how they are used. If you have been able to carry out most of the practical work in the specification, this kind of question is very straightforward, for example:

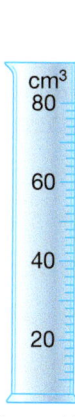

(a) What is the name of this piece of apparatus? *(1 mark)*
(b) Draw a line on the apparatus to show a volume of 30 cm^3. *(1 mark)*

Note: Use a pencil to draw any lines on diagrams, graphs etc. Then you can rub it out if you make a mistake. On this measuring cylinder, a horizontal line should be drawn half way between the 20 and 40 cm^3 marks – it's that easy!

3. RECALLING TESTS FOR SUBSTANCES

There are only six chemical tests in the specification. These are for:

- starch – using iodine solution
- reducing sugar (e.g. glucose) – heating with Benedict's solution
- protein – the biuret test
- lipid – the emulsion test using ethanol
- vitamin C – using DCPIP
- carbon dioxide in exhaled air – using limewater or hydrogencarbonate indicator.

The tests for starch, reducing sugar, protein and lipid are described on pages 35–36. The use of DCPIP to investigate the amount of vitamin C in food is on pages 59–60. The effects of carbon dioxide on limewater or hydrogencarbonate indicator are described on page 82.

Once you have learnt these tests, the questions are straightforward. For example:

John decided to test some milk for reducing sugar.
(a) Describe the test he would do. *(2 marks)*
(b) What result would he see if reducing sugar was present? *(1 mark)*
(c) Suggest how he might use the results to say how much reducing sugar was present. *(1 mark)*

Note: You might get 1 mark for 'Use Benedict's solution', one for 'heating' and one for the colour produced (orange or similar). In (c) the depth of colour would show how much reducing sugar was present.

4. MANIPULATING AND ANALYSING DATA

Some questions in an exam will involve manipulating data. This means you will be provided with some results from an experiment and you will be asked to process the data in various ways, such as:

- putting raw data from a notebook into an ordered table
- counting numbers of observations
- summing totals
- calculating an average
- calculating a missing value
- identifying anomalous results.

For example:

Kirsty monitored how her temperature changed during exercise. She took her temperature before she started the exercise, and every two minutes during the exercise. Here are her results:

Before exercise 36.4°C   At 12 minutes = 37.6°C
37.3°C after 10 mins, 2 mins = 36.8°C.
After 4 mins = 37.1°C, 6 min = 37.2°C, 8 min = 36.9°C

(a) Organise Kirsty's results into a table. *(4 marks)*
(a) Identify the time when an anomalous temperature reading was taken. *(1 mark)*

Marks for the answer are awarded like this:

Time / min	Temperature / °C
0	36.4
2	36.8
4	37.1
6	37.2
8	36.9
10	37.3
12	37.6

- Table with time and temperature headings (1 mark)
- Readings in order (1 mark)
- Units in header row (1 mark)
- Two columns (1 mark)

(b) The anomalous temperature reading was taken at 8 minutes. 'Anomalous' means that the result doesn't fit into the pattern of the other results. In this case, the temperature reading at 8 minutes is lower than expected. You might have to circle an anomalous result in a table, or on a graph – see below.

You may have to calculate totals, averages, percentages or missing values from a table. For example:

The table below contains results from an investigation into diffusion. The table shows the time taken for different cubes of agar jelly containing potassium permanganate to turn colourless when placed in beakers of dilute hydrochloric acid. Three groups of students carried out the same experiment. The table is incomplete.

Cube size (side length)	Time taken for cube to turn colourless / minutes				Mean time / minutes
	Group 1	Group 2	Group 3	Total for three groups	
2 cm	52	45	49	?	?
1 cm	10	15	13	38	12.7
0.5 cm	6	5	?	19	6.3

(a) For the 2 cm cubes, calculate the missing value in the 'Total' column. *(1 mark)*
(b) For the 2 cm cubes, calculate the mean time for the three groups. *(1 mark)*
(c) For the 0.5 cm cubes, calculate the missing value for Group 3. *(1 mark)*

The answers are:
(a) Total = 52 + 45 + 49 = 146 minutes
(b) Mean time = total / 3 = 48.7 minutes
   (Use the same number of decimal places as in the given answers.)
(c) Missing value = 19 – (6 + 5) = 8

This is straightforward arithmetic – just be careful!

If you have to find a percentage change, this is calculated using the equation: $\frac{\text{(final value – starting value)}}{\text{starting value}} \times 100$

For example:

After exercise, a boy's heart rate increased from 75 beats per minute to 97 beats per minute. What is the percentage increase in his heart rate?

The answer is: % increase = $\frac{(97 - 75)}{75} \times 100 = 29.3\%$

5. PLOTTING AND INTERPRETING GRAPHS

Take the following example.

The table below shows the activity of the enzyme amylase at different temperatures. The activity is measured as cm^3 of starch suspension digested per minute.

Plot a graph of these results.

Temperature / °C	Activity of amylase / cm^3 per minute
10	0.14
20	0.53
30	1.75
40	1.95
50	1.43
60	0.59
70	0.13

A graph of the data looks like this.

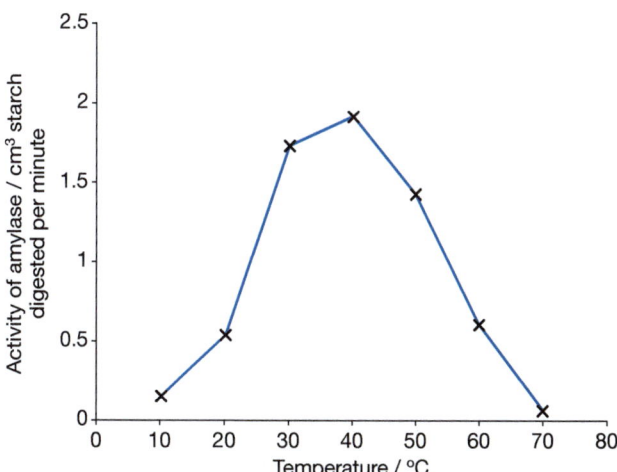

It is important to use a ruler and a sharp pencil for drawing graphs. A pen is no good, and if you make a mistake with a pen you can't erase it. Remember to label the axes and include units.

The variable that is set by the person designing the experiment is temperature. This is called the **independent variable**, and goes on the horizontal or *x*-axis. The **dependent variable** is the activity of the enzyme; this goes on the vertical or *y*-axis.

You will probably have to describe the pattern shown by the graph, or identify a point on the graph where something happens. For instance, in the graph above, the rate of reaction (activity of the enzyme) increases with temperature from 10 °C to 40 °C. Above 40 °C, the rate decreases again until 70 °C, where the reaction has almost stopped. The optimum temperature for this enzyme is about 40 °C.

A question like this would go on to ask about the biology behind the graph – this one is all about the effect of temperature on enzyme activity (see pages 37–38 and 40–41).

Note that in many graphs the line doesn't pass through the (0,0) coordinates.

You may also have to draw a bar chart from data you are given.

6. EXPERIMENTAL DESIGN

Expect questions on various aspects of experimental design, especially:

- suggesting a suitable Control for an experiment
- the meaning of controlled variables and how to control them
- how to improve accuracy and precision
- how to improve reliability.

(Note that here we are using a capital 'C' for Control, to distinguish it from 'controlled variables'.)

In a controlled experiment, only one factor is changed at a time. For example, imagine you are carrying out an investigation to compare the energy content of three different foods (pasta, biscuit and chocolate), using this apparatus:

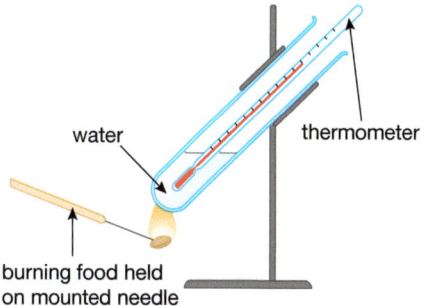

The factor that you are changing is the type of food. It is important that all other factors are kept constant. For example, the following should all be the same for each food tested:

- volume of water in the test tube
- starting temperature of the water
- type of test tube and thermometer
- distance of the burning food from the tube
- the way the food is ignited.

Ideally, the mass of each food should also be the same, although it is easier to weigh each sample and correct for any difference in mass by calculating the energy content per gram of food.

The factors above are called **controlled variables** because you, the experimenter, are controlling them (keeping them from changing) so that they shouldn't affect the results.

Some biological experiments involve the use of a **Control**. This is an experimental set-up where the key factor is missing. For example, if you were investigating the effect of the enzyme amylase on starch (as in Section 5 above) the experiment would involve mixing amylase with starch to see if the amylase breaks down the starch. It is possible (but unlikely) that the starch might break down on its own, so a Control would be a tube containing only starch, to check whether this happens. A better Control would be starch with *boiled* amylase; this is as near as possible to the experimental set-up but the amylase would not be able to act as an enzyme.

Hopefully, findings from experiments will be *accurate, precise* and *reliable*. Students often confuse these terms.

Accuracy means how close an experimental result is to its true value. Accuracy of the results depends mainly on the methods you use to obtain them.

For example, the starch–amylase experiment measures amylase activity by looking at how quickly the amylase breaks down the starch, as shown by a colour change with iodine solution (see Activity 4 on pages 40–41). As the starch is broken down, the colour of the iodine changes from blue-black to brown and finally to yellow. This is a gradual process, and the experimenter has to judge the colour change by eye. It may not be possible to do this better than to the nearest half minute, as in the table on page 41 so the accuracy is limited by the method used. (It is possible to measure the colour change using a machine called a colorimeter. This would increase the accuracy but is beyond what you would be expected to do for an experiment at International GCSE level.)

Sometimes, you have to accept that the accuracy of your results will be limited by the apparatus available. In an exam, however, you might need to comment on how accuracy could be improved, for example, by using more sensitive measuring equipment.

Precision relates to the smallest division on the scale of the measuring instrument you are using. For example, a thermometer that reads to 0.1 °C is more precise than one that only reads to 1 °C. However, you have to be sensible when choosing suitable measuring instruments to use. If you are using a water bath consisting of a beaker of water heated by a Bunsen burner, a thermometer that reads to 0.1 °C is too precise. You wouldn't expect to be able to maintain the temperature any better than ±1 °C,

so it is good enough to use a thermometer that reads to that degree of precision. Precision is also shown by the number of digits you give in a measurement, e.g. 9.998 g is more precise than 10 g.

Reliability is a measure of how similar the results are, if you carry out the same experiment several times. It tells you how confident you can be that your results are correct. For example, imagine you measured the activity of an enzyme at 40 °C eight times, using two different sets of apparatus, and obtained these results:

Measurement	Activity / arbitrary units*	
	Apparatus 1	Apparatus 2
1	26	14
2	25	19
3	27	52
4	22	44
5	28	12
6	25	10
7	26	37
8	26	21
Mean	25.6	26.1

* 'arbitrary units' means that the actual units are not important, but they are the same in both columns, so comparisons can be made.

Which mean value in the table is a more reliable measure of the activity? You can tell that the mean for Apparatus 1 is more reliable, because the individual readings are closer together. The only way you can tell if results are reliable is to carry out repeats.

You should make sure you understand the differences between accuracy, precision and reliability.

7. PLANNING AN INVESTIGATION

You may have to give a short description of how you would carry out an investigation. This may be one based on an experiment in the book, or it might involve an unfamiliar situation. Take this example:

Describe an investigation you could carry out to find out the effect of changing the concentration of the enzyme amylase on the rate of starch digestion. *(6 marks)*

This is similar to the experiment described on pages 40–41 in this book, but involves changing the *concentration* of amylase, rather than the *temperature*. In any question like this, you have to answer these points:

- What are you going to change?
- What are you going to keep constant (control) during the experiment?
- What are you going to measure and how?
- How are you going to check that the results are reliable?

A good answer would be:

I will make up 5 different concentrations of amylase solution (e.g. 1%, 2%, 3%, 4% and 5%) using the same source of enzyme, such as fungal amylase. I will mix 10 cm³ of each concentration with 10 cm³ of 1% starch suspension, and time how long it takes for the amylase to digest the starch. To tell when the starch has been digested, I will remove a sample of the mixture at 30-second intervals and add one drop of iodine solution to it. The iodine solution will start off blue-black when the starch is present, but change colour to yellow when the starch has been digested, so I will time how long it takes for this to happen. I will use a water bath to keep the temperature of the reaction mixture at 40 °C. To ensure reliability, I will repeat the experiment three times at each concentration and compare the results.

The marks (up to a maximum of 6) would be given for points like these:

- using two or more concentrations of amylase (the independent variable)
- using equal volumes of amylase (a controlled variable)
- using the same source of amylase, or explaining its source (a controlled variable)
- using the same concentration or volume of starch (a controlled variable)
- keeping the temperature constant (a controlled variable)
- use of iodine solution to show a colour change from blue-black to yellow
- measuring the time taken for the change (the dependent variable)
- repeating tests at each concentration (for reliability).

APPENDIX B: COMMAND WORDS

Command word	Definition
Add/Label	Requires the addition or labelling of a stimulus material given in the question, for example labelling a diagram or adding units to a table.
Calculate	Obtain a numerical answer, showing relevant working.
Comment on	Requires the synthesis of a number of variables from data/information to form a judgement.
Complete	Requires the completion of a table/diagram.
Deduce	Draw/reach conclusion(s) from the information provided.
Describe	To give an account of something. Statements in the response need to be developed, as they are often linked but do not need to include a justification or reason.
Determine	The answer must have an element that is quantitative from the stimulus provided, or must show how the answer can be reached quantitatively. To gain maximum marks, there must be a quantitative element to the answer.
Design	Plan or invent a procedure from existing principles/ideas.
Discuss	■ Identify the issue/situation/problem/argument that is being assessed within the question. ■ Explore all aspects of an issue/situation/problem/argument. ■ Investigate the issue/situation etc. by reasoning or argument.
Draw	Produce a diagram either using a ruler or freehand.
Estimate	Find an approximate value, number or quantity from a diagram/given data or through a calculation.
Evaluate	Review information (e.g. data, methods) then bring it together to form a conclusion, drawing on evidence including strengths, weaknesses, alternative actions, relevant data or information. Come to a supported judgement of a subject's quality and relate it to its context.

Command word	Definition
Explain	An explanation requires a justification/exemplification of a point. The answer must contain some element of reasoning/justification – this can include mathematical explanations.
Give/State/Name	All of these command words are really synonyms. They generally all require recall of one or more pieces of information.
Give a reason/reasons	When a statement has been made and the requirement is only to give the reason(s) why.
Identify	Usually requires some key information to be selected from a given stimulus/resource.
Justify	Give evidence to support (either the statement given in the question or an earlier answer).
Plot	Produce a graph by marking points accurately on a grid from data that is provided and then draw a line of best fit through these points. A suitable scale and appropriately labelled axes must be included if these are not provided in the question.
Predict	Give an expected result.
Show that	Verify the statement given in the question.
Sketch	Produce a freehand drawing. For a graph, this would need a line and labelled axes with important features indicated. The axes are not scaled.
State what is meant by	When the meaning of a term is expected but there are different ways for how these can be described.
Suggest	Use your knowledge to propose a solution to a problem in a novel context.
Verb proceeding a command word	
Analyse the data/graph to explain	Examine the data/graph in detail to provide an explanation.
Multiple choice questions	
What, Why	Direct command words used for multiple-choice questions.

GLOSSARY

accommodation Changes taking place in the eye which allow it to focus on objects at different distances

accuracy (of experimental results) Closeness of an experimental result to its true value

ACE inhibitors (angiotensin converting enzyme inhibitors) Substances that inhibit an enzyme in the renin–angiotensin system in order to control blood pressure in a person with hypertension

acquired immune deficiency syndrome (AIDS) Set of symptoms (syndrome) that develops as a result of an infection by the HIV virus, due to damage to the immune system

activated sludge method Method of sewage treatment where sludge from sewage settling tanks is mixed with incoming waste water and oxygen. Aerobic bacteria break down organic matter in the sewage

active immunity Production of antibodies by the body in response to an antigen

active site Area on the surface of an enzyme where the substrate attaches and products are formed

active transport Movement of molecules or ions against a concentration gradient, using energy from respiration

adaptation Feature of an organism that suits its structure to its function

adenosine triphosphate (ATP) Chemical present in all cells which acts as an energy 'currency'. ATP is made by respiration and used up by any process that needs a supply of energy. ATP is broken down into adenosine diphosphate (ADP) and phosphate

ADH (See antidiuretic hormone)

adrenal glands Pair of endocrine glands situated above the kidneys. Secrete adrenaline

adrenaline Hormone secreted by the adrenal glands. Stimulates several organs in the 'fight or flight' response. Also acts as a neurotransmitter

adult stem cell Stem cell found in certain adult tissues such as bone marrow and skin. Can differentiate to form a limited number of specialised tissues

aerobic respiration Chemical reaction that releases energy from food. Uses oxygen and produces carbon dioxide and water

agar Jelly-like substance used as a culture medium for growing microorganisms

agglutinate / agglutination Sticking together in clumps, as with an antigen–antibody reaction

AI (See artificial insemination)

AIDS (See acquired immune deficiency syndrome)

alginate Substance that forms jelly-like beads used for immobilising enzymes

alleles Different forms of a gene

alveoli (singular = alveolus) Microscopic air sacs in the lungs where gas exchange takes place

Alzheimer's disease Disease of the brain that causes dementia in elderly people. Caused by the build-up of certain proteins in brain cells

amino acid One of 20 different molecules that form the building blocks of proteins

amnion Membrane enclosing the embryo during pregnancy

amniotic fluid Fluid secreted by the amnion that protects the embryo by acting as a shock absorber

amylase Enzyme that digests starch into maltose

anaerobic respiration Reaction that releases energy from food, without using oxygen. Produces lactate (lactic acid) in humans

angina Chest pain caused by blockage of the coronary arteries

angioplasty Medical procedure using a catheter and stent to widen the coronary arteries

antagonistic pairs Pairs of muscles working together to move a part of the body. One muscle contracts while the other relaxes

antibiotic Chemical produced by a microorganism (bacteria or fungi) to kill other microorganisms. Used in medicine to treat bacterial infections

antibody Protein produced by lymphocytes that binds with foreign antigens as part of the immune response

anticodon Group of three bases on a tRNA molecule that is complementary to a codon on the mRNA molecule

antidiuretic hormone (ADH) Hormone released from the pituitary gland. Controls the water content of the blood by increasing reabsorption of water from the collecting ducts of the kidneys into the blood

antigen Chemical 'marker' on the surface of a cell that identifies the cell as 'self' or 'non-self'

anus Outlet of the gut where faeces is expelled from the body

aorta The main artery. Blood vessel carrying oxygenated blood from the heart to the body

appendicular skeleton Bones attached to the axial skeleton – scapulas, clavicles, pelvis and limb bones

arteriole Small artery

artery Blood vessel with a thick muscular wall and a narrow lumen, carrying blood away from the heart

artificial heart Synthetic device designed to carry out the functions of a heart, usually in the time leading up to a heart transplant

artificial immunity Immunity gained by vaccination or by injection of antibodies

artificial insemination (AI) Medical procedure where sperm is placed in a woman's uterus at the time of ovulation, to increase the chances of fertilisation

asexual reproduction Reproduction that does not involve fusion of gametes. New organisms are produced by part of an organism separating from a single parent

assimilation Manufacture of new substances in cells using the products of digestion

astigmatism Defect of the eye where the surface of the cornea or lens is not perfectly spherical but rounder in one plane than in another, so that an image is in focus in one direction but not the other

atherosclerosis Hardening of the arteries caused by a build-up of fatty deposits

athlete's foot Skin disease caused by a fungus. Symptoms include sore, red patches on the skin between the toes

ATP (See adenosine triphosphate)

atria (singular = atrium) Two upper chambers of the heart where blood enters the heart from the vena cava (right atrium) and pulmonary vein (left atrium)

axial skeleton Central axis of bones in the body – vertebral column, cranium, ribcage and sternum

axon Long extension of a neurone that carries nerve impulses in a direction away from the cell body

bacteria (singular = bacterium) Small single-celled organisms with no nucleus

balanced diet Diet containing all the necessary food types and in the correct amounts and proportions to keep the body healthy

ball and socket joint Joint that allows movement in all three directions, e.g. shoulder or hip

barrier methods (of contraception) Methods of contraception involving some kind of barrier to prevent sperm reaching the egg, e.g. a condom

basal metabolic rate (BMR) The rate of oxygen consumption of a person at rest

base (in DNA) One of four nitrogen-containing groups in the DNA molecule, called adenine, thymine, cytosine and guanine. Bases form complementary pairs linking the chains of the double helix

basement membrane (in Bowman's capsule) Membrane in the wall of the Bowman's capsule of a kidney tubule that acts as a molecular filter during ultrafiltration in the kidney

GLOSSARY

beta blockers Drugs that block the action of adrenaline. Used to manage unusual heart rhythms

biceps Muscle at the front of the upper arm, responsible for flexing the arm at the elbow

bicuspid valve Valve in the heart between the left atrium and left ventricle. Prevents backflow of blood when the ventricle contracts

bile Green liquid made by the liver and stored in the gall bladder. Causes lipids in the gut to form an emulsion, increasing their surface area for easier digestion by enzymes

bile duct Tube carrying bile from the gall bladder to the duodenum

binary fission Method of asexual reproduction in bacteria where the cell divides into two

biosensor Instrument that measures concentrations of substances in biological fluids, e.g. glucose in the blood

bladder Muscular bag that stores urine before its removal from the body

blind spot Area of the retina where the optic nerve leaves the eye. Contains no light-sensitive cells, so an image cannot be detected

blood Transport tissue made of various types of red and white blood cells in a liquid called plasma

BMI (See body mass index)

BMR (See basal metabolic rate)

body mass index (BMI) Person's weight in kilograms divided by their height in metres, squared. BMI = weight (kg) / [height (m) × height (m)]

bone Skeletal tissue composed of cells that secrete a hard material made of calcium phosphate

bone marrow Soft material in central cavities of larger bones. Makes blood cells

Bowman's capsule Structure consisting of a hollow cup of cells at the start of a kidney tubule. The site of ultrafiltration

bronchi (singular = bronchus) Tubes leading from the trachea to the lungs

bronchial tree Branching network of air passages in the lungs

bronchioles Small air passages leading from the bronchi to the alveoli

bronchitis Lung disease caused by irritation of the bronchial tree and infection by bacteria, resulting in breathing difficulties

canines Pointed teeth near the front of the mouth behind the incisors

cannabis Drug, illegal in most countries, that produces psychoactive effects

cap (contraceptive) (See diaphragm)

capillary Microscopic blood vessel that carries blood through organs and allows exchange of substances between the blood and the cells of the organ

capsule (of bacteria) Slime layer covering some bacterial cells. Protects the bacterium and stops it drying out

capsule (of joint) Tough, fibrous structure surrounding a joint

carbohydrase Enzyme that digests carbohydrates

carbohydrate Organic compound composed of one or more sugar molecules

carbon monoxide Poisonous gas present in cigarette smoke

carboxyhaemoglobin Substance formed when carbon monoxide combines with haemoglobin, displacing oxygen from the haemoglobin

carcinogen Something that causes cancer, e.g. a chemical or certain types of radiation

cardiac centre Region in the medulla of the brain that controls heart rate

cardiac cycle Sequence of events taking place in the heart during one heartbeat

cardiac muscle Specialised muscle making up the heart wall. Able to contract rhythmically without fatiguing

cardiovascular system Heart and blood vessels

carotid artery Artery leading from the aorta to the head and neck

carrier (in genetics) Person who has inherited a recessive allele for a genetic condition but is heterozygous. The carrier does not display the condition but can pass the recessive allele on to their offspring

cartilage Tough tissue present in several places in the body, such as rings in the trachea and between the bones at a joint

catalyst Chemical that increases the rate of a reaction but remains unchanged at the end of the reaction

cataract Eye condition where the lens becomes cloudy and opaque, so that the person is unable to see

cell Basic structural unit of living organisms

cell membrane Thin surface layer around the cytoplasm of a cell. Forms a partially permeable barrier between the cell contents and the outside of the cell

cell wall Non-living layer outside the cell membrane of certain types of cell, including plants, bacteria and fungi

cellulose Polysaccharide of glucose that forms plant cell walls. The main component of dietary fibre

central nervous system (CNS) Brain and spinal cord

cerebellum Region of the brain behind the cerebrum, involved with coordinating the contraction of sets of muscles and maintaining balance

cerebral cortex Outer, folded part of the cerebrum of the brain

cerebral hemispheres (See cerebrum)

cerebrum Largest part of the brain, made of two cerebral hemispheres. The source of all conscious thoughts

cervix 'Neck' of the uterus

CF (See cystic fibrosis)

CHD (See coronary heart disease)

cholera Disease caused by the bacterium *Vibrio cholerae*. Symptoms include severe diarrhoea, weakness, muscle pain, low blood pressure and rapid heart rate

cholesterol Lipid substance present in the blood and linked to coronary heart disease

chondrocytes Cells in cartilage

choroid Dark layer of tissue below the sclera of the eye. Contains blood vessels and pigment cells

chromatid One of two thread-like strands of a replicated chromosome. Each chromatid contains an exact copy of the double helix of DNA. Chromatids become visible at the start of mitosis and meiosis

chromosome Thread-like structure found in the nucleus of a cell, made of DNA and protein. Contains the genetic information (genes)

cilia (singular = cilium) Microscopic hair-like projections from the surface of some cells, such as those lining the trachea and bronchi. Waving of cilia moves mucus and trapped particles towards the mouth

ciliary muscle Ring of muscle around the lens of the eye that alters the shape of the lens during accommodation

ciliated epithelium Epithelium composed of cells with cilia

cirrhosis Liver disease caused by a person drinking large amounts of alcohol over a long period of time

clone Group of cells, or organisms, that are genetically identical

CNS (See central nervous system)

cocaine Drug with limited uses as a local anaesthetic. Also used illegally as a powerful stimulant. Highly addictive

cochlea Spiral-shaped structure of the inner ear, which contains the organ of Corti

codominance / codominant Pattern of inheritance where neither allele of a gene is dominant over the other so that both alleles are expressed in the phenotype

codon Triplet of bases on the mRNA molecule. Different triplets code for different amino acids in a protein

collecting duct Last part of a kidney tubule, where water is reabsorbed before the final urine is produced

colon First part of the large intestine, where water is absorbed from the waste material in the gut

compact bone Outer, harder part of a bone

competitive inhibitor Molecule with a shape that is similar to the shape of the substrate of an enzyme. Fits into the active site of the enzyme, stopping the substrate from entering and slowing the reaction

condom Barrier method of contraception. Thin rubber sheath that is placed over the penis before sexual intercourse to prevent semen entering the vagina

GLOSSARY

cone (cell) Cell in the retina of the eye that is sensitive to different wavelengths of light and results in colour vision

contraception Methods used to prevent a woman becoming pregnant

contraceptive pill Chemical means of contraception. Pill containing female hormones that prevent the production of FSH and LH from the pituitary gland, stopping ovulation

Control Part of an experiment which is set up to show that other variables are not having an effect on the outcome of the experiment

controlled variables Variables in an experiment other than the independent variable, which are kept constant by the person carrying out the experiment so that they do not affect the results

cornea Transparent 'window' at the front of the eye that allows light to enter the eye. Also (along with the lens) refracts the light as it enters the eye

coronary arteries Small arteries supplying blood to the heart muscle

coronary heart disease (CHD) Disease caused by a blockage of the coronary arteries due to a build-up of fatty material. Can cut off the blood supply to the heart and result in a heart attack

coronary veins Small veins carrying blood away from the heart muscle

corpus luteum Remains of an ovarian follicle after ovulation. Secretes progesterone

cortex (of kidney) Outer part of the kidney, containing kidney tubules and blood vessels

cranium Bones of the skull

cupula Structure in the inner ear consisting of a mass of jelly with hair cells embedded in it. Detects movement

cystic fibrosis (CF) Disease caused by a mutated gene that results in the over-production of mucus in the airways leading to the lungs, in the lining of the gut and in the reproductive organs

cytoplasm Jelly-like material that makes up most of a cell

decomposers Organisms that feed by breaking down the dead remains of other organisms, e.g. some bacteria and fungi

denatured / denaturing Condition or process where the structure of a protein is damaged by high temperatures (becomes denatured). If the protein is an enzyme, it will no longer catalyse its reaction

dendrites Fine extensions of the dendrons of a neurone

dendron Extension of the cytoplasm of a neurone that carries impulses towards the cell body

dental caries Tooth decay

dentine Hard, living material underneath the enamel of a tooth

deoxyribonucleic acid (DNA) Chemical of which genes are made. Double helix composed of deoxyribose sugar, phosphates and four bases

dependent variable Variable in an experiment that changes as a result of changes in the independent variable

depression Mental illness when a person has intense feelings of sadness and other symptoms that last for long periods of time

dermis Middle layer of the skin containing many sensory receptors

diabetes Disease where the blood glucose concentration cannot be properly controlled. Caused by a lack of insulin

diaphragm (contraceptive) Barrier method of contraception. Dome-shaped piece of rubber that a woman inserts into her vagina before intercourse. Also called a cap

diaphragm (in chest) Muscular sheet separating the thorax from the abdomen. Involved in the mechanism that ventilates the lungs

diastole Phase during the heart cycle when the heart muscle is relaxing

dietary fibre Indigestible plant material, mainly cellulose, in the diet. Helps to prevent constipation and bowel diseases

differentiation Process taking place during the development of an embryo, where cells become specialised to carry out particular functions

diffusion Movement of molecules or ions down a concentration gradient

digestion Process by which food is broken down into simpler molecules that can be absorbed

diploid (number / cells) Number of chromosomes found in body cells. Diploid cells contain both chromosomes of each homologous pair

disaccharide Sugar made up of two monosaccharides, e.g. sucrose is a disaccharide of glucose and fructose

DNA (See deoxyribonucleic acid)

DNA polymerase Enzyme involved in the replication of DNA

dominant (allele) Allele of a gene that is expressed in the heterozygote

dopamine Neurotransmitter in the brain. Low levels of dopamine are the cause of Parkinson's disease

dorsal root Part of a spinal nerve that emerges from the dorsal (back) side of the spinal cord

dorsal root ganglion Swelling in the spinal nerve that contains the cell bodies of sensory neurones

double circulation Blood circulatory system in mammals, where the blood passes through the heart to the lungs and returns to the heart before passing to the rest of the body

drug Chemical that affects the normal chemical reactions taking place in a person's body

duct Tube leading from an exocrine gland, through which the product of the gland is secreted

duodenum First part of the small intestine following the stomach

ear ossicles Three small bones in the middle ear that pass vibrations from the eardrum to the cochlea

eardrum Membrane in the middle ear that vibrates in response to sound waves

Ebola Severe, often fatal disease caused by a virus. Symptoms include fever, muscle pain, vomiting, headache and a sore throat, followed by vomiting, diarrhoea, a rash, and damaged liver and kidneys, leading to coma and death

effector Organ that brings about a response (a muscle or gland)

egestion Process of removing faeces from the intestine

ejaculation Release of semen during sexual intercourse

embryo Multicellular structure formed by division of a zygote

embryonic stem cell Stem cell found in the early stage of development of the embryo, which can differentiate to form any type of cell

emphysema Lung disease where the walls of the alveoli break down and fuse together again, forming air spaces with a reduced surface area. Results in breathing difficulties

enamel Hard outer non-living material of a tooth

endocrine gland Gland secreting a hormone into the bloodstream

endoplasmic reticulum (ER) Network of membranes throughout the cytoplasm of a cell. In places covered with ribosomes, where protein is made. Spaces between the membranes allow for transport of proteins and other materials around the cell

enzyme Protein that acts as a biological catalyst

epidermis Outer layer of skin, consisting of dead cells

epithelium (plural = epithelia) Tissue forming the internal or external lining to organs

ER (See endoplasmic reticulum)

erythrocyte Red blood cell

Eustachian tube Tube connecting the middle ear with the throat, allowing the air pressure to be equalised either side of the eardrum

excretion Removal from the body of the waste products of metabolism

exocrine gland Gland secreting a product through a duct

F1 generation Offspring formed from breeding the parent organisms

F2 generation Offspring formed from breeding individuals from the F_1 generation

faeces Semi-solid indigestible waste that passes out of the gut via the anus

fatty acid Type of molecule that, with glycerol, is one of the building blocks of lipids

femidom Barrier method of contraception. Thin rubber sheath that is placed inside the vagina before sexual intercourse to prevent semen entering the vagina

femur Thigh bone

fermenter A vessel used to grow microorganisms

fertilisation Fusion of male and female gametes to form a zygote

fetus Unborn offspring of a mammal, in particular an unborn human embryo more than 2 months after fertilisation, when it shows recognisably human features

fibrin Protein formed from fibrinogen during blood clotting

fibrinogen Protein in blood plasma that forms insoluble fibres of fibrin during blood clotting

flagellum (plural = flagella) In animal cells such as sperm: a tail-like structure that beats from side-to-side, producing movement. In some bacteria: a structure with a similar function but much smaller and with quite a different structure

fluoride Mineral that is added to water supplies to reduce tooth decay

follicle (in ovary) Structure in the mammalian ovary that contains a single developing egg cell

follicle stimulating hormone (FSH) Hormone made by the pituitary gland. Stimulates the maturation of eggs in the ovary and sperm production in the testes

fovea Region at the centre of the retina of the eye where there is a high concentration of light-sensitive receptor cells

fructose Monosaccharide sugar found in fruits

FSH (See follicle stimulating hormone)

fungi Group of organisms that includes mushrooms, moulds and yeasts. Mostly free-living decomposers

gall bladder Organ that stores bile from the liver

gametes Male and female sex cells, formed by meiosis

gene Part of a chromosome, the basic unit of inheritance. A length of DNA that controls a characteristic of an organism by coding for the production of a specific protein

gene therapy Medical procedure by which a patient is treated for a genetic disorder by having a functioning gene inserted into their DNA

genotype Alleles present in an organism for a certain characteristic

glomerular filtrate Fluid that passes through the Bowman's capsule at the start of a kidney tubule

glomerulus Ball of capillaries surrounded by the Bowman's capsule at the start of a kidney tubule

glucagon Hormone released by the pancreas which causes an increase in the concentration of glucose in the blood

glucose Monosaccharide sugar, the main 'fuel' for respiration

glycerol Molecule that, along with fatty acids, is a component of lipids

glycogen Polysaccharide of glucose that acts as a storage carbohydrate in the liver and muscles

gonorrhoea Sexually transmitted disease caused by the bacterium *Neisseria gonorrhoeae*

grey matter Tissue in the middle of the spinal cord and outer part of the brain. Consists mainly of nerve cell bodies

haemoglobin Chemical present in red blood cells that combines with oxygen and carries it around the body

haemophilia Blood disorder where the blood does not clot properly. Caused by a faulty gene on the X chromosome

hair erector muscle Muscle attached to the base of each hair in the skin. Contracts to pull the hair upright

haploid (number / cells) Number of chromosomes found in gametes. Haploid cells contain one chromosome from each homologous pair

Haversian systems (in bone) Osteocytes arranged in rings around central canals containing nerves and blood vessels

heart attack Sudden and sometimes fatal stoppage of the heart contractions caused by the heart muscle being deprived of oxygen

heart transplant Surgical procedure where a functioning heart from a donor is put into a patient whose heart has failed

hepatic portal vein Blood vessel transporting the products of digestion from the ileum to the liver

heroin Drug used to treat severe pain. Also used illegally as a highly addictive narcotic drug

heterozygous Genotype with different alleles of a gene, e.g. Aa

hinge joint Joint that allows movement in one direction, e.g. knee or elbow

histone Protein associated with the DNA in a chromosome

HIV (See human immunodeficiency virus)

homeostasis Maintaining constant conditions in the body. Maintaining a constant internal environment

homeotherm Animal that maintains a constant body temperature by physiological means (mammals and birds)

homologous pair Pair of chromosomes that carries genes controlling the same features at the same positions on each chromosome. The members of each homologous pair are the same size and shape

homozygous Genotype with the same alleles of a gene, e.g. AA or aa

hormonal methods (of contraception) Methods of contraception using the contraceptive pill, or hormone implants or injections

hormone Chemical 'messenger' that travels in the blood. Produced by endocrine glands

human immunodeficiency virus (HIV) Virus that causes AIDS

hybridoma cells Cells made by fusing together antibody-producing lymphocytes with cancer cells. Used to make monoclonal antibodies

hypertension Long-term high blood pressure

hypodermis Layer of skin below the dermis, containing fatty tissue

hypothalamus Region at the base of the brain above the pituitary gland. Secretes hormones and monitors various 'drives' such as hunger and thirst

ileum Last part of the small intestine, where the products of digestion are absorbed into the blood

immobilised enzyme Enzyme attached to, or trapped within, an insoluble material. Used for industrial or medical purposes

immune response Mechanism by which the body recognises and deals with exposure to a pathogenic microorganism. Involves the production of memory cells that respond to a subsequent infection by dividing to produce many antibody-producing cells

in vitro fertilisation (IVF) Medical procedure where sperm is added to an egg in a Petri dish or similar container, outside the woman's body, to monitor fertilisation and growth of the embryo

incisors Sharp, chisel-shaped teeth at the front of the mouth

incubation period Time between when a person is first infected with a pathogen and when they first show symptoms of the disease

incus (anvil) The second of the three ear ossicles

independent variable Variable in an experiment that is set by the person designing the experiment. Results in a change in the dependent variable

infection Transfer of a pathogen to a person

infertility Condition where a couple have difficulty in achieving a pregnancy

inhibitor Substance that reduces the rate of an enzyme-catalysed reaction by interfering with the action of the enzyme

insertion (of a muscle) Place where a muscle is attached to a bone that moves when the muscle contracts

insulin Hormone produced by the pancreas which causes a decrease in the concentration of glucose in the blood

intercostal muscles Two sets of antagonistic muscles lying between the ribs. Contract and relax to move the ribs in order to ventilate the lungs

internal environment Blood and tissue fluid

intrauterine device (IUD) Method of contraception. Small plastic or copper device that is inserted into the uterus and prevents a fertilised egg from implanting in the wall of the uterus

invertase Enzyme that breaks down the disaccharide sucrose into glucose and fructose. Made by yeast cells

invertebrate Animal without a vertebral column (backbone)

involuntary muscle (smooth muscle) Muscle whose contraction is not under conscious control by the brain. Present in walls of organs such as blood vessels, bladder and gut

iris Coloured part of the eye visible from the front. Muscles in the iris change the size of the pupil

IUD (See intrauterine device)

IVF (See in vitro fertilisation)

GLOSSARY

joint Point where two bones meet, allowing movement of one bone relative to the other

kidney transplant Surgical procedure where a functioning kidney from a donor is put into a patient whose kidneys have failed

knee-jerk reflex Sudden involuntary forward movement of the leg produced by a tap on the tendon just below the kneecap

kwashiorkor Disease caused by a lack of protein. Usually due to starvation, resulting in body proteins being metabolised

lactase Enzyme that breaks down the disaccharide lactose into glucose and galactose

lactate Waste product of anaerobic respiration in muscle cells

lacteal Structure in the middle of a villus, containing lymph and forming part of the lymphatic system. Absorbs products of lipid digestion

lactose Sugar found in milk. Disaccharide of glucose and galactose

LH (See luteinising hormone)

ligaments Fibrous structures with a high tensile strength and some elasticity. Positioned across a joint to hold bones together while allowing movement

ligase Enzyme used to join pieces of DNA in genetic engineering

limiting factor Factor in a process or reaction that limits (holds back) the process, e.g. a nutrient that is in short supply

lipase Enzyme that digests lipids

lipid Fats and oils. Most lipids are composed of fatty acids and glycerol

liver Large organ in the abdomen that has many functions, including the storage of glycogen, manufacture of bile, and breakdown of amino acids

lock and key model Model of enzyme action where the substrate is the 'key', fitting into the 'lock', which is the active site of the enzyme

long sight Defect of vision where either the lens is not convex enough or the eyeball is too short from front to back, so that light rays from a nearby object are out of focus behind the retina

loop of Henlé U-shaped part in the middle of a kidney tubule. Involved in concentrating the fluid in the tubule

lumen Space in the middle of a tube such as an artery or the gut

lung capacity Maximum total volume of the lungs, equal to the vital capacity plus the residual volume

luteinising hormone (LH) Hormone released from the pituitary gland, which stimulates the release of a mature egg from the ovary (ovulation). Also stimulates the ovary to make oestrogen

lymph Fluid formed from the blood plasma, which drains away through the lymphatic system back into the bloodstream

lymph nodes Swellings in the vessels of the lymphatic system that contain white blood cells

lymphatic system System of vessels throughout the body containing a fluid called lymph

lymphocyte Type of white blood cell that produces antibodies

malaria Disease caused by the protozoan parasite *Plasmodium*, passed to humans by mosquitoes

malleus (hammer) The first of the three ear ossicles

mammary glands Exocrine glands in a woman's breasts, which secrete milk

medulla (of brain) Part of brain that controls basic body functions such as heart rate and breathing rate

medulla (of kidney) Middle part of the kidney containing blood vessels, loops of Henlé and collecting ducts

meiosis Type of cell division that produces haploid cells (gametes)

memory cells Cells formed from lymphocytes during the immune response. Remain in the blood for many years, producing long-lasting immunity to a disease

menstrual cycle Monthly cycle of events preparing a woman's uterus for the possible implantation of a fertilised egg. Controlled by hormones from the pituitary gland

mental illness Health condition that involves abnormal changes in a person's thoughts, emotions or behaviour

messenger RNA (mRNA) Type of RNA that forms a copy of the template strand of the DNA during transcription

metabolism Chemical reactions taking place inside cells

methicillin-resistant Staphylococcus aureus (MRSA) Pathogenic bacterium that is resistant to many antibiotics. In hospitals, MRSA is responsible for many difficult-to-treat infections

microorganism Organism that can only be seen through a microscope. Include bacteria, viruses, protozoa and some fungi

microvilli Minute projections from the surface membrane of some cells to increase the surface area

minerals Elements needed by the body and gained from food but not present in carbohydrates, lipids or proteins

mitochondrion (plural = mitochondria) Organelle that carries out aerobic respiration, releasing energy for the cell. Place where most of the cell's ATP is made

mitosis Type of cell division that produces diploid body cells for growth and repair of tissues

molars Teeth at the back of the mouth behind the premolars. Have flattened crowns and are used for chewing

monoclonal antibodies Antibodies made in the laboratory from a single clone of hybridoma cells. Used for the diagnosis and treatment of disease, and for medical research

monohybrid cross Genetic cross involving one gene

monosaccharide 'Single' sugar such as glucose, which cannot be broken down into a simpler sugar

motor neurone Nerve cell that transmits impulses from the central nervous system to an effector organ

mRNA (See messenger RNA)

MRSA (See methicillin-resistant *Staphylococcus aureus*)

mucus Sticky liquid secreted by certain cells, e.g. in the lining of the trachea and bronchi to trap dust and bacteria

multiple alleles More than two alleles of a single gene

mutation Change in the structure of a gene or chromosome

myelin sheath Covering made of a lipid material that surrounds an axon. Nerve cells that have a myelin sheath are described as myelinated

natural immunity Immunity developed naturally; either in response to an infection, through antibodies received by a baby in its mother's milk, or by a fetus across the placenta

negative feedback Process where a change in the body is detected and brings about events that return conditions to normal

nephron Kidney tubule, the functional unit of a kidney

nerve impulse Tiny electrical signal that passes down a nerve cell. Caused by movements of ions into and out of the axon

nervous tissue Tissue made of cells called neurones, which carry electrical signals called nerve impulses

neuromuscular junction Synapse of a nerve cell on a muscle

neurone Nerve cell

neurotransmitter Chemical that carries a nerve impulse across a synapse

nicotine Addictive drug present in tobacco and cigarette smoke

noise-induced hearing loss (NIHL) Deafness caused by prolonged exposure to loud noise

non-competitive inhibitor Molecule that attaches to an enzyme at a point away from the active site. It changes the shape of the enzyme molecule, including the active site, slowing the reaction

nucleotide Molecule that is the building block of DNA. Consists of a phosphate group, the sugar deoxyribose and one of four nitrogenous bases

nucleus Cell organelle that contains chromosomes. Controls the activities of the cell

obesity Being severely overweight

oesophagus Part of the alimentary canal between the mouth and the stomach

oestrogen Female sex hormone secreted by the ovaries. Controls the development of the female sex characteristics and the repair of the uterine lining during the menstrual cycle

optic nerve Nerve carrying impulses from the retina of the eye to the brain

oral rehydration therapy Treatment for diarrhoea involving feeding a patient a solution of glucose and salts in sterile water

organ Structure in the body that is a collection of different tissues working together to perform a function

organ of Corti Structure in the cochlea of the ear that produces nerve impulses in response to sound vibrations

organ rejection Rejection of a transplanted organ by the immune system of the recipient

organelle Part of a cell with a particular function, e.g. the nucleus

origin (of a muscle) Place where a muscle is attached to a bone that remains stationary when the muscle contracts

osmoregulation Regulation of salt and water balance in the body

osmosis Net diffusion of water molecules across a partially permeable membrane from a solution with a high water potential to a solution with a low water potential

ossification Process of replacement of cartilage by bone during the growth of a fetus and baby

osteocytes Bone cells that secrete the bone matrix, made of calcium phosphate

otolith Structure within the inner ear involved in the sense of balance. Consists of a mass of jelly containing calcium carbonate crystals. The weight of the otolith pulls on the hairs of sensory cells, giving information about the position of the head

oval window Membrane-covered opening that leads from the middle ear to the cochlea of the inner ear

ovaries (singular = ovary) Female reproductive organs that produce eggs (ova)

oviduct Tube leading from the ovary to the uterus

ovulation Release of an ovum (egg cell) from a follicle in the ovary

ovum (plural = ova) Female gamete (egg cell)

oxygen debt The volume of oxygen that is needed to completely oxidise the lactate built up during a period of anaerobic respiration

oxyhaemoglobin Haemoglobin bound to oxygen

oxytocin Hormone from the pituitary gland that stimulates the release of milk by the mammary glands

pancreas Gland discharging into the duodenum. Makes digestive enzymes and is also an endocrine organ, secreting the hormones insulin and glucagon

paracetamol Drug used to treat common aches and pains such as headache and toothache

parasite Animal or plant that lives in or on another organism (called the host) and gets nutrients from the host

Parkinson's disease Disease of the brain in middle-aged or elderly people that results in muscle tremor, muscular rigidity and slow, imprecise movements. Caused by a deficiency in the neurotransmitter dopamine

partially permeable Structure (e.g. the cell membrane) that is permeable to some molecules but not permeable to others

passive immunity Immunity gained as a result of antibodies supplied to the body from somewhere else, e.g. to a fetus across the placenta, to a baby in its mother's milk or by injection

pathogen Organism that causes disease, e.g. some bacteria

pedigree Diagram showing a family tree for an inherited characteristic

pelvis (of kidney) Funnel-like part of the kidney leading to the ureter

pelvis (of skeleton) Basin-shaped structure between the spine and the upper leg bones, made of several fused bones

penicillin Antibiotic obtained from the mould *Penicillium*

pepsin Protease enzyme made in the stomach

percolating filter method Method of sewage treatment where sewage is passed through a filter bed, where aerobic microorganisms break down organic matter

period Common name for loss of blood during the menstrual cycle

periodontal disease Gum disease

periosteum Tough membrane covering the outer surface of a bone

peristalsis Waves of muscular contraction that push food along the gut

phagocyte Cell capable of phagocytosis, e.g. some white blood cells

phagocytosis Process by which cells engulf and digest material, e.g. white blood cells engulfing bacteria

phenotype How a gene is expressed. The 'appearance' of an organism resulting from its genotype

photosynthesis Process carried out by plants, where light energy is used to drive reactions in which carbon dioxide and water are used to make glucose and oxygen

physiology Branch of biology that deals with how the body works, e.g. how muscles contract or how nerves carry impulses

pit latrine Type of composting toilet where the user squats over a trench or pit. Relies on microorganisms in the pit to break down urine and faeces

pituitary gland Gland at the base of the brain which secretes a number of hormones and substances that control the release of hormones from other endocrine glands

placenta Organ in mammals which contains blood vessels of the embryo in close proximity to blood vessels of the mother. Allows exchange of gases, nutrients, waste products and other substances between the maternal blood and the embryo's blood

plaque Invisible layer of bacteria on the surface of teeth

plasma Liquid part of the blood

plasmid Small circular piece of DNA found in bacteria and used in genetic engineering

platelets Small fragments of cells in blood. Responsible for releasing chemicals involved in blood clotting

pleural cavity Space between the pleural membranes

pleural fluid Thin layer of liquid filling the pleural cavity

pleural membranes Two layers of membrane forming a continuous envelope around the lungs

polysaccharide Carbohydrate made of many sugar units, e.g. starch is a polysaccharide of glucose

precision (of experimental results) Smallest increment that can be usefully measured, i.e. the smallest division on the scale of any measuring instrument being used

premolars Teeth between the canines and molars. Have flattened crowns and are used for chewing

progesterone Female sex hormone made by the ovaries, corpus luteum, and later by the placenta. Causes thickening of the uterus lining during the menstrual cycle and maintenance of the lining during pregnancy. Drop in levels stimulates menstruation

prolactin Hormone from the pituitary gland that stimulates the production of milk by the mammary glands

protease Enzyme that digests proteins

protein Organic substance made of chains of amino acids

prothrombin A soluble protein in plasma that is involved in blood clotting

protozoa Single-celled organisms with 'animal-like' cells. Some species are pathogenic

puberty Time when developmental changes take place in boys and girls that lead to sexual maturity

pulmonary arteries Blood vessels carrying deoxygenated blood from the heart to the lungs

pulmonary circulation Circulation of blood from the heart to the lungs via the pulmonary artery and back to the heart via the pulmonary vein

pulmonary veins Blood vessels carrying oxygenated blood from the lungs to the heart

pulp cavity Middle part of a tooth, containing nerves and blood vessels

pulse Wave of stretching followed by constriction passing through the arteries and arterioles, caused by the surge of blood at each heartbeat

Punnett square Diagram showing a genetic cross, with the outcomes of the cross arranged in a square

pupil Hole in the centre of the iris that allows light to enter the eye

receptor Cell or organ that detects a stimulus

recessive Allele that is not expressed in the phenotype when a dominant allele of the gene is present (i.e. in the heterozygote)

GLOSSARY

recombinant DNA DNA made by genetic engineering, by combining DNA from two species of organism

rectum Last part of the large intestine, where faeces is stored

red-green colour blindness Medical condition where a person cannot distinguish between certain shades of red and green colours; caused by a faulty allele

reducing sugar A sugar such as glucose, which will reduce an alkaline solution of copper (II) sulfate to copper (I) oxide in the test using Benedict's solution

reflex action Rapid, automatic, involuntary response to a stimulus

reflex arc Nerve pathway of a reflex action

relay neurone Short neurone that connects a sensory neurone with a motor neurone in the CNS

reliability (of experimental results) Measure of the similarity of results when an experiment is carried out several times

renal (kidney) dialysis Filtration of a patient's blood through an artificial kidney machine, used when a person's kidneys have stopped working

renal artery Blood vessel that supplies blood to a kidney

renal vein Blood vessel that takes blood away from a kidney

renin An enzyme produced by the kidneys when blood pressure falls

renin–angiotensin system System of hormones and other chemicals that regulates blood pressure

replication (of DNA) Copying of DNA that takes place before cell division

residual volume Volume of air left in the lungs after a maximum exhalation

respiration Chemical reaction taking place in cells, where glucose is broken down to release energy

respiratory centre Region in the medulla of the brain that controls breathing

response Reaction by an organism to a change in its surroundings

restriction endonuclease (See restriction enzyme)

restriction enzyme Enzyme used in genetic engineering to cut out a section from a molecule of DNA

retina Inner, light-sensitive layer at the back of the eye

ribonucleic acid (RNA) Nucleic acid similar to DNA but made of a single strand, with ribose sugar and the base uracil instead of thymine. Involved in protein synthesis

ribosome Tiny structure in the cytoplasm of cells, the site of protein synthesis

RNA (See ribonucleic acid)

rod (cell) Light-sensitive cell in the retina which works in dim light, but cannot distinguish between different colours

round window Membrane-covered opening in the cochlea that vibrates with the opposite phase to vibrations entering the inner ear through the oval window. Allows for pressure changes between the middle ear and the cochlea

sacculus Structure in the inner ear (along with the utriculus) involved with the sense of balance, in particular detecting the position of the head

saliva Digestive juice secreted into the mouth by the salivary glands. Contains the enzyme amylase

salivary glands Glands in the mouth that secrete saliva

sclera Tough outer coat of the eye

secondary sexual characteristics Characteristics that develop in the bodies of boys and girls at puberty (e.g. growth of facial hair in males or breast development in females)

secretion Release of a fluid or substance from a cell or tissue

selective reabsorption Process taking place in a kidney tubule whereby different amounts of substances are absorbed from the filtrate into the blood

selectively permeable Structure (e.g. the cell membrane) that can control its permeability to molecules and ions

semen Mixture of sperm from the testes and fluids from glands. Ejaculated during sexual intercourse

semicircular canals Structures in the inner ear involved with the sense of balance, detecting movements of the head

semilunar valves Valves present at the start of the aorta and the pulmonary artery. Prevent backflow of blood when the ventricles relax

sensory neurone Nerve cell which carries impulses from a receptor into the CNS

sex chromosomes Pair of chromosomes that determine sex in humans. XX in females, XY in males

sex-linked genes Genes inherited on the sex chromosomes

sexual intercourse Insertion of the penis into the vagina, followed by release of semen

sexually transmitted disease (STD) Disease transmitted through sexual intercourse

shivering Rapid, involuntary contraction of skeletal muscles which generates heat when a person is cold

short sight Defect of vision where either the lens is too convex or the eyeball is too long from front to back, so that light rays from a distant object are out of focus in front of the retina

skeletal muscle (See voluntary muscle)

smooth muscle (See involuntary muscle)

sperm Male gamete of an animal, with a tail for swimming and a head containing the nucleus

sperm count Number of sperm per cm^3 of semen

spermicidal cream Contraceptive method. Cream that kills sperm, may be applied to barrier contraceptives

sphincter muscle Ring of muscle in the wall of an organ, which holds back its contents, e.g. the sphincter muscle at the outlet of the bladder

spirometer A piece of apparatus for measuring breathing volumes

spongy bone Middle part of a bone containing spaces like a sponge, filled with bone marrow

squamous epithelium Epithelium consisting of a single layer of thin, flattened cells

stanol esters Chemicals found in plants, which lower blood cholesterol if eaten, e.g. in dairy products

stapes (stirrup) The third of the three ear ossicles

starch Polysaccharide of glucose that acts as a storage carbohydrate in plants

statins Group of drugs that lower blood cholesterol

STD (See sexually transmitted disease)

stem cell Cell that can divide several times but remains undifferentiated. Present in the early embryo and in some adult tissues such as bone marrow

stem cell therapy Use of stem cells to treat or prevent a disease

stent Small inflatable balloon in an expandable mesh tube, used to perform an angioplasty

sterilisation Contraceptive method involving surgery carried out on men to prevent sperm passing to the penis, or on women to prevent eggs passing to the uterus

sternum Breastbone

stimulus Change in the environment of an organism that produces a response

substrate Molecule upon which an enzyme acts

sucrose Disaccharide made from glucose and fructose

suspensory ligaments Fibres between the lens and ciliary body of the eye that hold the lens in position

sweat gland Structure in the dermis of the skin that secretes sweat

symptoms Things that a patient notices that tell them they have a disease, e.g. pain, nausea or dizziness. Sometimes also used to describe measurable signs of a disease, such as a high temperature

synapse Junction between two neurones, crossed by a neurotransmitter

synovial fluid Lubricating fluid lining the space between bones in a movable joint

synovial membrane Membrane lining bones at a movable joint; secretes synovial fluid

systemic circulation Part of a double circulation that supplies blood to all parts of the body except the lungs

systole Phase during the heart cycle when the heart muscle is contracting

tartar Hard deposit that forms on teeth

template strand (of DNA) Strand of DNA that codes for the manufacture of proteins

tendon Structure attaching muscle to bone. High tensile strength and low elasticity

testes (singular = testis) Male reproductive organs that produce the male gametes (sperm)

testosterone Male sex hormone, made by the testes. Responsible for the development of the male secondary sexual characteristics

thermoregulatory centre Part of the brain that monitors core body temperature

thorax Chest, which includes the ribcage enclosing the heart and lungs

thrombin An enzyme involved in blood clotting. Thrombin causes fibrinogen to change into fibrin

thyroid (gland) Endocrine gland in the neck that secretes several hormones including thyroxine

thyroid-stimulating hormone Hormone released by the pituitary gland that stimulates secretion of thyroxine by the thyroid gland

thyrotropin-releasing hormone (TRH) Hormone released by the hypothalamus that stimulates the secretion of thyroid-stimulating hormone by the pituitary gland

thyroxine Hormone secreted by the thyroid gland, which speeds up the metabolic rate of the body

tidal volume Volume of air breathed in and out during normal breathing

tissue Collection of similar cells working together to perform a function

tissue fluid Watery solution of salts, glucose and other solutes. Surrounds all the cells of the body, forming a pathway for the transfer of nutrients between the blood and the cells

toxin Poisonous substance produced by organisms, e.g. by some pathogenic bacteria

trachea 'Windpipe' leading from the nose and mouth to the bronchi

transcription Process by which the information in the base sequence of a strand of DNA is copied into a molecule of mRNA

transfer RNA (tRNA) Type of RNA that carries amino acids to the ribosomes during translation of the mRNA

transgenic Organism that has been engineered with a gene from another species

translation Process by which the information in the base sequence of mRNA is used to produce the sequence of amino acids in a protein. Takes place at ribosomes

TRH (See thyrotropin-releasing hormone)

triceps Muscle at the back of the upper arm, responsible for extending the arm at the elbow

tricuspid valve Valve in the heart between the right atrium and right ventricle. Prevents backflow of blood when the ventricle contracts

tRNA (See transfer RNA)

TSH (See thyroid-stimulating hormone)

tubal ligation Method of female sterilisation involving cutting or tying the oviducts

tumour Mass of cells produced by mutation and uncontrolled cell division

typhoid Disease caused by the bacterium *Salmonella typhi*. Symptoms include fever, headaches, cough, stomach cramps and diarrhoea. Toxins from the bacteria can cause multiple organ failure

ultrafiltration Filtration of the blood taking place in the Bowman's capsule of a kidney tubule, where the filter separates differently sized molecules under pressure

umbilical cord Tubular structure leading from the placenta to the developing fetus. Contains blood vessels which supply the fetus with nutrients and take away waste products

urea Main nitrogenous excretory product of mammals

ureter Tube carrying urine from the kidney to the bladder

urethra Tube carrying urine from the bladder to the outside of the body

utriculus Structure in the inner ear (along with the sacculus) involved with the sense of balance, in particular detecting the position of the head

vaccination Artificially supplying antigens to a person (e.g. as an injection) in order to stimulate an immune response and protect them against a disease-causing pathogen

vascular dementia Disease of the brain that causes dementia in elderly people. Caused by diseased and damaged blood vessels in the brain

vasectomy Method of male sterilisation involving cutting or tying the sperm ducts

vasoconstriction Narrowing of blood vessels in the skin. Decreases the blood flowing through the skin to reduce heat loss

vasodilation Widening of blood vessels in the skin. Increases the blood flow through the skin to increase heat loss

vector (in genetic engineering) Structure which can be used to transfer genes in genetic engineering, e.g. a plasmid or a virus

vein Blood vessel with a thin muscular wall and a wide lumen, carrying blood towards the heart

vena cava (plural = vena cavae) The main vein. Blood vessel carrying deoxygenated blood from the body to the heart

ventilation Movement of air into and out of the lungs

ventral root Part of a spinal nerve that emerges from the ventral (front) side of the spinal cord

ventricles Two lower chambers of the heart, which pump blood out of the heart via the aorta (from the left ventricle) and pulmonary artery (from the right ventricle)

vertebrae (singular = vertebra) Individual bones of the vertebral column

vertebral column Backbone

vertebrate Animal with a vertebral column or 'backbone'

villi (singular = villus) Tiny projections from the lining of the ileum that increase the surface area for the absorption of the products of digestion

viruses Very small microorganisms that are not composed of cells. Virus particles consist of genetic material (DNA or RNA) surrounded by a protein coat

Visking tubing Artificial partially permeable membrane, used in kidney dialysis machines

vital capacity Difference in volume between the maximum breath in and the maximum breath out

vitamins Chemicals that are obtained from the diet and needed in small amounts to maintain health

voluntary action Movement of part of the body under conscious control by the brain

voluntary muscle (skeletal muscle) Muscle that contracts under conscious control by the brain. Attached to bones

water potential Measure of the ability of water molecules to move in a solution. Pure water has the highest water potential

white matter Tissue in the outer part of the spinal cord and the middle of the brain, consisting mainly of nerve cell axons

zygote Single cell resulting from fusion of a male and female gamete

INDEX

A

ABO blood groups 104–6, 218–19
accelerator nerve 97
accommodation 141–2
accuracy 260
ACE inhibitors 111–12
action potential 168
activated sludge method 249
active immunity 244
active site 36
active transport 26
adenosine diphosphate (ADP) 75
adenosine triphosphate (ATP) 75
ADH 163, 178–9
adolescent growth spurt 208
ADP 75
adrenal glands 163, 166
adrenaline 110–11, 166
adult stem cells 16
aerobic exercise 109–10
aerobic respiration 74–5
afterbirth 200
agar 231
agglutination 105
Agrobacterium 19–20
AIDS 234–6
alcohol 159–60
alleles 195, 212–13
   codominant 217–18
   dominant 213, 214
   multiple 218
   recessive 213, 214
alveoli 78, 80–1
Alzheimer's disease 155–6
amino acids 7, 34, 56
ammonia 173
amnion 200
amniotic fluid 200
amylase 67
anaemia 57
anaerobic digestion 250
anaerobic exercise 109
anaerobic respiration 76

angina 106, 111
angioplasty 107
angiotensin I/II 111
antagonistic pairs 130
antibiotic-resistant bacteria 246
antibiotics 245–7
antibodies 103, 104, 243
   monoclonal 112–13
anticodons 9
antidiuretic hormone (ADH) 163, 178–9
antigens 104, 242–3
anus 70
aorta 93
apparatus 257
appendicular skeleton 126
arteries 94, 98
arterioles 98
artificial active immunity 244
artificial heart 108
artificial insemination (AI) 207
artificial passive immunity 244
asexual reproduction 13
assimilation 69
astigmatism 144
atherosclerosis 106, 110
athlete's foot 240
ATP 75
atria 96
axial skeleton 126
axon 150–1

B

B-lymphocytes 243
bacteria 229 230–2
   antibiotic-resistant 246
   food contamination 71
   genetically-modified 16–18
   tooth decay 64
bactericidal 246
bacteriostatic 246
balance 147
balanced diet 54–9

ball and socket joints 129
barrier methods of contraception 204, 206
basal metabolic rate 164
base 6
base-pairing rule 6
basement membrane 175
Benedict's solution 35
beta blockers 110–11
biceps 130–1
bile 68
bile duct 68
binary fission 231
biogas 250
biological substances, tests for 34–5, 257–8
biosensors 46
birth 200
biuret test 35
bladder 174
blind spot 140
blood 14, 93, 94
   composition 102–4
blood clot 104
blood glucose
   control 166–8
   tests 44, 46
blood groups 104–6, 218–19
blood pressure 111
blood transfusions 104–6
blood vessels 94
body mass index 70
body temperature 182–5
body water content 178–9
Bohr effect 114
bone 14, 126–8
   compact 127
   spongy 127
bone marrow 125, 127
   transplants 16
Bowman's capsule 175–6
brain 154–5
   diseases 155–7
brain stem 155

breast cancer 113
breastfeeding 200–1
breathing 74, 76
    exercise 82–3
    regulation 85
    volumes 83–4
broad spectrum antibiotics 246
bronchi 78
bronchial tree 77
bronchioles 78
bronchitis 86
buffer solutions 42

C

calorimeter 61
cancer
    breast 113
    lung 87–8
canines 64
cannabis 160
capillaries 94, 99
capillary loops 184
caps 204
capsule
    bacteria 230
    joint 129
carbohydrases 66
carbohydrates
    balanced diet 54–5
    structure 32–3
carbon cycle 247–8
carbon dioxide, inhaled and exhaled air 82
carbon monoxide 89
carboxyhaemoglobin 89
carcinogens 88
cardiac centre 97
cardiac cycle 95–6
cardiac muscle 14, 95, 133
cardiovascular system 108–10
caries 64
carrier 220
cartilage 78, 127, 129
catalysts 36
cataracts 144
cell differentiation 13–16
cell division 11–12

cell membrane 4
cell structure 4–5
cell surface membrane 4
cell wall 230
cellulose 55
central nervous system (CNS) 149–53
cerebellum 155
cerebral cortex 154
cerebral hemispheres 154
cerebrum 154
cervix 200
chemical coordination 161–8
chemical digestion 65
chemical tests 34–5, 257–8
childbirth 200
cholera 238
cholesterol 56
chondrocytes 127
choroid 139
chromatid 195
chromosomes 4, 5, 11
    sex (X and Y) 219
cilia 78
ciliary muscle 141
ciliated epithelium 14
circulatory system 93–4
cirrhosis 160
clone 12, 13
clot 104
CNS 149–53
cocaine 161
cochlea 145
codominance 217–18
codons 9
cognitive behavioral therapy (CBT) 158
coil 205, 206
collecting duct 176
colon 70
colour blindness 221–2
combined pill 205, 206
command words 262
compact bone 127
competitive inhibitors 39
complementary bases 6
composting toilets 250

concentration gradient 24, 25
condoms 204, 206
cones 139
contraception 203–6
contraceptive pill 205, 206
control 260
controlled variables 260
cornea 139
    transplants 144
coronary arteries 96, 106
coronary heart disease 106–7
coronary veins 96
corpus luteum 203
cortex
    cerebral 154
    kidney 174
cranium 125
cupula 147
cystic fibrosis 222–3
cytoplasm 4

D

data manipulation and analysis 258–9
daughter cells 11–12, 195
decelerator nerve 97
decomposers 230, 247–50
deep brain stimulation 157
deletion 11
dementia
    Alzheimer's disease 155–6
    vascular 156
denatured 37
dendrites 150
dendrons 150
dental caries 64
dental hygiene 64–5
dentine 64
dentition 63
deoxyribonucleic acid (DNA) 5, 6–7, 8, 16
dependent variable 259
depolarised 168
depression 158–9
dermis 183
development 207–9
    hormones 165

diabetes 44, 107, 167–8
dialysis machines 180–1
diamorphine 160
diaphragm
   chest 76, 79, 80
   contraceptive 204, 206
diastole 96
diet
   balanced 54–9
   coronary heart disease 107
differentiation 13–16
diffusion 24–6
digestion 65–70
digestive system 66–8
dihybrid inheritance 224–5
diploid cells 11, 194
disaccharide 33
disease 228–9, 232–4
diuresis 178
DNA 5, 6–7, 8, 16
DNA polymerase 6
dominant allele 213, 214
dominant eye 141
dopamine 156, 158
dorsal root 152
dorsal root ganglion 152
double circulation 93–4
drug abuse 158
drugs 159–61
duct 162
ductless glands 162
duodenum 66, 68
duplication 11

E

ear 145–7
ear ossicles 145
eardrum 145
Ebola 236–7
effector organ 136
egestion 70
egg 194, 196
ejaculation 198
electron microscopy 21
embryo 13, 199
embryonic stem cells 16
emphysema 86–7

enamel 64
endemic 234
endocrine glands 161, 162, 163–4
endocrine system 161–8
endoplasmic reticulum (ER) 4
endotherms 182
energy
   from food 60–3
   needs 61
   respiration 74–6
environment 223–4
enzymes 4, 36–46
   digestion 65, 66
   immobilised 43–6
   intracellular and extracellular 36
epidemic 234
epidermis 183
epithelium 14, 69
   ciliated 14
   squamous 14
erythrocytes 102–3
essential amino acids 56
Eustachian tube 145
exchange surfaces 29
excretion 171, 173
exercise
   breathing rate 82–3
   cardiovascular system 107, 108–10
   pulse rate 101
exhalation 80
exocrine glands 162
experimental design 259–60
expiration 80
external environment 136
extracellular enzymes 36
eye 138–44

F

F1/F2 generation 214
faeces 66, 70
family intervention therapy 158
fats
   balanced diet 55–6
   structure 33–4
   test for 35
fatty acids 33

femidom 204
fermenters 18
fertilisation 11, 193, 196–8
fertility treatment 206–7
fetus 199
fibre, dietary 55, 65
fibrinogen 104
'fight or flight' response 98, 166
fixed joints 129, 130
flagella 230
fluoride 65
follicle 203
follicle stimulating hormone 201, 203
food
   balanced diet 54–9
   energy from 60–3
   hygiene 71
   preservation 71–2
fovea 139
freely moveable joints 129
fructose 32
fungi 229

G

gall bladder 68
gametes 11, 193, 194–8
gas exchange system 76–85
gene gun 20
gene therapy 223
genes 4, 5, 207, 212–13
   environment and 223–4
   mutations 10–11
   sex-linked 219–22
genetic code 7
genetic crosses 213–18
genetic diagrams 214–15
genetic engineering 16–21
genetic variation 195, 196–7
genotype 214
glands 162
glomerular filtrate 176
glomerulus 175
glucagon 164, 166–7
glucose 32, 54, 74
   blood glucose control 166–8
   blood glucose testing 44, 46

test for 35
glycerol 33
glycogen 33, 55, 166
GM bacteria and plants 16–21
goitre 164
golden rice 20
gonadotrophic hormones 201
gonorrhoea 239–40
graphs 259
grey matter 153, 154
growth 165, 207–9
growth hormone 19, 165

H

haemoglobin 89, 102–3, 114
haemophilia 220–1
hair erector muscles 184
haploid cells 11, 194
hashish 160
Haversian systems 127
hearing range 146
heart
    artificial 108
    attack 106
    rate 97–8
    structure and function 94–6
    transplant 108
heavy metals 39
hepatic portal vein 69, 94
hepatitis B vaccine 19
HER2 113
Herceptin™ 113
heredity 107, 158
heroin 160–1
heterozygous 214
high blood pressure 107, 111–12
hinge joints 129
histones 5
HIV 234–6
homeostasis 171–2
homeotherms 182
homologous pairs 11
homozygous 214
hormonal methods of contraception 205, 206
hormones 155, 161, 162, 163
    breastfeeding 200
    growth and development 165
    menstrual cycle 202–3
    reproduction 201
    see also specific hormones
hybridoma cells 112
hypertension 107, 111–12
hypodermis 183
hypothalamus 155, 163, 183

I

identical twins 223
ileum 66, 68–9
illegal drugs 160–1
immobilised enzyme 43–6
immoveable joints 129, 130
immunity 104, 242–5
in vitro fertilisation (IVF) 207
incisors 64
incubation period 232, 233
incus 145
independent variable 259
infant growth spurt 208
infection 232–3
infertility 206
inhalation 79
inhibitors 38–9
inspiration 79
insulin 19, 164, 166–7
    injections 168
    resistance 167
intelligence 224
intercostal muscles 76, 79, 80
internal environment 136, 172
intracellular enzymes 36
intrauterine devices (IUDs) 205, 206
inversion 11
invertase 44
investigation planning 260–1
involuntary muscle 14
iris 139, 140
islets of Langerhans 167
IUDs 205, 206
IVF 207

J

joints 128–30

K

Kaposi's sarcoma 235
kidney 173, 174–8
    dialysis machine 180–1
    failure 179–81
    transplants 179–80
kilojoules (kJ) 60
kinetic energy 37
knee-jerk reflex 153
kwashiorkor 56

L

lactase 44, 45–6
lactate 76, 109
lacteals 69, 99
lactose 33
lactose-free milk 44, 45–6
large intestine 66, 70
latency period 235
legal drugs 159–60
lens 141
Lewy bodies 156
ligaments 129
ligases 17
limiting factors 231
lipases 66
lipids
    balanced diet 55–6
    structure 33–4
    test for 35
liver 68, 69, 173
    cirrhosis 160
lock and key model 36
long sight 142–3
loop of Henlé 176, 178
lungs 76, 78
    cancer 87–8
    capacity 84–5
    ventilation 79–80
luteinising hormone 201
lymph 69, 99
lymph nodes 101
lymphatic capillaries 99
lymphatic system 69, 99–101
lymphocytes 102, 104, 242–3

M

malaria 241–2
malleus 145
mammary glands 200
marijuana 160
mechanical digestion 65
medulla
   brain 97
   kidney 174
megacities 239
meiosis 11, 194–6
memory cells 104, 243–4
menstrual cycle 202–3
mental illnesses 157–9
messenger RNA (mRNA) 8
metabolic rate 164–5
metabolism 36
methanogens 250
methanol 39
methicillin-resistant Staphylococcus aureus 246
microorganisms 228–32
microvilli 69
milk, lactose-free 44, 45–6
milk teeth 64
minerals 57
mini-pill 205, 206
mitochondria 5
mitosis 11–12, 196
molars 64
monoclonal antibodies 112–13
monohybrid cross 214
monosaccharide 32
motor neurones 150–1
MRSA 246
mucus 78
multiple alleles 218
multiple births 207
muscle fibres 95, 132–3
muscles 14
   anaerobic respiration 76
   cardiac 14, 95, 133
   ciliary 141
   hair erector 184
   insertion 131
   intercostal 76, 79, 80
   involuntary 14
   origin 131
   skeletal 130–3
   smooth 133
   sphincter 68, 174, 185
   voluntary 14
mutations 10–11
myelin sheath 151

N

natural active immunity 244
natural methods of contraception 203–4, 206
natural passive immunity 244
nature/nurture argument 224
negative feedback 164–5, 167, 179, 183
nephrons 174, 175–8
nerve impulse 149, 168
nervous tissue 14
neuromuscular junction 151
neurones 149, 150–1
   motor 150–1
   relay 152
   sensory 150, 151
nicotine 88
nitrates 248
nitrogenous waste 173
noise-induced hearing loss 147
non-competitive inhibitors 39
nucleotides 6
nucleus 4
nutrient deficiency disease 56

O

obesity 70
oesophagus 67
oestrogen 165, 201, 203
oils
   balanced diet 55–6
   structure 33–4
   test for 35
optic nerve 139
optimum pH 37–8
optimum temperature 37
oral rehydration therapy 237
organ of Corti 145, 146
organ rejection 108
organelles 4–5
organs 14, 15
osmosis 26–9
ossicles 145
ossification 127
osteocytes 126, 127
osteoporosis 128
otolith 147
oval window 145
ovaries 163, 164, 194
oviduct 199
ovulation 199, 202
ovum 193
oxygen debt 76
oxygen dissociation curve 114
oxyhaemoglobin 102–3
oxytocin 200

P

pancreas 66, 68, 163, 164, 166, 167
pandemic 234
paracetamol 159
parasites 232
Parkinson's disease 156–7
partial pressure 114
partially moveable joints 129, 130
partially permeable membrane 4, 27
passive immunity 244
pedigrees 215–17
pelvis 174
penicillin 245
pepsin 38, 68
percolating filter method 249
period 202
periodontal disease 64
periosteum 127
peristalsis 65–6
pH 37–8, 42–3
phagocytes 102, 103
phagocytosis 103
phenotype 214
photosynthesis 247
physical fitness 101
pit latrines 250
pituitary gland 155, 163, 165
placenta 199, 200

plants
- genetically-modified 19–21
- stanol esters 110

plaque
- arteries 106
- teeth 64

plasma 102
plasmids 16–17, 230
platelets 102, 104
pleural cavity 78
pleural fluid 78
pleural membranes 78
polysaccharide 33
positive feedback 200
precision 260
pregnancy 199–200
- smoking 89

premolars 64
primary sex characteristics 201
progesterone 200, 203
programmed cell death 243
prolactin 200
prostaglandins 159
proteases 66
protein
- balanced diet 56
- structure 34
- synthesis 7–10
- test for 35

prothrombin 104
protozoa 229
pseudopodia 103
puberty 201
pulmonary arteries 93
pulmonary circulation 93
pulmonary veins 93
pulp cavity 64
pulse 98
- rate 101

Punnett square 214–15
pupil 139, 140

R

receptors 137
recessive allele 213, 214
recombinant DNA 16
recovery period 109
rectum 70
red blood cells 102–3
red-green colour blindness 221–2
reflex actions 97, 140, 152–3
reflex arc 152–3
refraction 139
relay neurones 152
reliability 260
renal artery 174
renal dialysis 180–1
renal vein 174
renin 111
renin–angiotensin system 111–12
replication 6
repolarised 168
reproduction
- asexual 13
- sexual 193

reproductive system 198–200
- hormones 201

residual volume 84
respiration 5, 74–6
respiratory centre 85
response 136–7
restriction endonucleases (enzymes) 17
retina 139
ribonucleic acid (RNA) 7, 8
ribosomal RNA (rRNA) 8
ribosomes 4
rickets 57
RNA 7, 8
RNA nucleotides 8
rods 139
round window 146

S

sacculus 147
'safe period' method 203, 206
safety precautions 257
saliva 67
salivary glands 66
saturated fats 56
sausage test 141
schizophrenia 157–8
sclera 139
scurvy 58
secondary sexual characteristics 201
secretion 162
selective reabsorption 177
selectively permeable membrane 4, 27
semen 198
semicircular canals 147
semilunar valves 99
sensory neurones 150, 151
sewage treatment 248–50
sex cells see gametes
sex chromosomes 219
sex determination 219
sex-linked genes 219–22
sexual intercourse 198
sexual reproduction 193
shivering 185
short sight 142–3
signs of disease 233
skeletal muscles 130–3
skeleton 125–30
skin
- temperature control 183–4
- touch receptors 137–8

small intestine 66, 68–9
smallpox 245
smoking 86–9, 107
smooth muscle 133
specific immune responses 242
sperm 193, 196, 201
sperm count 206
spermicidal cream 204
sphincter muscles 68, 174, 185
spirometer 83–4
spongy bone 127
squamous epithelium 14
stanol esters 110
stapes 145
starch 33, 55
- test for 35

statins 110
stem cell therapy 16
stem cells 16
stent 107
stereoscopic vision 141
sterilisation 205, 206

stimulus 136–7
stomach 67–8
stress 107, 158
stroke 110
substitution 11
substrate 36, 38
sucrose 33, 44, 54
suspensory ligaments 141
sweat glands 184
symptoms of disease 233
synapses 150, 151–2
synovial fluid 129
synovial joints 129
synovial membrane 129
systemic circulation 93
systole 96

T

T-helper cell 234
T-lymphocytes 243
talking treatments 158
target organs 162
tartar 65
teeth 63–5
temperature
    bodily control 182–5
    diffusion 25
    enzyme action 37, 40–1
template strand 7
tendon 130
testes 163, 164, 194, 201
testosterone 165
thermoregulatory centre 183
thermostat 183
thrombin 104
thyroid 163, 164–5
thyroid-stimulating hormone (TSH) 164
thyrotropin releasing hormone (TRH) 164
thyroxine 164
tidal volume 84
tinnitus 147
tissue fluid 99, 172
tissues 14
tooth decay 64
toothbrushes 64–5

touch receptors 137–8
trachea 78
transcription 8
transfer RNA (tRNA) 8
transfusions 104–6
transgenic organisms 16–21
translation 9–10
transplant
    bone marrow 16
    cornea 144
    heart 108
    kidney 179–80
triceps 130–1
triplet code 7
tubal ligation 205, 206
twins 223
typhoid 238–9

U

ultrafiltration 176
umbilical cord 199
universal code 7
unsaturated fats 56
urea 173
ureters 174
urethra 174
urinary system 174–8
urine 173
    glucose test 44
utriculus 147

V

vaccines and vaccination 19, 21, 245
variation 195, 196–7, 223–4
vascular dementia 156
vasectomy 205, 206
vasoconstriction 185
vasodilation 184
vector 18
veins 94, 98–9
vena cava 93
ventilated improved pit latrine (VIP) 250
ventilation 79–80
ventral root 152

ventricle 96
vertebrae 125, 131, 132
vertebral column 125, 131
villi 68–9
viruses 229, 232
vision defects 142–4
Visking tubing 27, 180, 181–2
vital capacity 84
vitamins 57–60
voluntary muscle 14

W

waste elimination 70
water content 178–9
water potential 28
white blood cells 102, 103–4, 242–3
white matter 153, 154
'withdrawal' method 204, 206

X

X chromosome 219

Y

Y chromosome 219

Z

zygote 11, 193, 196, 197